Handbook of Food Science and Technology 3

Series Editor
Jack Legrand & Gilles Trystram

Handbook of Food Science and Technology 3

Food Biochemistry and Technology

Edited by

Romain Jeantet
Thomas Croguennec
Pierre Schuck
Gérard Brulé

WILEY

First published 2016 in Great Britain and the United States by ISTE Ltd and John Wiley & Sons, Inc.
Translated by Geraldine Brodkorb from "Science des aliments" © Tec & Doc Lavoisier 2006.

Apart from any fair dealing for the purposes of research or private study, or criticism or review, as permitted under the Copyright, Designs and Patents Act 1988, this publication may only be reproduced, stored or transmitted, in any form or by any means, with the prior permission in writing of the publishers, or in the case of reprographic reproduction in accordance with the terms and licenses issued by the CLA. Enquiries concerning reproduction outside these terms should be sent to the publishers at the undermentioned address:

ISTE Ltd
27-37 St George's Road
London SW19 4EU
UK

www.iste.co.uk

John Wiley & Sons, Inc.
111 River Street
Hoboken, NJ 07030
USA

www.wiley.com

© ISTE Ltd 2016
The rights of Romain Jeantet, Thomas Croguennec, Pierre Schuck and Gérard Brulé to be identified as the authors of this work have been asserted by them in accordance with the Copyright, Designs and Patents Act 1988.

Library of Congress Control Number: 2016936913

British Library Cataloguing-in-Publication Data
A CIP record for this book is available from the British Library
ISBN 978-1-84821-934-2

Contents

Introduction xi
Gérard BRULÉ

Part 1. Food from Animal Sources 1

Chapter 1. From Milk to Dairy Products 3
Thomas CROGUENNEC, Romain JEANTET and Pierre SCHUCK

    1.1. The biochemistry and physical chemistry of milk 3
        1.1.1. Milk fat 4
        1.1.2. Carbohydrates 8
        1.1.3. Proteins 10
        1.1.4. Milk minerals 15
    1.2. Biological and physicochemical aspects of milk processing 17
        1.2.1. The stability of fat globules 17
        1.2.2. Protein stability 19
    1.3. Dairy product technology 25
        1.3.1. Liquid milk 25
        1.3.2. Fermented milk products 29
        1.3.3. Milk powder 32
        1.3.4. Cheese 39
        1.3.5. Cream and butter 58

Chapter 2. From Muscle to Meat and Meat Products 65
Catherine GUÉRIN

    2.1. The biochemistry of muscle (land animals and fish) 65
        2.1.1. The structure and composition of meat and fish muscle 66
        2.1.2. Muscle structure 73
        2.1.3. Proteins 78

  2.1.4. Carbohydrates 88
  2.1.5. Vitamins and minerals 88
 2.2. Biological and physicochemical changes in muscle 89
  2.2.1. Muscle contraction 89
  2.2.2. Changes in muscle after death 91
 2.3. Meat and fish processing technology 102
  2.3.1. Meat processing technology 102
  2.3.2. Fish processing technology 109

Chapter 3. From Eggs to Egg Products 115
Marc ANTON, Valérie LECHEVALIER and Françoise NAU

 3.1. Chicken egg – raw material in the egg industry 117
  3.1.1. Structure and composition 117
  3.1.2. Biochemical and physicochemical properties
  of the protein and lipid fractions of egg 120
 3.2. Physicochemical properties of the different egg fractions 125
  3.2.1. Interfacial properties 125
  3.2.2. Gelling properties 131
 3.3. The egg industry: technology and products 136
  3.3.1. Decontamination of shells 138
  3.3.2. Breaking and separation of the egg white and yolk 138
  3.3.3. Primary processing of egg products – decontamination
  and stabilization 139
  3.3.4. Secondary processing of egg products 142
  3.3.5. Egg extracts 143

Part 2. Food from Plant Sources 145

Chapter 4. From Wheat to Bread and Pasta 147
Hubert CHIRON and Philippe ROUSSEL

 4.1. Biochemistry and physical chemistry of wheat 150
  4.1.1. Overall composition 150
  4.1.2. Structure and properties of the constituents 154
 4.2. Biological and physicochemical factors
 of wheat processing 163
  4.2.1. Development of texture 164
  4.2.2. Development of color and flavor 170
 4.3. The technology of milling, bread making and pasta making 172
  4.3.1. Processing of wheat into flour and semolina 172
  4.3.2. Bread making 180
  4.3.3. Pasta making 195

Chapter 5. From Barley to Beer 205
Romain JEANTET and Ludivine PERROCHEAU

5.1. Biochemistry and structure of barley and malt 205
   5.1.1. Morphology of barley grain. 206
   5.1.2. Biochemical composition of barley 207
   5.1.3. Composition and structure of starch and protein 208
   5.1.4. Effect of malting 209
5.2. Biological and physicochemical factors of processing 213
   5.2.1. Enzymatic degradation of starch and protein. 214
   5.2.2. Fermentability of the wort. 220
5.3. Brewing technology. 221
   5.3.1. Stages of malting 221
   5.3.2. Stages of beer production 224

Chapter 6. From Fruit to Fruit Juice
and Fermented Products 231
Alain BARON, Mohammad TURK and Jean-Michel Le QUÉRÉ

6.1. Fruit development. 231
   6.1.1. Stages of development. 231
   6.1.2. Fruit ripening 233
6.2. Biochemistry of fruit juice 237
   6.2.1. Pectins 238
   6.2.2. Pectinolytic enzymes. 241
   6.2.3. Bitter and astringent compounds 245
6.3. Fruit juice processing 249
   6.3.1. Preparation of fruit 249
   6.3.2. Pre-treatment 250
   6.3.3. Pressing. 250
   6.3.4. Treatment of fruit juice 253
   6.3.5. Pasteurization, high-pressure treatment,
   pulsed electric fields and concentration. 262
6.4. Cider. 264
   6.4.1. French cider 264
   6.4.2. Fermentation process. 265
   6.4.3. Action of microorganisms. 267
   6.4.4. Fermentation and post-fermentation. 271

Chapter 7. From Grape to Wine 275
Thomas CROGUENNEC

7.1. Raw materials 276
   7.1.1. Grape variety 276
   7.1.2. Composition of grapes. 276

7.2. Winemaking techniques 280
   7.2.1. State of the harvest and adjustments. 281
   7.2.2. Physicochemical processes involved
   in winemaking 282
   7.2.3. Biological processes involved in
   winemaking: fermentation 285
7.3. Stabilization and maturation of wine 289
   7.3.1. Biological stabilization 289
   7.3.2. Physicochemical stabilization 290
   7.3.3. Maturation of wine 291
7.4. Specific technology 292
   7.4.1. Sparkling wines (traditional method) 292
   7.4.2. Sweet wines 293

Chapter 8. From Fruit and Vegetables to
Fresh-Cut Products 297
Florence CHARLES and Patrick VAROQUAUX

8.1. Respiratory activity of plants 298
   8.1.1. Measurement and modeling of
   respiratory activity 299
   8.1.2. Control of respiratory activity 301
8.2. Enzymatic browning 302
   8.2.1. Mechanism and evaluation 302
   8.2.2. Prevention of enzymatic browning 303
8.3. Unit operations in the production of fresh-cut
products: main scientific and technical challenges 304
   8.3.1. Raw materials: selection of varieties and
   cultivation methods 306
   8.3.2. Raw material quality control: grading 307
   8.3.3. Trimming and mixing 307
   8.3.4. Cutting 308
   8.3.5. Washing and disinfection 309
   8.3.6. Draining and drying 312
   8.3.7. Weighing 313
   8.3.8. Bagging 313
8.4. Modified atmosphere packaging 314
   8.4.1. Diffusion of gases through packaging 315
   8.4.2. Change in gas content in modified
   atmosphere packaging 317
8.5. Conclusion 319

Part 3. Food Ingredients ... 321

Chapter 9. Functional Properties of Ingredients ... 323
Gérard BRULÉ and Thomas CROGUENNEC

9.1. Interactions with water: hydration and
thickening properties ... 324
  9.1.1. Types of interaction ... 324
  9.1.2. Influence of hydrophilic components on
  water availability and mobility ... 325
  9.1.3. Influence of hydration on the solubilization,
  structure and mobility of compounds ... 325
  9.1.4. Effect of the hydration of components
  on rheological properties ... 326
9.2. Intermolecular interactions: texture properties ... 326
  9.2.1. Aggregation/gelation by destabilization of
  macromolecules or particles ... 326
  9.2.2. Aggregation/gelation by covalent cross-linking ... 327
  9.2.3. Sol–gel transitions ... 329
  9.2.4. Influence of denaturation kinetics and
  molecular interactions ... 329
9.3. Interfacial properties: foaming and emulsification ... 330
  9.3.1. Interfacial tension ... 330
  9.3.2. Surfactants ... 332
  9.3.3. Emulsification and foaming ... 332

Chapter 10. Separation Techniques ... 335
Thomas CROGUENNEC and Valérie LECHEVALIER

10.1. Proteins and peptides ... 335
  10.1.1. Milk proteins and peptides ... 335
  10.1.2. Extraction of lysozyme from egg white ... 346
  10.1.3. Extraction of gelatin ... 348
  10.1.4. Plant proteins ... 349
10.2. Carbohydrates ... 351
  10.2.1. Sucrose ... 351
  10.2.2. Lactose ... 364
  10.2.3. Polysaccharides ... 369
10.3. Lipids ... 378
  10.3.1. Production of vegetable oils ... 379
  10.3.2. Lipid modification ... 383
10.4. Pigments and flavorings ... 391
  10.4.1. Types of pigments and flavorings ... 391

10.4.2. Extraction/concentration of
colorings and flavors. 397
10.4.3. Formulation 400

Bibliography. 403

List of Authors 417

Index 419

Introduction

The processing into food of raw materials from hunting, gathering, fishing and subsequently arable and livestock farming has always had two objectives: to preserve nutrients in order to defer the time and place of consumption, and develop products with a wide variety of textures and flavors to satisfy the sensory needs of consumers. The development of arable and livestock farming has facilitated an improved control of supply, even though the provision of agricultural products has long remained very irregular due to climatic or health risks and the seasonality of certain products. Furthermore, the importance of stabilization and/or processing has significantly increased with the rural exodus, which has led to a distancing of production from consumption areas.

The production of certain foods that still form the basis of our diet today dates back several centuries or even millennia, as in the case of bread, cheese and wine for example. These products, particularly those derived from fermentation, were developed based on empirical observations, with no knowledge of the raw materials or phenomena involved in their processing. It was not until the work of Pasteur in the 19th Century that microorganisms gained a key role in the development and processing of agricultural products.

The agri-food industry has undergone a major change over the past few decades in order to better meet the quality requirements of consumers; while traditional food is the result of a series of increasingly understood and controlled biological and physicochemical phenomena, this is not the case for a number of new products designed to meet market expectations. These

---

Introduction written by Gérard BRULÉ.

products are the result of an assembly of various ingredients (Figure I.1), the control of which is a real challenge for food technologists and engineers.

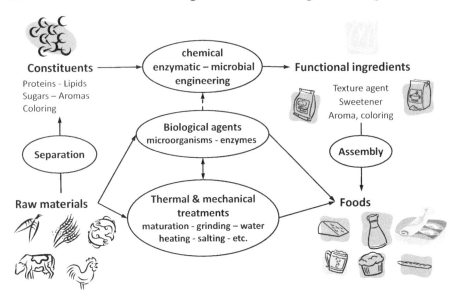

Figure I.1. *New products and assembly technology*

I.1. From empiricism to rational technology

The oldest forms of processing (milk to cheese, grain to bread or beer, grapes to wine, muscle to meat, etc.) were based on biological phenomena that could occur naturally under specific water content and temperature conditions, since the biological agents (enzymes or microorganisms) responsible for processing and the reaction substrates and growth factors were present in the raw materials and/or immediate environment; it was enough to simply mature (milk, meat), crush, grind and sometimes hydrate (fruit, grain) in order for biological reactions to take place. This is why these types of products were able to develop based solely on the observation of natural processes.

The knowledge acquired since the end of the 19th Century in the field of microbiology and the early 20th Century in the field of enzymology has gradually helped explain the biological phenomena involved in the development of certain food products. Based on this knowledge, the food industry has sought to control these processes rather than witness them, which is how the fermentation and then an enzyme industry arose, producing and marketing biological agents for each type of processing. The use of fermentation and in some cases enzymes has become indispensable given the existing food safety requirements, which include increasingly stringent hygiene conditions in production and processing, and technological treatments to eliminate potential pathogenic microorganisms (microfiltration, heat treatment). This change has led to a reduction in the endogenous biological potential, which needs to be replenished by adding fermentation steps and enzymes. Reconstituting microbial ecosystems through the assembly of exogenous flora requires the identification of the endogenous flora and their role in the characteristic features of the food; progress in the field of molecular biology should enable significant progress in this area.

Over the past 40 years, many teams have focused on the study of food science, which has resulted in a better understanding of the composition of various raw materials and the biological and physicochemical mechanisms involved in the development of texture, flavor and aroma; this work has allowed the food industry to better identify the key technological tools in the development of quality, and to rely less on empiricism and more on technology.

I.2. From traditional foods to assembly technology

The quality requirements of consumers are increasingly specific and segmented. Food must be absolutely safe (no pathogens, toxins, residues or contaminants), have the closest possible nutritional profile to that recommended by nutritionists, meet sensory needs, integrate practicality (ease of storage and use) and convey social values (fair trade, environmental protection and animal welfare) while remaining at an affordable price. These market expectations are identified by the marketing services of the food industry for specific target consumers whose needs depend on several factors (gender, age, activity, health, metabolic disorders, food trends, etc.). These services, in consultation with nutrition and health specialists, identify which nutrients and micronutrients (minerals and vitamins) to assemble and define the structure, sensory characteristics and practicality of the food based on consumer research. The path from conception to completion can sometimes be difficult, since food is a complex and thermodynamically-unstable system, which can be defined as a continuous, usually aqueous phase, a three-dimensional protein and/or polysaccharide network and dispersed elements (gas, fat globules, solids); the aim of technologists is to stabilize this system throughout the marketing period while taking into account mechanical (transport) and thermal (refrigeration, freezing, thawing) constraints.

The progress made in recent years in food science has provided insight into the key role of various biological components and particularly their structure, whether native or modified by technological treatments, in the development of texture, the thermodynamics of dispersed systems and the role of interfaces. This knowledge allows us to better understand and control the instability of food using technological processes or functional ingredients; the industry has a very large range of functional ingredients that is used to create texture and stabilize complex multiphase systems. It is therefore possible, by assembly, to create new foods that meet the quality requirements of the market.

PART 1

Food from Animal Sources

1

From Milk to Dairy Products

Secreted from the mammary glands of mammals, milk is a complete food designed to provide newborns with energy, compounds necessary for growth, immunological protection, and so forth, which are all vital in the early stages of life. From a physicochemical point of view, milk is complex in terms of its structure, the interactions between its various components and its variability in composition based on species, breed, diet and lactation period. It is a dynamic system due to the presence of endogenous enzymes and microorganisms as well as ionic equilibria, which depend on pH and temperature, and determine the stability of dispersed elements. These physical, physicochemical and biological changes lead to instability in milk, which can be exploited during processing into a variety of dairy products, such as fermented products, cheese, cream, butter, and so on.

1.1. The biochemistry and physical chemistry of milk

Milk is a natural emulsion. Fat, which represents approximately 4% of the overall composition of cow's milk (w/w), is present in the form of fat globules dispersed in the skimmed milk phase.

The non-fat phase of cow's milk (skimmed milk) is composed mainly of water (90% (w/w) of the overall composition) in which the following are dispersed or dissolved:

– lactose (4.8 – 5% (w/w) of overall composition);

– protein (3.2 – 3.5% (w/w));

Chapter written by Thomas CROGUENNEC, Romain JEANTET and Pierre SCHUCK.

– non-protein nitrogen (NPN) consisting of urea, amino acids and peptides, representing about 5% of the nitrogen fraction of milk;

– inorganic minerals (calcium, phosphate, chloride, potassium, sodium) and organic acids (mainly citric acid in fresh milk);

– water-soluble vitamins.

1.1.1. *Milk fat*

The fat content of cow's milk varies between about 3.3 and 4.7% (w/w) depending on breed, lactation stage, season, and so forth. Milk fat is mostly present in the form of fat globules measuring between 0.2 and 15 μm in diameter. Around 75% of fat globules are smaller than 1 μm, but they represent less than 10% of the total volume of milk fat. Similarly, there are very few fat globules larger than 8 μm; they represent less than 3% of the overall volume. Thus, almost 90% of milk fat is in the form of milk globules measuring between 1 and 8 μm in diameter. The average diameter of fat globules is approximately 4 μm. The core of the fat globule almost exclusively consists of neutral lipids, while the fat globule membrane is composed of complex lipids and proteins. The amphiphilic properties of these complex lipids and proteins facilitate the creation of interfaces and help keep the fat in the dispersed state (Figure 1.1).

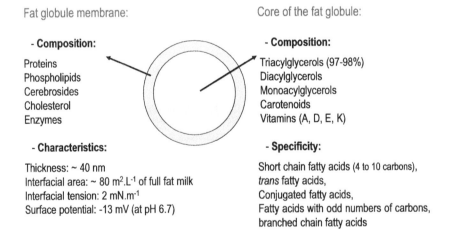

Figure 1.1. *Composition and main characteristics of milk fat globules*

1.1.1.1. Composition and characteristics of milk fat

Table 1.1 shows the average lipid composition of cow's milk. Triacylglycerols represent approximately 97.5% of the total lipids. Diacylglycerols, monoacylglycerols and free fatty acids are naturally present in small amounts but their proportion can increase with lipolysis. The many other compounds (cholesterol, steroid hormones, vitamins, flavorings and flavor substrates, etc.), even though low in number, play a crucial nutritional and sensory role.

Class of lipids	Percentage of total lipids (w/w)
Triacylglycerols	97.5
Diacylglycerols	0.36
Monoacylglycerols	0.027
Free fatty acids	0.1
Cholesterol	0.31
Hydrocarbons	Traces
Carotenoids	0.008
Phospholipids	0.6

Table 1.1. *Average lipid composition of cow's milk (Source: [CHR 95])*

Milk triacylglycerols are made up of more than 400 different fatty acids, which makes milk fat a very complex lipid source, as each fatty acid can be esterified to one of the three hydroxyl groups of glycerol (Table 1.2). However, only 12 fatty acids are present in quantities of more than 1% (mol/mol). Fatty acids are either synthesized in the secretory cells in the udder or taken from the bloodstream (body fat or food origin). Thus, milk fat varies depending on the season, the cow's diet and the energy level of the food intake, which could determine the ratio of *de novo* synthesis with regard to plasma uptake.

Fatty acids	Symbol	% mol	Distribution on the glycerol sites (% mol)			Melting point (°C)
			Sn1	Sn2	Sn3	
Butyric	4:0	4.8	-	-	35.4	-7.9
Caproic	6:0	2.2	-	0.9	12.9	-1.5
Caprylic	8:0	1.3	1.4	0.7	3.6	+16.5
Capric	10:0	2.9	1.9	3.0	6.2	+31.4
Lauric	12:0	3.3	4.9	6.2	0.6	+43.6
Myristic	14:0	10.8	9.7	17.5	6.4	+53.8
Palmitic	16:0	26.1	34.0	32.3	5.4	+62.6
Palmitoleic	16:1	1.4	2.8	3.6	1.4	-0.5
Stearic	18:0	10.8	10.3	9.5	1.2	+69.3
Oleic	18:1	24.1	30.0	18.9	23.1	+14.0
Linoleic	18:2	2.4	1.7	2.5	2.3	-5.0
Linolenic	18:3	1.1	-	-	-	-11.0

Table 1.2. *Fatty acid composition of milk and distribution on the three positions of glycerol (adapted from [CHR 95])*

Milk fat is characterized by:

– a high proportion of short-chain fatty acids (chain lengths of four to ten carbons) synthesized from acetate and β-hydroxybutyrate produced by microorganisms during cellulose degradation in the rumen. These fatty acids are preferentially in the Sn3 position of triacylglycerols. They are easily released by the action of microbial or milk lipases, and are actively involved in the flavor of dairy products due to their volatility at acidic pH;

– a high proportion of saturated fatty acids (with 14, 16 and 18 carbon atoms), some of which come from the hydrogenation in the rumen of unsaturated fatty acids originating from food;

– unsaturated fatty acids from either the diet or the desaturation of saturated fatty acids by Δ9-desaturase in epithelial cells;

– unsaturated fatty acids whose double bonds are in *trans* configuration and/or are conjugated resulting from the hydrogenation of fatty acids in food by microorganisms;

– the presence of bacterial fatty acids (fatty acids with odd numbers of carbons, branched-chain fatty acids).

Fatty acids determine the physical properties of fat (melting point, crystallization properties) by the length of their carbon chain, their level of unsaturation and their position on the glycerol molecule. Milk fat has a broad melting profile that varies throughout the year, mainly due to diet. At -30°C, milk fat is completely solid and at 40°C it is completely liquid. Between these two temperatures, liquid fat, located mainly in the core of the globule, and solid fat, forming a solid shell located at the periphery of the globule, coexist.

1.1.1.2. *Milk fat globule membrane*

The milk fat globule membrane (MFGM) accounts for $1 - 2\%$ (w/w) of total lipids. It is primarily composed of proteins (butyrophilin, xanthine oxidase, several enzymes, etc.), phospholipids (phosphatidylethanolamine, phosphatidylinositol, phosphatidyl-serine, phosphatidylcholine, and sphingomyelin), neutral lipids (triacylglycerol) and a small proportion of other components (cholesterol, cerebrosides, β-carotene, etc.). Its structure is closely linked to the mechanisms involved in the formation of lipid droplets in secretory cells and to their method of secretion in the alveolus of the mammary gland. It is composed of an inner layer of proteins and polar lipids from the endoplasmic reticulum, allowing the lipid fraction to be dispersed as lipid droplets in the cytoplasm of secretory cells. These lipid droplets, when secreted, are surrounded by the phospholipid bilayer membrane of secretory cells (Figure 1.2). A portion of the cytoplasm from the secretory cells can be trapped in the MFGM. The membrane is typically around 40 nm thick. The MFGM is composed of lipid rafts, consisting of rigid domains rich in sphingomyelin, which move in a continuous bilayer made of the other phospholipids. The size of the lipid rafts depends on the temperature and time of milk fat globule handling.

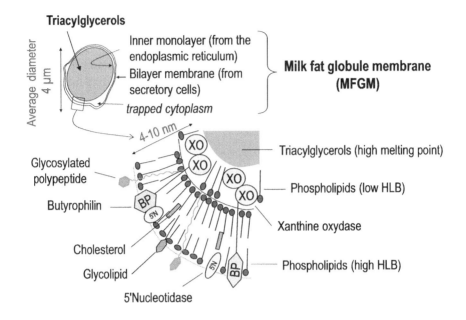

Figure 1.2. *Diagram of the structure of a native milk fat globule membrane (Source: [MIC 01]). For a color version of this figure, see www.iste.co.uk/jeantet/foodscience.zip*

Due to the composition of the MFGM, the interfacial tension between the fat phase and skimmed milk is low at around 2 mN m^{-1}, which makes it very sensitive to local perturbations. The total surface of the MFGM is around 80 m^2 L^{-1} of fresh milk. This can, however, be considerably increased during processing (agitation, homogenization, etc.). The surface electrostatic potential of the fat globule, which is close to –13 mV at the natural pH of milk, contributes to stability by limiting the risk of flocculation and coalescence.

1.1.2. *Carbohydrates*

Milk contains free carbohydrates, the main one being lactose, and carbohydrates bound to proteins. The lactose concentration in mammalian milk is inversely proportional to the mineral content, both of which contribute to the balance of osmotic pressure. The lactose content of cow's milk varies from 4.8 to 5% (w/w) and represents 97% of total carbohydrates. Lactose is a disaccharide composed of a galactose and a glucose unit (Figure 1.3). It is made from blood glucose in the presence of galactosyltransferase and α-lactalbumin. For absorption, lactose should be hydrolyzed by β-galactosidase (lactase) secreted by enterocytes in the small intestine. The low hydrolysis rate of lactose provides young mammals with prolonged energy and a constant blood glucose level between feedings. Lactase-deficient individuals cannot digest lactose as it provokes intestinal problems (diarrhoea, bloating) when ingested. Galactose and its amino derivative galactosamine contribute to the formation of several glycoproteins and/or glycolipids.

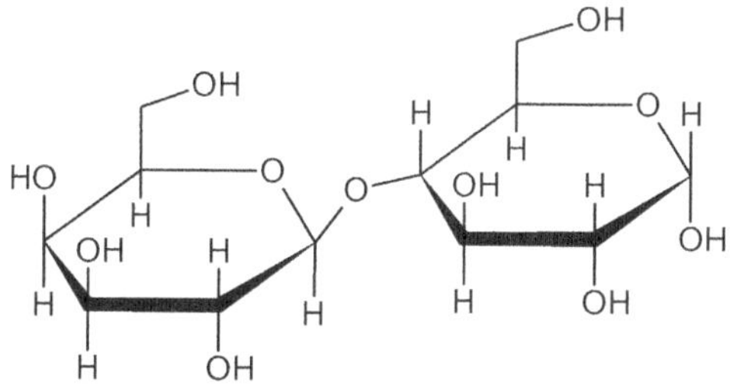

Figure 1.3. *Chemical structure of lactose*

Lactose has a low solubility (around 18 g/100 g of water at 20°C) compared to other carbohydrates: it can crystallize when concentrated in the aqueous phase of milk or derivatives (evaporation, freezing, storage in powder form). Lactose has a high melting point for a disaccharide (over 200°C). It has a low sweetness level (0.3 with reference to sucrose, which has a sweetness level of 1). Lactose has one reducing function per molecule, carried by the glucose unit. It is thus prone to non-enzymatic browning, which changes the flavor and color of foods (Maillard reaction). Enzymatic hydrolysis by β-galactosidase combats lactose intolerance, improves the sweetness of milk and doubles its reducing power, which promotes non-enzymatic browning. Lactose is the main substrate for lactic acid bacteria. The transformation of lactose to lactic acid lowers the pH of milk and destabilizes the dispersed elements, which is the basis of the production of fermented dairy products.

1.1.3. *Proteins*

Cow's milk contains 3.2 – 3.5% (w/w) protein, which can be divided into two separate fractions:

– caseins that precipitate at pH 4.6, representing 80% of total protein;

– whey proteins, soluble at pH 4.6, representing 20% of total protein.

Differential protein precipitation is used industrially in the preparation of acid casein. Casein exists as micelles comprising colloidal minerals mostly in the form of calcium phosphate as described later.

1.1.3.1. *Caseins*

Caseins (α_{S1}, α_{S2}, β, κ), present in cow's milk in proportions of 37, 10, 35 and 12%, respectively (w/w), are synthesized from four different genes. Protein diversity is increased by the presence of numerous variants resulting from genetic polymorphism and differences in post-translational modifications (phosphorylation, glycosylation).

Caseins are small proteins with a molecular weight of 19 – 25 kDa. They have a high proportion of non-uniformly distributed charged amino acids and non-polar amino acids (Table 1.3), giving them amphiphilic properties.

Due to the presence of a large number of proline residues, caseins have a low level of secondary structure (α-helices or β-sheets). Thus, caseins can withstand intense heat treatment but are very sensitive to enzymatic action, in particular digestive enzymes (pepsin and trypsin). β-casein is particularly sensitive to plasmin, an endogenous milk protease found on the surface of casein micelles. The hydrolysis of β-casein by plasmin generates hydrophilic peptides from the N-terminal fragments of β-casein and hydrophobic γ-caseins, which precipitates at pH 4.6, similar to other caseins.

Caseins are rich in lysine, an essential amino acid that in the presence of a reducing sugar is heavily involved in non-enzymatic browning. However, the large number of acidic amino acids gives caseins an isoelectric point of close to 4.6 at the ionic strength of milk.

Amino acid	Caseins			
	α_{S1} (B)	α_{S2} (A)	β (A)	κ
Asp	7	4	4	4
Glu	24	25	18	13
Asn	8	14	5	7
Gln	15	15	21	14
Thr	5	15	9	14
Ser	8	6	11	12
SerP	8	11	5	1
Pro	17	10	35	20
Gly	9	2	5	2
Ala	9	8	5	15
Val	11	14	19	11
Ile	11	11	10	13
Leu	17	13	22	8
Phe	8	6	9	4
Tyr	10	12	4	9
Met	5	4	6	2
Cys	0	0	0	0 or 2
Cystine/2	0	2	0	2 or 0
Lys	14	24	11	9
His	5	3	5	3
Arg	6	6	4	5
Trp	2	2	1	1
Total	199	207	209	169

Table 1.3. *Amino acid composition of cow's milk casein (α_{S1} (variant B), α_{S2} (variant A), β (variant A), casein κ)*

The isoelectric point of casein is closely linked to the phosphoserine content. Caseins, with the exception of κ-casein, contain a high proportion of phosphorylated serine predominantly arranged in clusters (sequence of phosphoserines in the primary structure). α_{S1}-casein mostly has eight phosphoserines. α_{s2}-casein mainly has ten to 13 phosphoserines in almost equivalent proportions. β-casein contains five phosphoserines whereas κ-casein mainly has one. However, κ-casein is the only protein that can sometimes be glycosylated. Phosphoserines, arranged in clusters, display a strong affinity for divalent or polyvalent cations, which depending on their type can make casein insoluble. The sensitivity of casein to calcium increases with the rate of phosphorylation. κ-casein does not precipitate in the presence of calcium.

Caseins have few sulfuric amino acids, which limits their nutritional value. α_{s2}- and κ-caseins each have two cysteines involved in intermolecular disulphide bonds. While α_{s2}-casein is mostly present as covalent homodimers, κ-casein forms polymers of up to 15 κ-casein units.

1.1.3.2. Structure of the casein micelle

Casein micelles are spherical particles formed by the aggregation of different caseins (α_{S1}, α_{S2}, β and κ), some peptide fragments resulting from the proteolysis of β-casein by plasmin (γ-casein) and salt components, the main ones being calcium and phosphate. Table 1.4 shows the average composition of casein micelles.

Caseins		Salt components	
α_{S1}	33	Calcium	2.9
α_{S2}	11	Magnesium	0.2
β	33	Inorganic	4.3
κ	11	Citrate	0.5
γ	4		
Total caseins	92	Total salt components	8.0

Table 1.4. *Average composition of casein micelles in % (w/w)*

The composition of casein micelles varies slightly depending on their diameter, which varies between 50 and 600 nm for an average diameter of about 150 nm. Regardless of the size of the micelle, the proportion of α_{S1} and α_{S2} casein varies marginally, whereas the ratio of β- to κ-caseins increases with the size of the micelle. Micelle organization, or the arrangement and distribution of the various micelle components and their types of association, is still a subject of intense debate. The non-charged regions of caseins form rigid structures maintained by hydrophobic associations and hydrogen bonds; colloidal calcium phosphate, in the form of nanoclusters, shields the negative charges of phosphoserine clusters and allows the association of casein micelles. κ-casein is structured into inhomogeneous clusters almost exclusively located on the micelle surface. Without phosphoserine clusters, κ-casein remains associated to the casein micelle by its hydrophobic N-terminus, but prevents further micelle growth. Its charged hydrophilic C-terminus protrudes about 5 – 10 nm into the solvent phase, making the micelle appear hairy.

Table 1.5 shows some of the properties of casein micelles. Their composition and physicochemical properties are highly dependent on the solvent phase.

Property	*Values*
Average diameter (nm)	150
Area (cm^2)	8×10^{-10}
Volume (mL)	2.1×10^{-15}
Density (hydrated)	1.0632
Hydration (g H_2O g^{-1} of proteins)	3.7
Voluminosity (mL g^{-1} of proteins)	4.4
Molecular weight (hydrated) (Da)	1.3×10^9
Molecular weight (dehydrated) (Da)	5×10^8
Water content (%, w/w)	63
Number of caseins per micelle	2×10^4
Number of nanoclusters of calcium phosphate per micelle	3×10^3
Number of micelles per L of milk	$10^{17} - 10^{19}$
Mean free distance between micelles (nm)	240
Zeta potential (mV)	−13

Table 1.5. *Average physicochemical properties of casein micelles at 20°C and pH 6.7 (modified from [MCM 84])*

1.1.3.3. Whey proteins

Whey proteins are defined as the protein fraction that remains soluble at pH 4.6. β-lactoglobulin, α-lactalbumin, bovine serum albumin (BSA), immunoglobulins and lactoferrin represent more than 90% of all whey proteins. They are mostly globular proteins with a high sensitivity to heat treatment. They are generally rich in sulfuric amino acids and tryptophan residues making them highly nutritious.

β-lactoglobulin has a molecular weight of 18.3 kDa and its concentration in cow's milk ranges from 0.2 to 0.4% (w/w). Its biological function is still unknown. There are several genetic variants of β-lactoglobulin, but types A and B are the most common. Its secondary structure consists primarily of two perpendicular β-sheets forming a central hydrophobic cavity held in place by two disulphide bridges and partially closed by an α-helix. The cavity can hold a small hydrophobic molecule, which can be a fatty acid, retinol or an aromatic molecule. In addition, β-lactoglobulin has a free cysteine residue naturally buried in the protein core, which upon input of energy (e.g. heat) is exposed to the solvent and can initiate intermolecular exchange reactions. β-lactoglobulin has a pI of 5.2 and its quaternary structure varies depending on pH. Under physiological conditions (pH 6.8), β-lactoglobulin exists mainly in the form of non-covalent dimers.

α-lactalbumin has a molecular weight of 14.1 kDa and a pI of 4.5. Its concentration in cow's milk ranges from 0.1 to 0.15% (w/w). The secondary structure of α-lactalbumin consists of four α-helices and a β-sheet; its tertiary structure is stabilized by four disulphide bridges and the presence of one calcium ion at a specific site on the protein. The affinity of α-lactalbumin for calcium and its conformation are highly dependent on pH. A drop in pH below 4 induces protonation of carboxylic groups involved in the coordination of calcium, which results in the release of calcium. α-lactalbumin contributes to the regulation of galactosyltransferase activity in the synthesis of lactose.

BSA is present in cow's milk at a concentration of between 0.01 and 0.04% (w/w). Its molecular weight is 66 kDa and it has the distinction of having 35 cysteine residues, 34 of which are involved in intramolecular disulphide bridges. It has an ellipsoidal shape and its surface is comprised of hydrophobic pockets allowing the attachment of long-chain fatty acids.

Immunoglobulins are present in cow's milk at a concentration of 0.06 – 0.1% (w/w). Their pI is within a pH range of 5 – 8. They are glycoproteins derived from blood and have antibody properties. They are synthesized in response to stimulation by antigens. Immunoglobulins are comprised of two types of polypeptide chains, a light chain with a molecular weight of about 28 kDa and a heavy chain of about 50 – 70 kDa. The basic structure of immunoglobulins, the molecular weight of which is close to 160 kDa, consists of four subunits linked by disulphide bonds. Each subunit differs in its amino acid sequence at the N-terminus, which gives the subunits immunological specificity.

Lactoferrin has a molecular weight of 75 kDa and a pI of 8.5. Its tertiary structure is stabilized by 16 disulphide bridges. It has two free cysteine residues. A molecule of lactoferrin has the ability to bind two iron ions in the presence of a synergistic anion (carbonate under physiological conditions). Affinity to iron is high at neutral or basic pH (stabilization of iron in basic medium), but iron is quickly released in acid medium. Its iron chelating property gives lactoferrin antimicrobial activity.

1.1.4. Milk minerals

Although the mineral fraction of milk is relatively small, it is very important from a structural, nutritional and technological point of view. Calcium phosphate nanoclusters, associated with casein phosphoserins α_{S1}, α_{S2} and β, contribute to the structure and stability of casein micelles (see section 1.1.3.2). The solubilization of colloidal calcium phosphate in the presence of a calcium complexing agent such as EDTA (ethylene diamine tetra-acteic acid) results in disintegration of the micelle. Milk and milk derivatives are the main supply of calcium and phosphorus in the diet. In cheese-making, the rheological properties of cheese strongly depend on the retention of these elements in the curd. Table 1.6 shows the average concentrations of the main minerals in cow's milk.

Minerals	Concentration ($mg\ kg^{-1}$)	Concentration ($mmol\ kg^{-1}$)
Calcium	1,043–1,283	26–32
Magnesium	97–146	4–6
Inorganic phosphate	1,805–2,185	19–23
– (total phosphorus)	*930–992*	*30–32*
Citrate	1,323–2,079	7–11
Sodium	391–644	17–28
Potassium	1,212–1,681	31–43
Chloride	772–1,207	22–34

Table 1.6. *Mineral composition of cow's milk (according to [GAU 05])*

Milk also contains many trace elements. The concentration of mineral elements is not influenced by diet, even though differences are observed for citrate. However, there are more variations during lactation or with pathological conditions (mastitis).

Mineral elements are distributed differently between the soluble phase and the colloidal phase, depending on their respective affinities for proteins and organic solutes. Monovalent ions (sodium, chloride, potassium) are found exclusively in the soluble phase of milk, while divalent or polyvalent ions are distributed between both phases. The mineral balance between the colloidal phase and the soluble phase is rather complex, since many different types of minerals are involved. Figure 1.4 shows the main mineral balances of milk.

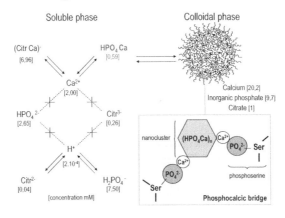

Figure 1.4. *Main mineral balances of milk (salt concentrations shown are for milk under physiological conditions)*

Calcium phosphate is poorly soluble and is saturated in the soluble phase of milk (0.59 mM) over a wide pH range. In milk at pH 6.7, the natural content of calcium phosphate is far greater than its solubility limit. Micelles increase the solubility of calcium phosphate by the integration of calcium phosphate nanoclusters with a core-shell structure: calcium phosphate clusters stabilized by the phosphorylated caseins α_{S1}, α_{S2} and β. In addition, a fraction of calcium is directly linked to casein phosphoserines. Under physiological conditions, approximately two-thirds of calcium and half of the inorganic phosphate are associated with micelles and are in equilibrium with the serum phase. Any physicochemical change in milk will affect the concentration of minerals in the soluble phase of milk, causing a shift in the mineral balance and an alteration in the structure and stability of micelles (see section 1.2.2).

1.2. Biological and physicochemical aspects of milk processing

1.2.1. *The stability of fat globules*

1.2.1.1. *Native fat globules*

Native fat globules have a natural tendency to rise to the surface (creaming), the rate of which depends on globule size, the temperature affecting the viscosity of the continuous phase, the difference in density between the continuous and dispersed phases and the gravitational acceleration (Stokes' law, equation [1.1], Volume 2). These characteristics are taken into account in the preparation of milk and cream for consumption. Flocculation and the coalescence of fat globules accelerate creaming by increasing particle size. In milk, flocculation or agglutination of fat globules is mainly due to the IgM class of immunoglobulins that result in the formation of aggregates of up to 1mm (up to 10^6 fat globules). These aggregates form at cold temperatures and can be separated by stirring or reheating milk above 37°C. Fat globules of heat-treated milk are less prone to agglutination due to IgM denaturation. Coalescence of fat globules does not occur in milk emulsions due to the electrostatic barrier generated by the charge of the native MFGM (surface potential of –13mV at the natural pH of milk) and the steric barrier formed by hydrophilic carbohydrate chains of glycoproteins in the membrane. Lowering pH that reduces the MFGM surface potential or increasing the ionic strength that reduces the thickness of the electrical double layer, decreases emulsion stability by promoting flocculation and the coalescence of fat globules. In addition, coalescence of fat globules can be obtained by the vigorous stirring of cream (churning) or by a series of freeze–thaw cycles, which have a destabilizing effect on the membrane.

Partial coalescence occurs when fat globules, the fat of which is partially crystallized, aggregate but keep their shape after contact despite perforation of their membrane by lipid crystals. The mechanical rigidity provided by fat crystals on the surface of the fat globules prevents complete fusion. Partial coalescence is favored by the low interfacial tension and low viscoelasticity of the native MFGM.

Fats are also susceptible to biological (lipolysis) or chemical (oxidation) degradation. In fresh cow's milk, lipolysis and oxidation of fat are virtually non-existent despite the natural presence of lipoprotein lipase (lipolysis catalyst), oxygen and oxidation catalysts dissolved in the non-fat phase. The native MFGM, although relatively weak due to low interfacial tension (around

2 mN m^{-1}) and the large radius of curvature, forms a protective layer against such reactions. However, any change to the native MFGM increases the risk of lipolysis and oxidation of milk fat.

1.2.1.2. Homogenized fat globules

Changing the physicochemical characteristics of an emulsion affects its properties. Homogenization, for example, improves the physical stability of emulsions by reducing the average diameter of the fat globules, and thus lowers the rate of creaming. The membrane becomes thicker and more viscoelastic due to the adsorption of casein micelles and whey proteins to the newly-formed interface (Figure 1.5); this limits the possibility of penetration of fat crystals and therefore reduces the risk of partial coalescence of the fat globules. However, the strong increase in the interfacial area and the change in the nature of the membrane due to homogenization alter its protective properties against oxidation and lipolysis.

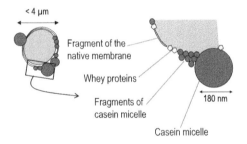

Figure 1.5. *Structure of the milk fat globule membrane after homogenization. For a color version of this figure, see www.iste.co.uk/jeantet/foodscience.zip*

In addition, homogenization changes the color of milk emulsions as well as the participation of fat globules in the formation of coagulum (cheese, yoghurt). Native fat globules cannot participate in the formation of the protein network. Moreover, if the diameter of the fat globules is greater than 1 μm, they even hinder the network formation. On the other hand, homogenized fat globules are involved in the formation of the protein network via the casein micelles incorporated into the interface created during homogenization. In low-fat products, homogenization (one stage) can be a means to increase the viscosity of dairy emulsions; in such systems, linear aggregates of flocculated fat globules are formed. Homogenization also causes an increase in the interfacial tension between the lipid and aqueous phases, which, together with a reduction in fat globule size makes the interface more resistant to mechanical processing and phase inversion. Greater stability of fat globules resulting from changes to the interface may have an adverse effect on the rheological, sensory and culinary properties of cheese (altered melting properties after homogenization).

1.2.2. Protein stability

Casein micelles and whey proteins differ in their resistance and technological stability during the processing of milk. Maintaining the micellar structure primarily depends on colloidal calcium phosphate, which acts within the micelles. On the other hand, the C-terminal part of κ-casein forms highly hydrated and negatively-charged protrusions on the micelle structure, which hinder casein micelle self-association. Many technological processes adversely affect stability by modifying the physicochemical properties of the micelle surface. Figure 1.6 shows the impact of the main physicochemical factors on the structure and stability of casein micelles.

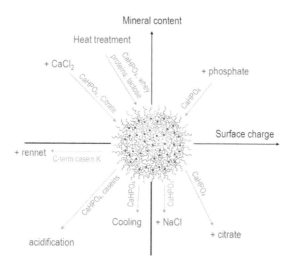

Figure 1.6. *Impact of the main physicochemical factors on casein micelle structure and stability*

The stability of whey proteins is governed by a set of low-energy bonds (hydrogen bonds, hydrophobic interactions, electrostatic bonds, salt bridges) and covalent bonds (disulphide bridges).

1.2.2.1. *Effect of temperature*

Temperature affects the solubility of calcium phosphate (inverse solubility salt) as well as the state of association of milk proteins. Both refrigeration and heat treatment alter the technological properties of the casein micelle, but the underlying mechanisms are different.

Partial solubilization of colloidal calcium phosphate (about 10%) during the cooling of milk (4°C) is reversible upon heating. In addition, the disassociation of β-casein from the micellar structure occurs at low temperature due to a reduction in hydrophobic interactions; once in the soluble phase, it can be hydrolyzed by plasmin (endogenous milk enzyme), which results in a decrease in cheese yield (see section 1.3.4). β-casein and/or hydrophobic fragments resulting from its hydrolysis by plasmin associate with the micelles during the heating of refrigerated milk. As a consequence, the rennet coagulation of such milk is altered. Indeed, it is likely that β-casein, mostly located in the center of the native micelle, moves on its surface during the refrigeration–heating cycle of milk. Its presence on the micelle surface could reduce chymosin site accessibility on κ casein.

Unlike whey proteins, casein micelles are relatively stable to heat treatment. Heat treatment decreases the solubility of calcium phosphate, which either insolubilizes inside the casein micelle or precipitates on the exchanger surface. The latter fraction is not recovered during the cooling of milk. If the heat treatment is below 95°C for a few seconds, the calcium phosphate insolubilized inside the micelle remains in equilibrium with the soluble phase of milk and resolubilizes during cooling. For more intense heat treatment (e.g. 120°C for 20 minutes), irreversible changes take place in the structure and distribution of salts between the micelle and the soluble fraction. At temperatures above 70°C, whey proteins denature and can interact with each other in the soluble phase of milk (formation of soluble aggregates) or with κ-casein (formation of stable aggregates on the micelle surface). The distribution of aggregates between the soluble phase or the micelle surface depends on pH and determines the heat stability of milk. The heat treatment of milk with a pH above 6.7 promotes the release of κ-casein, which decreases micelle stability. When heat treatment is carried out at a pH below 6.6, a large proportion of whey proteins remain associated with the casein micelle. Thus, the stability of heat-treated milk is greatest when the heat treatment is carried out between pH 6.6 and 6.7. Aggregation of whey proteins on the casein micelle surface makes them stable to chymosin hydrolysis by masking the cleavage site on κ-casein.

In addition, heat treatment applied to milk (e.g. 95°C for a few minutes) has a positive effect on the texture of the gels obtained after slow acidification (yoghurt).

On another level, the interaction of lactose with proteins during heat treatment (Maillard reaction) may alter their functional characteristics.

1.2.2.2. Effect of concentration

The concentration of milk by evaporation increases the colloidal calcium phosphate content of casein micelles. It also increases ionic strength and decreases the pH of milk, resulting in the shielding of the negative charges on the C-terminal portion of κ-casein. In contrast, the concentration of milk by ultrafiltration does not alter the mineral concentration of the soluble phase and therefore does not affect the structure and stability of casein micelles.

1.2.2.3. Effect of ionic environment

Calcium (generally calcium chloride) is widely used in cheese technology to offset the adverse effects of heat treatment and to improve the rheological properties of curd. It induces major changes in the distribution of salts between the soluble and the colloidal phase. It leads to the formation of calcium phosphate ($CaHPO_4$), which, given its low solubility, mainly insolubilizes inside casein micelles. In addition, some of the calcium ions reduce the zeta potential of the micelle and its thermal stability. At the same time, they cause a decrease in the level of hydration in the micelle.

The addition of sodium chloride causes an increase in ionic strength and a decrease in the activity coefficient of ions in the soluble phase. This results in the solubilization of colloidal calcium phosphate. Hydration of the casein micelles increases without any change in its size and its surface potential.

Citrate is a commonly used complexing agent of calcium. Its addition to milk causes a shift in equilibrium, which results in a dissociation of calcium phosphate and the solubilization of colloidal calcium phosphate. Depending on the amount added, citrate can cause the disintegration of the casein micelles and the release of free caseins. Unlike citrate, phosphate addition increases the calcium phosphate content of the casein micelles. By reducing the amount of ionic calcium, phosphate and citrate increases the thermal stability of milk.

1.2.2.4. *Effect of acidification*

The acidification of milk causes major physicochemical changes to both the casein micelle and serum. Rapid acidification of milk (concentrated organic or inorganic acid) causes destabilization of the casein micelle surface and flocculation of casein micelles in the form of a precipitate of varying granular size dispersed in whey. Slow acidification (lactic acid bacteria, glucono-delta-lactone) causes a greater rearrangement of casein micelles leading to the formation of a homogeneous gel throughout the entire milk volume. During slow acidification (Figure 1.7), the surface potential of casein micelles decreases gradually. At the same time, the protonation of citrate and phosphate causes the dissociation of soluble calcium salts (mainly calcium phosphate and calcium citrate) and a shift in the mineral balance of milk resulting in the solubilization of colloidal calcium phosphate and in the release of some caseins from the casein micelle. Up to a pH of 5.4, the solubilization of colloidal calcium phosphate has little impact on the organization of the micelle. At a pH below 5.4, the release of calcium bound to phosphoserines causes a gradual disintegration of the micelle, which loses its spherical shape. In addition, the amount of soluble caseins (mostly β casein) reaches a maximum between pH 5.5 and 5.2 (10 – 30%, depending on temperature). When the surface charge of the micelles is zero (pH 5.2), their distribution, homogenous until then, becomes inhomogeneous. The disintegrated casein micelles form aggregates of a few μm dispersed in the whey, which are progressively connected by the solubilized caseins. This results in the formation of a gel network containing the entire aqueous phase, which contracts continuously when the pH decreases from pH 5.0 to approximately 4.4 [HEE 85].

1.2.2.5. *Effect of renneting*

Rennet, a mixture of chymosin and pepsin, is the coagulating enzyme of casein and is widely used in cheese technology. The destabilization of the casein micelle by rennet resulting in the formation of a gel can be divided into three stages (Figure 1.8):

– enzymatic hydrolysis of κ-casein;

– aggregation of hydrolyzed casein micelles;

– reorganization of the aggregated casein micelles and formation of a gel network.

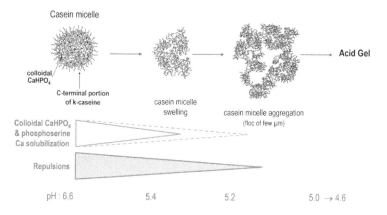

Figure 1.7. *Change in micelle structure during acidification*

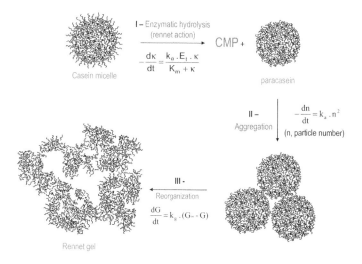

Figure 1.8. *Change in micelle structure during rennet coagulation (CMP = caseinomacropeptide)*

Hydrolysis of the Phe_{105}–Met_{106} bond in κ-casein is accompanied by the release of the hydrophilic and negatively-charged C-terminal segment (caseinomacropeptide CMP) in whey, whereas the hydrophobic and basic N-terminal remains associated with the micelle. The release of CMP destabilizes the colloidal complex by reducing the surface potential of casein micelles. The rate of hydrolysis is considered to be Michaelis–Menten type kinetics (see Volume 2, Chapter 7) in which the Michaelis constant (K_m) is much greater than the concentration of κ casein (κ):

$$-\frac{d\kappa}{dt} = \frac{k_\theta E_t \kappa}{K_m + \kappa} \quad [1.1]$$

with $-\frac{d\kappa}{dt}$ being the rate of hydrolysis of κ casein, k_θ (s^{-1}) the rate constant of the enzymatic reaction and E_t the total enzyme concentration.

When the repulsive forces (electrostatic and steric) responsible for colloidal stability are neutralized, which is achieved at 80% hydrolyzed κ-casein, close or adjacent casein micelles aggregate (second-order kinetics):

$$\frac{dn}{dt} = -k_a n^2 \quad [1.2]$$

where n denotes the number of casein micelles at time t, and k_a the aggregation constant.

The aggregation of destabilized casein micelles is driven by electrostatic interactions between oppositely-charged residues, hydrophobic bonds and presumably calcium bridges. The rate of aggregation quickly increases with the increasing rate of hydrolysis of κ-casein between 80 and 100% through the rapid increase in the aggregation constant (k_a):

$$k_a = k_{a0}\, e^{\frac{\psi(1-\chi)}{kT}} \quad [1.3]$$

where k_{a0} is the aggregation rate constant of uncharged particles (mol^{-1} s^{-1}), ψ is the repulsion potential energy between casein micelles (J), χ is the rate of hydrolysis (%), k is the Boltzman constant (J K^{-1}) and T is temperature (K).

During aggregation, the equilibration of soluble calcium with colloidal calcium phosphate leads to a major reorganization of casein micelles and the formation of a gel. The rate of change of the rheological properties of the gel is highly dependent on the concentration and the availability of calcium.

1.3. Dairy product technology

The processing of milk into dairy products is based on the influence of biochemical (composition), physicochemical (pH, ionic strength, intensity [time/temperature] of the heat treatment) and biological factors (enzyme or flora action) on the stability of milk. A distinction can be made between:

– products for which a high level of biological and physicochemical stability is desired (liquid milk, milk powder);

– products, such as fermented milks, which associate the physicochemical destabilization of the milk during acid coagulation and its biological stability up to consumption (risk associated with the presence of pathogens, exudative phenomena and post-acidification);

– products resulting from the separation and concentration of all or part of the more valuable fractions of milk (protein and/or fat) by exploiting their instability (butter, cheese).

1.3.1. *Liquid milk*

The changes in technological processing, preservation techniques and distribution have allowed the development of a wide range of liquid milk (i.e. drinking milk) that differs in its composition, nutritional and sensory quality and shelf-life. Global market trends show a strong decrease in the consumption of whole milk (3.6% fat (w/w)) in favor of semi-skimmed milk (1.5 – 1.8% fat (w/w)), skimmed milk (less than 0.3% fat (w/w)) and "special" milks (infant formula milk, milk fortified with vitamins, calcium, phosphorus, magnesium and/or fiber, organic milk, growth milk, flavored milk, lactose-free milk, etc.).

Milk for human consumption can be currently classified into three categories:

– untreated raw milk;

– heat-treated milk;

– microfiltered milk.

These milks are only subject to physical treatment such as fat and/or protein, mineral and vitamin standardization, homogenization to avoid creaming, and heating or cross-flow microfiltration to reduce microorganisms.

1.3.1.1. Raw milk

The production and sale of raw milk must be highly controlled due to potential health risks. Milk should come from:

– registered healthy animals free from brucellosis and tuberculosis;

– registered farms, subject to strict veterinary control;

– a process (milking, packaging, storage) that is carried out under good hygienic conditions.

Authorities specify the conditions of production and the microbiological quality standards of raw milk.

1.3.1.2. Heat-treated milk

Depending on the intensity of the heat treatment (see Chapter 4, Volume 2), a distinction can be made between:

– pasteurized milk;

– long-life milk.

Pasteurized milk

Pasteurization is used to destroy all pathogenic microorganisms in milk (Figure 1.9). The destruction of tubercle bacillus is often taken as a reference for the choice of pasteurization level. Pasteurization levels are defined by equivalent temperature/time relationships based on a z value of 5°C: the time is reduced by a factor of 10 for a temperature increase of 5°C.

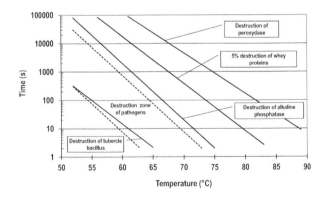

Figure 1.9. *Time–temperature diagram of pasteurization*

Two types of heat treatment are generally used for milk:

– *High temperature short time (HTST) pasteurization* (71–72°C/15–40 s): This is used for high-quality raw milk. From a sensory and nutritional perspective, high pasteurization has little impact: alkaline phosphatase is inhibited but peroxidase remains active. The use-by date of HTST pasteurized milk is 7 days after packaging (glass bottle, carton, polyethylene or aluminium container).

– *Flash pasteurization* (85 – 90°C/1 – 2 s): This is used for poor quality raw milk. Phosphatase and peroxidase are inhibited.

Long-life milk

Long-life milk has undergone sterilization, the purpose of which is to destroy all microorganisms; in return, the sensory and nutritional quality is altered compared to pasteurized milk. Sterilization levels are defined based on a 12 decimal reduction of *Clostridium botulinum*. In long-life milk, shelf-life is limited by slow time-dependent physicochemical changes in the product (precipitation, gelation, etc.).

Sterilized milk

Milk is pre-sterilized (135–150°C/3–10 s) after homogenization (in the case of milk containing fat). It is then cooled to 70–80°C and bottled (high-density polyethylene) before undergoing a second sterilization (115°C/15–20 min) followed by rapid cooling. This has a negative impact on color and flavor due to the Maillard reaction. The shelf life is around 150 days. In order to prevent lipid oxidation, such milk is stored away from light, generally in opaque containers. From a nutritional perspective, such heat treatments lead to a loss of thiamine and vitamins B_{12} and B_6.

Ultra high temperature (UHT) milk

Milk is heated to 135–150°C for 1–6 s. This process helps to preserve the original nutritional and sensory qualities of the milk because the z value of the Maillard reaction is greater than that of microbial inactivation. Its shelf life is around 120 days. This limit is imposed to ensure physicochemical stability against precipitation, flocculation and gelation due to the partial proteolysis of casein by residual plasmin or heat-resistant bacterial proteases.

UHT treatment is either direct or indirect, depending on the materials used:

– in the case of direct UHT treatment, food-grade steam is injected into milk preheated to 80°C, where it condenses releasing the latent heat of evaporation. The resulting dilution is corrected during cooling by expansion of the mixture in a partial vacuum chamber;

– in the case of indirect treatment, there is no contact between the milk and the steam. The treatment is carried out with plate or tubular heat exchangers. The limiting factor of the process is the gradual fouling caused by the precipitation of protein/mineral complexes on the walls of the exchanger:

- homogenization is carried out in either the rising or the falling phase; in the latter case, it is necessary to ensure sterilization of the homogenizer,

- the intensity of heat treatments applied is related to the quantity of lactulose in UHT milk.

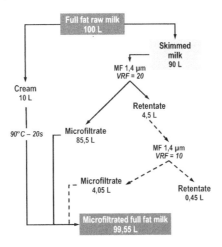

Figure 1.10. *Processing diagram for the production of microfiltered whole milk (VRF = Volume reduction factor)*

1.3.1.3. Microfiltered milk

Microfiltration (1.4 μm) is used to obtain liquid milk with the original flavor intact and a shelf life of around 21 days. Dispersed elements and microorganisms are concentrated at temperatures of around 50°C in the retentate (often called "bacterial retentate"), while all other constituents are transferred to the permeate (microfiltrate; Figure 1.10). In order to increase yield and the number of decimal reductions obtained, double filtration is generally performed to achieve a volume reduction factor (VRF) of 200. Milk is pre-skimmed and the cream is reincorporated to the microfiltrate after a specific heat treatment.

Combined with a moderate heat treatment, the shelf life of microfiltered milk can range from 35 days (treatment of 20 s at 72°C) to six months (treatment of 6 s at 96°C).

1.3.2. Fermented milk products

The bacterial conversion of lactose forms the basis of a wide variety of fermented products (yoghurt, kefir, kumis, etc.) and is one of the oldest methods used for stabilizing milk. Fermentation causes the formation of an acidic (or alcoholic) gel consisting of a network of proteins and fat globules trapped in the aqueous phase. Yoghurt is the most popular fermented milk product and is obtained exclusively by the growth of lactic acid bacteria *Streptococcus salivarius* subsp *thermophilus* and *Lactobacillus delbrueckii* subsp. *bulgaricus*, which should be inoculated simultaneously. Any products containing bacteria other than these cannot be called yoghurts but are fermented milk. In yoghurt, lactic acid bacteria should be viable, active and present in abundant quantities ($\sim 10^7$ bacteria g^{-1}); the lactic acid content must not be less than 0.7 % (w/w) in products sold to consumers [MAH 00]. Many molecules generated during fermentation, other than lactic acid, contribute to the sensory (diacetyl, acetaldehyde, etc.) and health (bioactive peptides, β-galactosidase, etc.) qualities of fermented products.

There are two types of yoghurts: set and stirred yoghurt. In the case of set yoghurt, fermentation occurs directly in the container; these are usually natural or flavored yoghurts. In the case of stirred yoghurt, fermentation occurs in tanks prior to stirring, smoothing (up to total liquefaction of the gel in the case of drinking yoghurts) and packaging; these are generally smooth natural or fruit yoghurts.

1.3.2.1. Standardization of milk

The standardization of milk in the production of yoghurt helps to achieve the qualitative requirements of the finished product. It mainly concerns total solids as well as the protein and fat content. Total solids are generally higher for set yoghurts than for stirred yoghurts. Enriching milk with proteins (to around 5 g per kg^{-1}) contributes to the firmness of the gel and prevents the risk of phase separation. This is achieved either by the addition of powder (skimmed milk powder, whey protein concentrate powder), evaporation or membrane technology (ultrafiltration, reverse osmosis).

In addition, carbohydrates such as sucrose or glucose are often added to sweetened or fruit yoghurts. Polysaccharides (pectin, xanthan, etc.) can also be used as stabilizers in fruit yoghurts.

1.3.2.2. Homogenization

The homogenization of milk used for fermentation has a number of objectives: it improves the firmness of the gels obtained after fermentation, increases their water retention capacity and reduces syneresis. It also prevents creaming during yoghurt production, in particular during the static incubation period in containers or fermentation tanks. Homogenization is usually carried out in the rising phase of pasteurization at a pressure of around 20 MPa and a temperature between 60 and 90°C. During homogenization, the lipid interface is covered with proteins (casein micelles, whey proteins). The protein coating of homogenized fat globules is involved in the formation of the protein network during acidification [LUC 98].

1.3.2.3. Heat treatment

By modifying the physicochemical properties of proteins, the heat treatment of milk (around 90°C/10 minutes) has a significant impact on the rheological properties of lactic gels. Through heat denaturation, whey proteins (more than 90%) form soluble covalent aggregates or aggregates bound to κ-casein on the surface of casein micelles. By changing the micelle surface, heat treatment causes an increase in the pH of acid gelation of milk, an increase in gel firmness and a reduction in its syneresis (Figure 1.11). Furthermore, heat treatment creates a favorable environment for the growth of lactic acid bacteria by destroying undesirable microorganisms and potential competitors to lactic acid bacteria, lowering the redox potential, contributing to the production of formic acid, and so forth.

1.3.2.4. Fermentation

After heat treatment, milk is cooled to between 40 and 45°C and inoculated with starter culture, resulting in acidification in either a tank or individual containers. In the case of yoghurt, starter cultures include *Streptococcus thermophilus* and *Lactobacillus bulgaricus*. They grow synergistically

(Figure 1.12) and can be differentiated by their optimum growth temperature, but also by their acidifying capacity and flavor production. Thus, the proportion of strains added during inoculation and the incubation temperature determine the sensory properties of the products. In addition, some strains release exopolysaccharides into the medium, which affect the rheological properties of the gel.

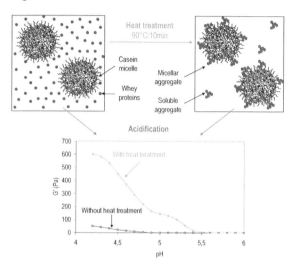

Figure 1.11. *Diagram of the influence of heat treatment on the protein constituents of milk and rheological properties (G') of the gels obtained*

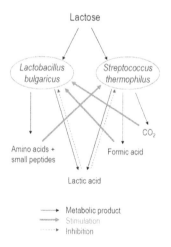

Figure 1.12. *Synergistic action of starter cultures in yoghurt*

During acidification, casein micelles covered with whey protein aggregates are destabilized and begin to associate when the pH of the medium drops below 5.5. This results in molecular rearrangement, leading to the formation of a gelled protein network that includes homogenized fat globules. The firmness of the network increases with the degree of acidification [TAM 99]. When the pH reaches 4.6, yoghurt is cooled to around 5°C in order to control the metabolic activity of the starter cultures. While set yoghurts (container fermentation) are cooled to 5°C in a single stage, stirred yoghurts (tank fermentation) are cooled in two stages. In the first stage, carried out in a plate heat exchanger, the yoghurt is cooled to 15–20°C. After stirring and smoothing, the yoghurt is then poured into containers and cooled to 5°C.

1.3.3. *Milk powder*

1.3.3.1. Drying of milk

After bactofugation to eliminate dispersed elements (butyric acid bacteria spores, casein fines), whole, skimmed or standardized milk is heat-treated before drying; it can also undergo different concentration operations (microfiltration, ultrafiltration, nanofiltration) that modify the ratio between milk components. Some components (polysaccharide, minerals, vitamins, etc.) can be dispersed in milk. After standardization, the liquid is homogenized, concentrated by vacuum evaporation and finally spray-dried or drum-dried (see Volume 2).

The concentration of milk and its derivatives by vacuum evaporation (the process of removing water by boiling) is based on lowering the boiling point of the liquid (and therefore the processing temperature) by reducing pressure. Vacuum is used for two main reasons: on the one hand, the temperature difference between the dairy product to be concentrated and the heating surface of the falling film evaporator is greater for a given heating steam pressure, which can reduce steam consumption by increasing the evaporation capacity and/or using more effects; on the other hand, it can evaporate heat-sensitive solutions. The most common apparatus in the dairy industry is the multiple effect evaporator, which incorporates a falling film evaporator equipped with thermal vapor recompression and mechanical vapor recompression systems. The energy cost of removing 1 ton of water is between 360 and 1080 kWh. The maximum boiling at the beginning of the cycle (first effect) is normally less than 70°C, corresponding to an absolute pressure of 30,664 Pa. The evaporation capacity of industrial evaporators varies from 10 to 30 tons h^{-1}. A concentration cycle lasts between 10 and 20 h. Theoretically, the average residence time of the product in an industrial evaporator is between 10 and 20 minutes: fouling of the evaporation tubes due to the precipitation of calcium phosphate causes a gradual increase in temperature throughout the entire evaporation unit of 10 – 15°C.

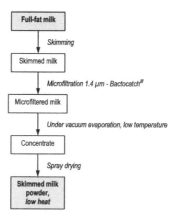

Figure 1.13. *Processing diagram for the production of low-heat skimmed milk powder*

The concentrate obtained by vacuum evaporation can be dried using various drying techniques, which are differentiated according to energy cost and powder quality; the most common method is spray drying, which involves spraying the product (liquid or suspension) into a hot gas stream so as to obtain a powder almost instantaneously. It is thus a form of entrainment drying where air acts both as a heater and a carrier for the water removed from the concentrate: the air is dry and hot when entering the drying tower and moist and cooler when it leaves. In Europe, industrial drying systems usually have an evaporation capacity of 0.5 – 4.5 tons h^{-1}, requiring air flows of 1×10^4 to 12×10^4 m^3 h^{-1}, and production cycles varying from 4 – 24 h on average. Operating a skimmed milk drying facility involves the control of many thermodynamic, physical and technological parameters, such as incoming and outgoing air temperatures (180 – 280°C and 80 – 90°C respectively), velocity and relative humidity of the air (5 – 10% for outgoing air), temperature and viscosity of the concentrate, spray nozzle type (high pressure, two-fluid or turbine), type of tower (1–3 stages) and so on. Under these conditions, it is possible to obtain a skimmed milk powder with a_w of 0.2 (25°C) and 4% residual moisture. Figures 1.13 and 1.14 are diagrams showing the production of low-heat skimmed milk powder and the production of reduced whey protein content skimmed milk powder prepared by combining microfiltration and ultrafiltration.

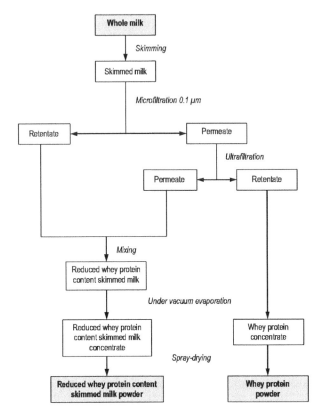

Figure 1.14. *Processing diagram for the production of skimmed milk powder by microfiltration and ultrafiltration (Primin® powder)*

1.3.3.2. *Physical properties of milk powders*

The quality of milk powder depends on many factors, including milk quality before drying and the implementation of the drying process itself. [PIS 81] and [MAS 91] classified the main properties of milk powders into two categories (Figure 1.15):

– properties inherent to the product (biochemical, microbiological, etc.);

– properties inherent to the process (functional properties and any defects).

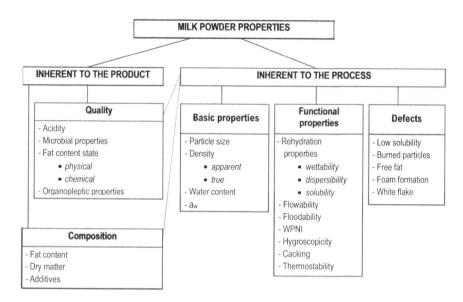

Figure 1.15. *Main properties of milk powders [PIS 81, MAS 91]*

Particle size

Particle size, determined by particle size analysis, determines several physical and functional properties (flow, density, solubility, wettability, etc.). Dry particle size is mainly influenced by the size of droplets during spraying.

Density

High-density powders can reduce transport costs. The density of milk powders is a complex property that depends on primary factors such as the true or absolute density of the product, the amount of occluded air in each particle and the amount of interstitial air between each particle. Bulk density is mainly influenced by the properties of the concentrate (dry matter, temperature, intensity of the heat treatment, foaming capacity), the drying air (thermodynamic properties at the entry and exit of the facility) and the powder (particle size, residual moisture content).

Hygroscopicity

The hygroscopicity of a powder is characterized by its final moisture content after equilibration with air of controlled relative humidity under defined temperature conditions. A powder is assumed to be non-hygroscopic if its hygroscopicity percentage is less than 10%. The hygroscopicity of a milk powder is determined by the hydrophilic nature of the components (mainly lactose and amorphous minerals). Reducing particle size, by increasing the powder surface in contact with air of controlled relative humidity, promotes water adsorption and the hygroscopicity of the powder.

Flowability and floodability

The ability of a powder to flow has a significant impact on storage, discharge, weighing, mixing, compression, transfer and so on. Carr's method [CAR 65] is used to determine two types of behavior: flowability and floodability. Flowability involves measuring the angle of repose, the angle of spatula, cohesion and compressibility. Floodability involves measuring the angle of fall, the angle of difference, dispersibility in the air and the value of the flowability index. The two main factors affecting the flowability of powders are particle size distribution and the state of the particle surface. Powders produced using spray nozzles have a higher level of flowability compared to powders produced using heated rollers. Two-stage drying also yields better flow results compared to single-stage drying. Other factors that improve flowability are powder agglomeration, a low level of fines, the addition of a flow agent (silica), the addition of hygroscopic compounds (carbohydrates, whey) and low levels of free fats.

1.3.3.3. *Technological properties of milk powders*

Rehydration properties

The ability of a milk powder to rehydrate in water is an essential property for industrial users of dried ingredients and can be characterized by three properties: wettability, dispersibility and solubility. They depend on powder composition and the affinity between these components and water, and the accessibility of water in terms of structure (porosity and capillarity) to the powder components.

Wettability, the ability of a powder to immerse itself once placed on the surface of water, reflects the capacity of powder to absorb the water on its surface. The swelling ability (swellability) of a powder is also linked to wettability. The structure of a powder disappears when the various components (in particular proteins) are dissolved or dispersed. Factors influencing wettability include:

– the presence of large primary particles, such as agglomerated particles: this is a desired effect with the granulation (with or without recycling fines) of milk powders;

– powder density;

– the presence of fat on the surface of powder particles (free fats);

– porosity and capillarity of powder particles as well as the presence of interstitial air.

Dispersibility is probably the best individual criterion for assessing the rehydration ability of a milk powder, since to a certain extent, it is influenced by wettability and solubility. Dispersibility is improved by:

– a decrease in protein content;

– an optimal particle size of 200 µm;

– drying at low temperatures (low heat powder).

The insoluble materials formed during the production of milk powder are usually due to the denaturation of soluble proteins and the precipitation of calcium phosphate. Thus, solubility is particularly influenced by heat treatment before drying, the viscosity and biochemical composition of the concentrate, the drying air temperature and the particle size of the powder.

Use of recombined milk in cheese processing

The use of recombined milk from powder is justified for several reasons: for economic, nutritional, dietary and geographical purposes, and also for sensory and technical purposes. It allows the transfer of cheese production to countries where milk production is insufficient, and where milk production has a high seasonality (in the case of goat's or sheep's milk).

Milk powders used in cheese production must have a level of microbiological quality that is in compliance with regulations and acceptable for cheese making. These factors depend on the initial quality of the milk used and the intensity of the heat treatments during processing into powder, which are the source of physicochemical changes resulting in reduced coagulation properties when milk powder is reconstituted with water. In order to meet these microbiological and technological requirements, HTST (pasteurization at 75°C for 20 s) is recommended prior to drying in order to ensure hygienic quality while maintaining a high level of coagulation. These recommendations only apply if the milk is of good microbiological quality. Otherwise, the intensity of the heat treatment must be higher, therefore compromising coagulation properties: in this case, the milk powders obtained cannot be used in the production of cheese.

Cross-flow microfiltration (1.4 μm) followed by vacuum evaporation at low temperature and spray drying is well suited for the production of an "ultra-low-heat" powder (low level of denaturation of soluble proteins); milk reconstituted from this powder, according to regulatory microbiological requirements, has the same level of rennet coagulation as the original raw milk ([SCH 94], Figure 1.16).

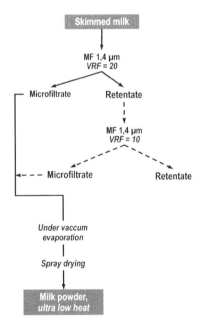

Figure 1.16. *Processing diagram for the production of ultra-low-heat skimmed milk powder (VRF = volume reduction factor)*

The technological quality of a powder intended for cheese production can be improved by reducing the soluble protein content of the milk. Native phosphocaseinate (NPC) powder can be produced using 0.1 μm cross-flow microfiltration (see section 9.1.). The rennet clotting time of NPC reconstituted to 3% from powder is reduced by 53% and the firmness of the rennet gel after 30 min is improved by 50% compared to raw milk at the identical casein concentration. Enriching milk with milk microfiltrate (0.1 µm) can significantly improve cheese yield, especially in the case of hard cheeses. In addition, the partial removal of soluble proteins that otherwise aggregate on the surface of casein micelles on heating, limits the negative effects of heat treatments on rennet coagulation. These factors have led to the development of technology for producing medium- or high-heat powder (e.g. Primin®), which has similar or even greater suitability for cheese production compared to raw milk (Figure 1.17).

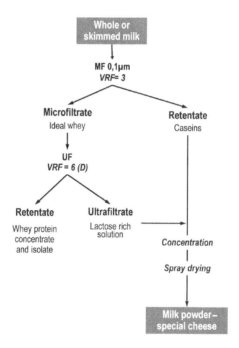

Figure 1.17. *Processing diagram for the production of Primin milk powder (VRF = volume reduction factor)*

1.3.4. Cheese

Cheese making is an ancient way of preserving milk (protein, fat and some calcium and phosphorus). Its nutritional and sensory qualities are valued in almost every part of the world.

The name "cheese" is reserved for a fermented or non-fermented, ripened or non-ripened product of exclusively dairy origin (milk, partially or fully skimmed milk, buttermilk) used alone or as a mixture. It is totally or partially coagulated before draining or after partial removal of water. Cheese may be considered a concentration of the major components of milk (protein, fat), produced by draining curd obtained by acidification and/or enzymatic action (usually rennet extracted from the stomach of a calf before weaning). Cheese production involves four phases: milk standardization, coagulation, draining and ripening (Figure 1.18).

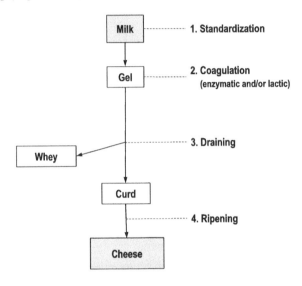

Figure 1.18. *Steps of cheese production*

The preparation of milk (standardization) for a given cheese relies on physicochemical and microbiological "standards". Transformation from the liquid state to the gel state (coagulation) differs depending on whether coagulation is induced by acidification and/or enzymatic (rennet) action. After phase separation (draining), the curd may undergo a ripening process specific to each type of cheese.

A wide variety of cheeses can be produced using traditional technologies depending on the type of milk used (cow's, goat's or sheep's milk – alone or mixed), the pH of coagulation and the relative kinetics of acidification and whey removal (draining) from the curd.

1.3.4.1. Physicochemical and biological standardization of milk

The quality of milk for cheese production can be defined by its suitability to form a coagulum resulting, after draining and eventually ripening, in a cheese with defined physicochemical properties and a satisfactory yield. Milk has a varied composition depending on the animal species, breed, individual, lactation stage and number, method and time of milking, season, climate, diet and so forth. Not all milk has the same suitability for cheese production since it differs in some characteristics such as casein content and composition, salt balance, lactose content, hygienic quality, pH and so on. These characteristics affect their ability to coagulate, which is necessary to pass from the liquid to the solid state, as well as the properties of the coagulum.

In order to avoid variations in the protein content of milk and improve coagulation properties, which affect cheese yield and quality, manufacturers are able to adjust the milk protein level to between about 30 and 42 g L^{-1} using various techniques: removal of water by evaporation or reverse osmosis, concentration by nanofiltration, ultrafiltration (most common), microfiltration or the addition of caseinates.

In order to adjust the "fat/dry matter" ratio that is specific to each type of cheese, manufacturers standardize the milk fat while taking into account the milk protein composition. Using a weight of standardized milk (w_{SM}) in terms of fat (F_{SM}) and protein (P_{SM}), and knowing the cheese yield and recovery coefficients of these constituents in the cheese (see section 1.3.5.3), it is possible to obtain a weight of cheese w_C with the desired characteristics (fat content F_C, protein content P_C; Figure 1.19).

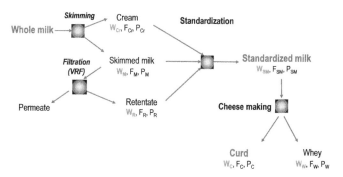

Figure 1.19. *Standardization of the fat and protein content of milk intended for cheese production*

To correct for variations in the calcium content of milk during the lactation stage or changes in the calcium balance between the soluble and colloidal phase due to the effects of refrigeration or heat treatment, manufacturers add $CaCl_2$ at a dose ranging usually from 80 to 200 mg L^{-1} of milk, which improves the coagulation properties of the milk.

To meet the processing time requirements (rennet clotting time, rate of curd formation, hardening time) and the desired mineral content in the curd, which depend on the type of cheese desired, manufacturers adjust the pH rennet added into milk by fermentation (lactic starters), the addition of glucono-δ-lactone, the injection of CO_2, or the addition of acid whey proteins.

For certain types of cheeses, the lactose content of milk is lowered by washing the curd (in medium-hard cheeses) or by ultrafiltration of the milk followed by diafiltration before coagulation (in "stabilized" soft cheeses). The partial removal of lactose slows down the activity of lactic bacteria and is a mean of controlling the pH of the curd at the end of the acidification.

Biological standardization of cheese milk consists of the elimination of the endogenous microorganisms in the milk that may be undesirable (psychrotrophs, pathogens) by heat treatment, bactofugation or microfiltration, followed by the addition of a controlled starter culture; prematuration at a low temperature (10–12°C, 12–18 h) by promoting the production of growth factors, improves the lactic fermentation process.

1.3.4.2. Coagulation

There are three types of coagulation (Figure 1.20).

Acid coagulation

Acid coagulation involves the precipitation of casein at its isoelectric point (pI = 4.6) by biological acidification using lactic acid bacteria to transform lactose to lactic acid, or by chemical acidification (injection of CO_2, addition of glucono-δ-lactone or addition of acid whey proteins). The chemical method (organic acid) is mainly used to standardize the pH of milk before renneting, while the addition of mineral acid is usually not permitted.

Gel formed by the solubilization of colloidal calcium phosphate during acidification has good permeability but high friability; the lack of structure in the network (low energy hydrophobic interactions) results in almost zero elasticity and plasticity as well as low resistance to mechanical treatment.

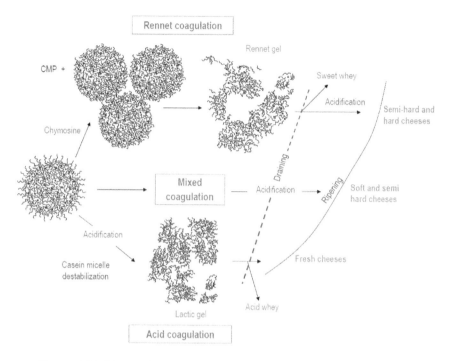

Figure 1.20. *Types of coagulation and categories of cheese. For a color version of this figure, see www.iste.co.uk/jeantet/foodscience.zip*

Enzymatic coagulation

Enzymatic coagulation is used to transform milk from a liquid to a gel state through the action of proteolytic enzymes, mostly of animal origin.

There are three phases:

– primary or enzymatic phase, corresponding to the hydrolysis of κ-casein at the bond between phenylalanine (105) and methionine (106);

– secondary phase or aggregation of hydrolyzed micelles, which, at pH 6.6, begins when 80 – 90% of the κ-casein is hydrolyzed;

– tertiary or cross-linking phase leading to gel formation.

Several factors influence coagulation, such as enzyme concentration, temperature, pH, calcium content, casein composition, micelle size and pre-treatment of milk such as cooling, heat treatment and homogenization.

The network formed at pH 6.6 is rich in minerals, given the numerous interactions between calcium and casein at this pH; this type of gel tends to contract, which leads to an expulsion of whey.

Mixed coagulation

Mixed coagulation results from the combined action of rennet and acidification. The variety of combinations, resulting in casein micelles with different mineral content when gelation occurs and whey is drawn off the curd, is the source of a wide range of soft cheeses, semi-soft cheeses and medium-hard cheeses.

1.3.4.3. Draining

Lactic acid and rennet gels

The draining stage involves removing some of the whey trapped in the gel network formed by acidification and/or enzymatic action. It begins in the coagulation tanks and continues in the moulds and finally in the cheese-ripening rooms. It is possible to express the whey flow rate based on Darcy's law (see Chapter 3, Volume 2):

$$\dot{V} = \frac{A}{R}\frac{\Delta P}{\eta} \qquad [1.4]$$

where ΔP is the differential pressure exerted on the gel (Pa), η is the viscosity of the whey (Pa s), R is the hydrodynamic resistance of the gel (m^{-1}) and A is the surface area of the gel (m^2).

As a result, whey removal depends on:

– the type of curd, which affects permeability in particular ($\frac{1}{R}$ term of [1.4]); gel porosity decreases during acidification (increase in R), but this is offset by the reduced water retention capacity of protein close to its pI;

– extent of the mechanical and thermal treatments applied on the curd in the tank, which involves slicing the curd (increase in A), subsequent stirring to avoid sticking (maintaining A) and heating (decrease in η; energy supply strengthens the protein network and contractability of the gel resulting in an increase in ΔP);

– the pressing stage after molding (increase in ΔP), which removes the remaining whey and strengthens the cohesion of the curd in the case of hard cheeses.

The kinetics of the removal of whey from the mould can be described by equation [6.9] (see Chapter 6, Volume 2). In this case, flow resistance increases due to the obstruction of the mould perforations by the curd grains, making it necessary to turn the cheese regularly so as to promote draining.

The natural drainage of a lactic gel is slow and limited. It results in a heterogeneous curd with a low dry matter and mineral content: the weakly cross-linked network contracts only slightly. Processes such as centrifugation or ultrafiltration of the curd can significantly increase drainage compared with traditional methods (strainer, bag or filter drainage). Subjecting milk and/or acid gel to intense heat treatment (80–95°C for several minutes) can increase cheese yield by denaturation and retention of whey proteins. However, heat treatment, as well as homogenization, limits the rate and intensity of drainage.

Rennet gel has strong cohesion, elasticity and porosity, but low permeability leading to limited natural drainage. As a result, it is necessary to carry out different operations in the tank (slicing, mixing, slow and steady heating up to 56°C for hard cheeses) to allow drainage of the gel. The higher the dry matter content required, the more intense these processing steps become; however, they also reduce cheese yield and the recovery coefficients of the cheese components.

Mass balance, cheese yield and recovery rate

When processing milk into cheese, it is possible to calculate the mass balance for a constituent X (protein, fat, etc.) using the following equation:

$$\begin{cases} w_M X_M = w_C X_C + w_W X_W \\ w_M = w_C + w_W \end{cases} \qquad [1.5]$$

where w_M, w_C and w_W, and X_M, X_C and X_W, respectively, are weight (kg) of milk, cheese, whey, and concentrations of the constituent X in milk, cheese and whey (g kg^{-1}). Cheese yield Y_C (dimensionless) is expressed in kg of cheese per 100 kg of milk used.

$$Y_C = \frac{w_c}{w_m} 100 \qquad [1.6]$$

By combining [1.5] and [1.6], we get

$$Y_C = \frac{X_M - X_W}{X_C - X_W} 100 \qquad [1.7]$$

For example, if 50 kg of milk with 32 g kg^{-1} of protein give 6.7 kg of cheese and 43 kg of whey with 185 g kg^{-1} and 8.5 g kg^{-1} of protein respectively, the cheese yield R_F is equal to 13.3%. To standardize the cheese yield calculation, technologists often calculate the yield of a reference cheese; this corrected yield (Y_{CC}) is:

$$Y_{CC} = \frac{X_M - X_W}{X_{ref} - X_W} 100 = Y_C \frac{X_C - X_W}{X_{ref} - X_W} 100 \qquad [1.8]$$

where X_{ref} is the concentration of constituent X in a reference cheese, representative of a given technology.

For a given manufacturing process, it is also possible to calculate the recovery rate R_X of a constituent X in cheese based on the following equation:

$$R_X = \frac{w_C X_C}{w_M X_M} 100 \qquad [1.9]$$

In the previous example (50 kg of milk with 32 g kg^{-1} of protein gives 6.7 kg of cheese with 185 g kg^{-1} of protein), the recovery rate of protein R_{prot} is therefore 77.5%.

Drainage and acidification kinetics: categories of cheese

The physicochemical properties of cheese during demolding (fat-free dry matter – FFDM, fat content, pH, moisture in non-fat substance – MNFS, calcium on a from basis – Ca/FFDM), which determine the ripening process by influencing microbial growth and biochemical and enzymatic reaction kinetics, depend on the intensity and relative position of the drainage and acidification stages (Figure 1.21).

MNFS (dimensionless), which expresses the availability of water in the curd, is calculated as follows:

$$\text{MNFS} = \frac{100 - \text{DM}_C}{100 - F_c} 100 \qquad [1.10]$$

where DM_C and F_C are the dry matter and fat content of cheese, respectively.

There are four main categories of cheese:

– *Acid curd*, such as fresh cheeses: This is high-moisture cheese; in this case acidification of the milk substrate precedes drainage. Whey is drawn off at acid pH (4.5–5), under conditions where more than 80% of calcium and phosphates are solubilized in the whey. This leads to a significant demineralization of the cheese, which accentuates its friability and crumbliness.

– *Rennet curd*, such as semi-hard and hard cheeses (cooked pressed cheese): Processing involves intense drainage after rennet coagulation. Drainage therefore precedes acidification that occurs in a lactose-depleted medium; the buffering capacity is largely due to the concentration of proteins and minerals (FFDM up to 30–35%). As a result, the pH of the cheese at the end of the acidification stage is generally between 5.2 and 5.4, and its calcium content is higher than other types of cheeses (2.9 < Ca/FFDM < 3.1%). These characteristics give an elastic and cohesive texture; the low level of MNFS results in a shelf life of several months.

– *Mixed curd with a predominantly acid nature*, such as traditional and industrial soft cheeses: These are high-moisture cheeses (MNFS around 75%), relatively acidic before ripening (pH 4.6–4.8) and depleted of minerals. The shelf life of these products is no more than a few week.

– *Mixed curd with a predominantly rennet nature*, such as semi-soft cheeses and medium-hard cheeses (stabilized soft cheese, uncooked or semi-cooked pressed cheese): There is a greater level of drainage compared with the previous category, which may involve a lactose-removal stage. The pH at the end of the acidification stage ranges from 4.8 to 5.2 and the cheese has a moderate mineral content. The shelf life is a number of weeks depending on the MNFS (60 – 72%).

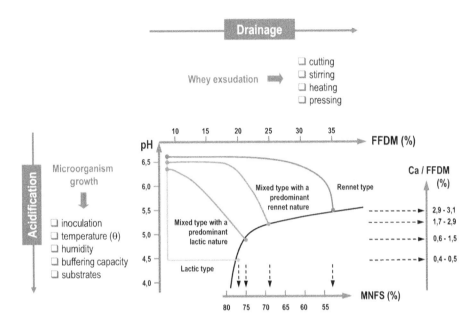

Figure 1.21. *Drainage and acidification kinetics and processing types (according to [MIE 91])*

The following Figures (1.22, 1.23, 1.24 and 1.25) give examples of processes associated with the production of acid curd (fresh cheese; Figure 1.22), mixed curd with a predominantly acid nature (industrial Camembert; Figure 1.23), mixed curd with a predominantly rennet nature (Saint-Paulin; Figure 1.24) and rennet curd (Beaufort; Figure 1.25).

Decoupling of drainage and acidification kinetics

It is possible to completely decouple drainage and acidification kinetics by concentrating milk proteins and fat by means of cross-flow filtration. Ultrafiltration (see Chapter 6, Volume 2) can concentrate all the milk proteins, whereas 0.1 µm microfiltration results in the total retention of caseins and the transmission of whey proteins to the permeate (transmission rate between 60 and 80%); as a result, 0.1 µm microfiltration is mainly carried out when the cheese manufacturer also looks for a high-quality permeate, which in this case is a "true whey" (absence of phospholipids, casein fines and any particles).

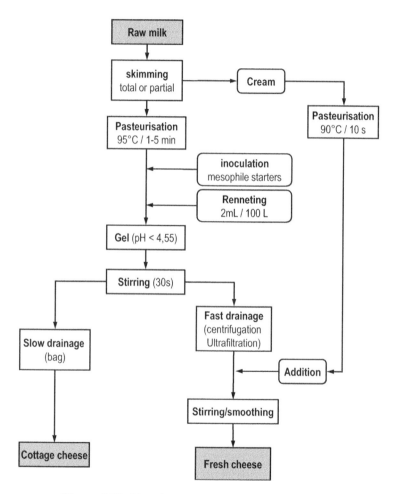

Figure 1.22. *Manufacturing process of fresh cheese*

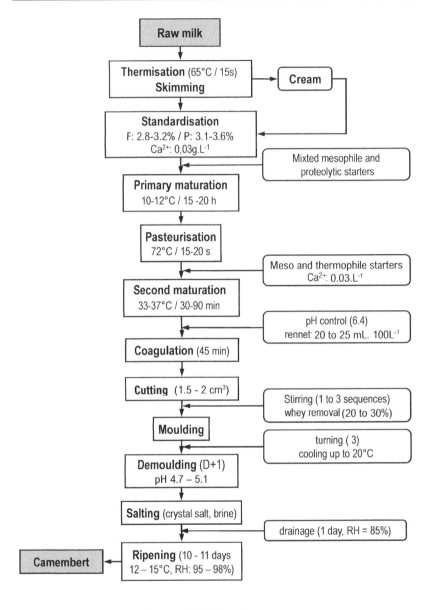

Figure 1.23. *Manufacturing process of industrial Camembert*

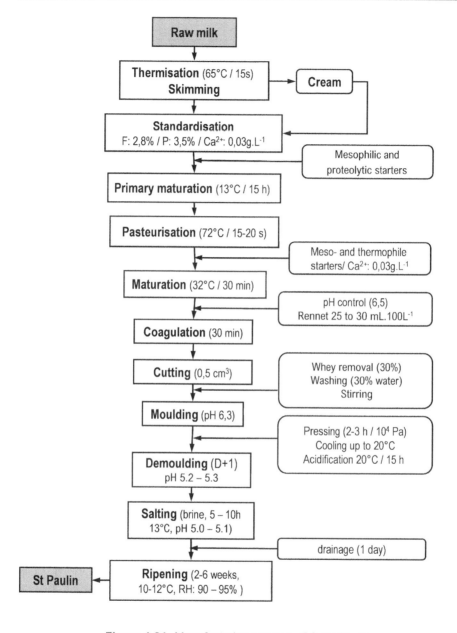

Figure 1.24. *Manufacturing process of Saint Paulin*

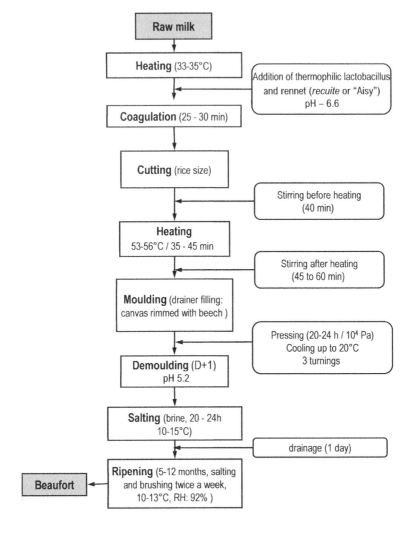

Figure 1.25. *Manufacturing process of Beaufort*

The ultrafiltration of milk results in a liquid pre-cheese, which has the composition of drained cheese [MAU 69]. This process eliminates drainage after coagulation and therefore allows, in some cases, molding directly into the containers for retail; it reduces the weight difference due to molding of the curd, increases yield (by 10 – 20% through whey protein retention in the cheese) and decreases rennet consumption (added after filtration). It has, however, the disadvantage of yielding cheeses with high lactate and lactic acid contents, due to the high buffering capacity of pre-cheeses resulting from the concentration of calcium phosphate associated with casein. To overcome these problems, it is possible to pre-acidify and salt the milk to solubilize some of the colloidal calcium phosphates, which are then removed during ultrafiltration. This method is limited, however, since pre-acidified milk is thermally less stable and it is more difficult to add value to the permeate.

1.3.4.4. Ripening

Ripening involves the enzymatic digestion of the protein and lipid components in the curd. It is a complex biochemical process for several reasons:

– the cheese matrix resulting from the coagulation and drainage of milk has a very high level of physicochemical heterogeneity;

– enzymes involved in ripening have several origins: they could be endogenous milk enzymes (plasmin, lipase, etc.), added to milk during manufacturing (coagulating enzymes, microorganisms) or produced during ripening by microbial synthesis (bacteria, yeasts, moulds).

The curd and biological agents constitute a complex ecosystem and a heterogeneous bioreactor, the parameters of which are not always well defined. Ripening is dominated by three major biochemical phenomena:

– the fermentation of residual lactose and the degradation of lactate;

– the hydrolysis of fat and proteins;

– the production of aroma from fatty acids and amino acids.

These transformations give the cheese its characteristics; they modify its appearance, composition and consistency, while at the same time flavor, aroma and texture develop.

Substrates

The physicochemical properties of the curd vary with the manufacturing process. While the composition of the curd is well defined, its structure is more complex. It is difficult to control microbial growth and enzymatic action in such a complex and heterogeneous medium.

The kinetics of ripening depends on the mobility of carbohydrates, proteins and lipid substrates, reaction products (lactate, amino acids, fatty acids) in the cheese matrix, and the rate of biological reactions, which depends on the pH of the matrix and its water availability. Reactions are generally faster when the pH of the curd is close to neutral (optimal activity pH of flora and enzymes) and the MNFS is high. The shelf life of the product depends on the buffering capacity of the cheese, which limits and regulates the increase in pH (alkalinization) of the cheese during ripening (rennet curd and mixed curd with a predominantly rennet nature).

Ripening agents

The enzymes involved in ripening have several origins: milk, coagulating agent and microorganisms in the cheese.

Milk enzymes

– *Plasmin*: heat resistant protease active in slow-ripened cooked and uncooked pressed cheeses.

– *Alkaline phosphatase*: denatured by pasteurization, it is active in raw milk cheeses only.

– *Lipase*: thermolabile enzyme active in raw milk cheeses only. It hydrolyzes short chain fatty acids in particular. Its action is more pronounced in goat's and sheep's milk because the proportion of short-chain fatty acids is higher and the fat globules are smaller than those of cow's milk.

Coagulating enzymes

Rennet (mixture of chymosin and pepsin), a coagulating agent added to milk, has a wide spectrum of proteolytic activities. Its action is dominant in uncooked pressed cheeses. The reaction products formed are mainly high-molecular-weight peptides.

Enzymes of microbial origin

These enzymes come from five main microbial groups:

– *Lactic acid bacteria*: present in the starter culture, convert lactose into lactic acid. They include:

- *lactococci*: dominant flora in uncooked soft and pressed cheeses; they produce lactic acid and exhibit proteolytic activity,

- *thermophilic streptococci and lactobacilli*: flora in cooked pressed cheeses; they exert acidifying and proteolytic activity,

– *Leuconostoc*: they produce aromatic components in addition to lactic acid and contribute to the open texture of blue cheese.

– *Propionic bacteria*: produce propionic acid from lactate, are responsible for the open texture of cooked pressed cheeses and contribute to the formation of flavor and aroma.

– *Surface bacteria*: the most common are micrococci and coryneform bacteria (*Bacterium linens*); they are present in washed-rind and smear-ripened soft cheeses. They exhibit proteolytic and lipolytic activity.

–*Yeasts*: the most common is *Geotrichum candidum*; it grows on the cheese surface by consuming lactic acid, producing ethanol and exhibiting lipolytic and proteolytic activity.

– *Molds*: the two most common are *Penicillium camemberti*, which is a surface mold on bloomy rind cheese, and *Penicillium roqueforti*, an internal mold in blue cheeses. They have the most lipolytic enzymes, are responsible for the formation of methyl ketones and secondary alcohols, and also exhibit proteolytic activity.

Influence of ripening on the flavor of cheese

The development of flavor and aroma in cheese is based on a number of changes that occur during ripening (Figure 1.26). Several components, from various classes, are involved in this process (acids, alcohols, esters, sulfur products, etc.). Most of these compounds are found in all cheeses but in varying quantities and proportions:

– in fresh cheeses, the flavor is based on acidity and acetaldehyde, which contributes to the fresh character of the cheese;

– in bloomy rind soft cheeses (Camembert), the main compounds are 1-Octen-3-ol, methyl ketones, secondary alcohols, phenolic compounds (phenylethanol and its esters), and various garlic-smelling volatile sulfur compounds;

– in washed-rind soft cheeses, (Limburger, Munster), surface bacteria (corynebacteria and micrococci) degrade amino acids and volatile fatty acids into sulphur compounds (methanethiol and thioesters);

– in blue cheeses, there is a high proportion of free fatty acids, methyl ketones, secondary alcohols and lactones;

– in semi-hard cheeses (cheddar), some authors attribute the aromatic base note to short-chain fatty acids (C_2 to C_6), methyl ketones and corresponding alcohols;

– in hard cheeses, where the level of proteolysis is high, the flavor is created by amino acids, acetic acid, propionic acid, alcohols, esters and sulfur products.

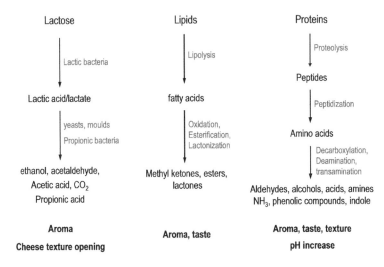

Figure 1.26. *Change in constituents during ripening*

1.3.4.5. *Errors and defects*

Given the diversity and complexity of cheese processing, manufacturers must face the risk of errors resulting in defects in the final product. These defects can be classified into two categories: coagulation and draining defects and ripening defects.

Coagulation and drainage defects

The growth of lactic acid bacteria plays a key role in cheese technology. The bacteria act as an acidifying agent, and therefore contribute to coagulation, drainage and the adjustment of the degree of mineralization in the cheese.

The ability of milk to allow the growth of lactic acid bacteria varies with the origin of the milk and the bacterial species. Milk contains a number of natural inhibitors (immunoglobulins, lactoperoxidase, lysozyme, lactoferrin, nisin, free fatty acids, leukocytes, etc.) and stimulants such as growth factors (group B vitamins, amino acids, nitrogenous bases, small peptides, proteose peptones). Subjecting milk to heat treatment may destroy natural inhibitors and growth factors, but it can also generate growth factors such as peptides, amino acids, formic acid and so on. Other exogenous factors such as bacteriophages, antibiotics or chemical residues can inhibit the growth of lactic acid bacteria.

Finally, coagulation defects (longer rennet clotting time, slower firming rate, formation of a soft gel with reduced cheese yield) may occur in milk due to its physicochemical and bacteriological composition (mastitis milk, milk from the beginning or end of lactation) or the type of treatment it has undergone (refrigeration, heat treatment, etc.).

Ripening defects

Ripening defects can be classified into three categories:

– *texture and swelling defects:* these defaults can be caused by processing (dry, oily or runny rind, split in the body of the cheese, untypical number, size and uniformity distribution of eye in certain semi-hard cheese, etc.) or microbiological reasons (early or late blowing, off-odor);

– *appearance defects (crust texture, undesirable mould growth)*: these can be caused by fungus on the cheese surface ("blue", "cat hair" or "toad skin" defects), or fungus and bacteria on the surface and inside the cheese (cheese rind rot, mottled appearance with flecks of orange, cream, pink, brown, white, red, etc.);

– *flavor and aroma defects*. These include:

- *bitter flavor*, frequently encountered in pressed, blue and soft cheeses. Caseins (mainly hydrophobic β-casein) are responsible for the formation of bitter peptides by the action of residual rennet, plasmin, penicillii, psychrotrophic bacteria and some starter cultures that acidify the curd rapidly,

- *rancid flavor*, which occurs with excessive lipolysis during ripening causing a large amount of short and medium chain-free fatty acids to form. The agents responsible are certain penicillii, psychrotrophic bacteria, natural or microbial lipases (contamination, heat-resistant enzymes, starter cultures, etc.),

- *other flavor defects* including cruciferous vegetable, mushroom, potato, or malt odors among others. The origins and mechanisms of their formation are diverse and difficult to establish.

In conclusion, the preparation of milk is an important step because it plays a key role in the production of cheese. With the increase in scientific knowledge over the past 30 years, the various stages of processing milk into cheese have become better controlled: the biological, biochemical, chemical, and physical properties of products are constantly changing, demonstrating a vast variety and complexity of reactions, in particular during ripening. However, greater understanding of the physicochemical and microbiological mechanisms involved in the various stages of cheese production is needed.

1.3.5. Cream and butter

Cream and butter are dairy products with a higher fat content than milk; except for low-fat products, the fat content of cream and butter, respectively, is greater than 30% and at least 80%. Thus, the production of cream and butter begins with the separation of the fat globules from the skimmed milk. This is achieved by centrifugal separation in hermetic separators with conical plates (see Chapter 3, Volume 2) at a temperature usually ranging between 45 and 55°C.

1.3.5.1. Cream

Cream can be liquid (whipping cream), thick (sour cream) or aerated (whipped cream). Liquid cream has either undergone pasteurization ("crème fraiche") or UHT sterilization. Thick cream is obtained after inoculating a pasteurized cream with certain starters. Whipped cream is obtained by introducing air into pasteurized or sterilized cream at low temperatures (usually between 4 and 10°C).

Homogenization and heat treatment

After separation, cream is homogenized to improve storage stability (creaming in liquid creams, release of whey in thick cream, etc.) or functionality (viscosity of cream, stability for cooking, foaming capacity, etc.). The concentration of available proteins in cream (casein, whey protein) for interface creation during homogenization is often limited. As a result, they are located at the interface of adjacent fat globules, leading to aggregate formation conducive to creaming. To limit this phenomenon, cream is homogenized using a two-stage homogenizer. The homogenization pressure of the first stage, between 13.5 and 20 MPa, is used to create the interface and decrease fat globule size; the second stage, set at a homogenization pressure of 10 – 20% of that of the first stage, separates the aggregates of fat globules formed during the first stage.

Cream is heat-treated in plate heat exchangers (before or after the homogenization); the exchange area is about three times larger than that used in the treatment of milk due to the lower heat transfer coefficient in cream. In addition, temperatures are higher generally due to higher microbial load in cream and to the high thermal resistance of microorganisms in the presence of fat: the intensity of the heat treatment increases with increased fat content. In the case of pasteurization, the time/temperature combination is approximately 50–10 s/80–100°C. Heat treatment is particularly challenging, since cream is a fragile emulsion and rapid temperature variations can significantly alter the properties of the emulsion.

Ripening

Ripened or sour cream is a thick cream. Pasteurized cream is inoculated with up to 0.5% starter culture consisting of a combination of acidifying, aromatic (*Lactococcus lactis* subsp. *Lactis cremoris*, *Lactococcus lactis* subsp *Lactis diacethylactis*, *Streptococcus thermophilus*) and sometimes thickening strains (*Leuconostoc*), which produce exopolysaccharides generating thick creams at less acidic pH. The ripening phase takes 12 – 18 h at temperatures between 12 and 22°C. Acidification causes the gradual destabilization of casein micelles, some of which are adsorbed on the surface of the homogenized fat globules, resulting in the formation of a network of proteins and fat globules, and a thickening of the cream. The most significant changes of texture occur at a pH below 5.0–5.2.

Whipping

Whipped cream is a foamed emulsion in which air bubbles are incorporated into a network of partially coalesced fat globules; in the presence of emulsifiers (mono- and diglycerides) and stabilizers (gelatine, carrageenan, etc.), this network ensures the rigidity and stability of the foam.

Cream intended for whipping is first homogenized, which increases the number of fat globules, and heat-treated before being refrigerated (4 – 10°C) for several hours (approximately 20 h) to promote fat globule crystallization.

During the aging of cream at low temperature, emulsifiers gradually displace adsorbed proteins from the homogenized fat globule surface, thereby reducing fat globule stability [GOF 17]. During whipping, the collision of destabilized fat globules promotes partial coalescence. Partially coalesced homogenized fat globules move to the air interface and form a network that stabilizes air bubbles. In addition, stabilizers increase the viscosity of the non-fat phase and limit drainage by interaction with proteins of the non-fat phase and adsorbed proteins on the fat globules.

1.3.5.2. Butter

Butter consists of a continuous liquid fat phase in which triglyceride crystals, small fat globules, aqueous phase droplets and air bubbles are dispersed (Figure 1.27). It is made from cream, typically pasteurized, containing 40 – 50% fat, which is traditionally ripened (cultured butter) and then churned to induce phase inversion. Ripening includes two combined operations:

– fat globule crystallization to develop the rheological properties of butter;

– cream fermentation to develop aroma and decrease the pH of cream.

These combined operations occur in cultured butters obtained by traditional batch churning or a continuous manufacturing process (Fritz process). The manufacture of cultured butters has gradually been replaced by the NIZO method, which is more flexible and economical, whereby fat globule crystallization and the production of flavor and acid are separate.

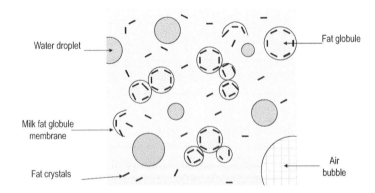

Figure 1.27. *Structure of butter*

Fat globule crystallization

Cream storage/aging at low temperature aims to induce the partial crystallization of fat, thereby promoting phase inversion. By strict control of the thermal cycle, it also adapts the consistency of butter to the seasonal and geographical variability in milk fat composition. In practice, there are two types of aging:

– low-temperature aging for winter cream, whereby the cream is immediately cooled to 6–7°C allowing the formation of many small fat crystals;

– high-temperature aging for summer cream whereby temperatures are adapted to obtain large fat crystals.

After aging, the solid and liquid fat ratio in cream is relatively stable.

Cream fermentation

Cream fermentation aims to acidify cream, allowing the development of a marked and typical aroma, promote phase inversion by decreasing the surface potential of fat globules at low pH, and ensure biological protection against microorganisms that can degrade butter. The major disadvantage of cream fermentation is that it generates an acidic and aromatic by-product after churning (buttermilk), which is difficult to stabilize and process further. This was the driving force behind the developments in butter technology and the introduction of the NIZO process.

Inoculation of cream with 3–5% lactic acid bacteria is achieved using a dosing pump. It can be performed either at the beginning of cream aging (before fat globule crystallization), resulting in pH values below 5.0, or after fat globule crystallization. Currently, the desired final acidity of butter is significantly less than it was in the past. Fermentation is usually carried out below 15°C for 10 – 12 hours. When the pH reaches a value close to 5.5–5.8, ripening is slowed by cooling the cream to 8°C. Butter produced in such a way has a storage pH ranging from 5.2 to 5.6.

Phase inversion

Phase inversion involves transforming ripened cream, an oil-in-water emulsion, into butter, a water-in-oil emulsion. Phase inversion is performed by churning, or vigorous agitation, at a temperature corresponding to the optimum ratio of crystallized and liquid fat (normally 10 – 13°C for churning winter cream and 7 – 10°C for summer cream). During churning, air bubbles are incorporated into the cream. The air bubble interface is first stabilized by (non-homogenized) fat globules. When they become insufficient in number to cover the newly-created interface, the foam collapses causing a rapid convergence of fat globules [VAN 01]. Coalescence of fat globules is promoted by the reduction of its surface charge depending on cream acidity and the presence of fat crystals deforming the fat globule surface. The release of liquid fat contained in the fat globules causes an agglomeration of fat crystals, intact fat globules and fat globule fragments in the form of granules. When a sufficient amount of liquid fat has been released, the granules are converted into butter grains in which droplets of buttermilk and small fat globules are dispersed. The emulsion is then rapidly reversed and the buttermilk is expelled. After washing and optional salting, the butter is kneaded to compact the butter grains and ensure homogenization by evenly distributing the aqueous phase and salt.

Conventional churning is carried out batch-wise in a barrel churn rotating about a horizontal axis. It is generally filled to 40–50% of its volume with ripened cream. Rotation ensures the incorporation of air into the cream and phase inversion. The churn has an outlet for releasing butter-cream and wash water. Continuous churns, or butter-making machines, operate according to the same principle as conventional churns, but without interruption. A butter-making machine consists of a cooled cylinder containing a rotating beater that incorporates air resulting in phase inversion and a tilted kneading cylinder containing two counter-rotating augers that compress and release the butter. The butter is generally washed and kneaded under vacuum to limit the risk of oxidation.

1.3.5.3. *NIZO butter*

The NIZO (Netherlands Institute for Dairy Research) method is used to prepare butter from cream that has not been fermented. Apart from this exception, all other processing stages remain the same as for the continuous manufacture of butter. Acidifying and flavoring agents (NIZO mixture) are added at a rate of approximately 0.8 – 1.25% to sweet butter after kneading. The NIZO mixture is prepared under aerobic conditions by vigorously mixing around 40% of a lactic acid concentrate (lactic acid content of the concentrate close to 18%) obtained by culturing *Lactobacillus helveticus* in whey, and around 60% of a mesophilic lactic starter (*Lactococcus lactis, Lactococcus cremoris* and *Lactococcus diacetylactis*). Intense oxygenation of the mixture is favorable for diacetyl production, characteristic of the butter aroma. The final mixture is very acidic and no longer contains live bacteria. This first injection is followed by a second injection of live bacterial culture consisting of acidifying strains of *Lactococcus lactis* and *Lactococcus cremoris* as well as a different aromatic strain of the previous culture, *Leuconostoc cremoris*. *Leuconostoc cremoris* is able to consume excess acetaldehyde, responsible for the "yoghurt" taste in the NIZO mixture (Figure 1.28).

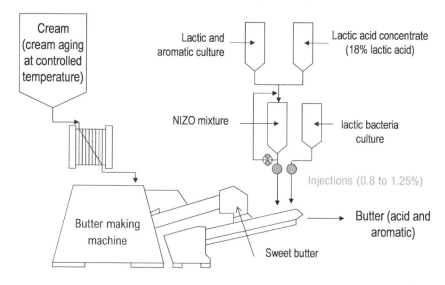

Figure 1.28. *Manufacture of butter based on the NIZO method*

There are several advantages to the NIZO method: the by-product generated (sweet buttermilk) is easier to use in further processing; the pH of butter can be adjusted upon injection of the lactic acid concentrate; and the control of starter culture conditions (controlled temperature and culture medium) ensures a high level of regularity in the production of aromas. Finally, the crystallization temperature of fat in cream is determined solely for the purpose of controlling the rheological properties of butter [MIL 98].

2

From Muscle to Meat and Meat Products

This chapter deals with the various technologies and products in the meat and fish industry. Meat production includes six main categories, three of which (pork, beef and chicken) represent over 90%. Fish products vary considerably, since they comprise almost 230 different species, including 150 species of fish, but also mollusks, crustaceans, cephalopods and algae; in addition, these products have highly variable compositions and characteristics.

It is also important to take into account production methods, which significantly impact the preservation and quality of products. In the meat industry, animals are mostly farmed; the composition and quality of raw materials is controlled by genetic selection, diet and slaughtering conditions. In the case of fish, the majority of production comes from fishing (71%), although aquaculture is growing rapidly: global production of farmed fish products rose from 3.9% in 1970 to about 42% in 2012. It is difficult to control the physiological stage, diet, history and capture of fish; these different factors affect muscle modification after the death of the animal as well as product quality.

2.1. The biochemistry of muscle (land animals and fish)

The muscle tissue of land animals and flesh of fish have similar biochemical compositions, especially with regard to protein, carbohydrates,

---

Chapter written by Catherine GUÉRIN.

minerals and vitamins. However, their lipid and water contents are quite different.

The distribution of fat varies significantly in fish and meat. Lipids in oily fish are primarily located in the muscle tissue, whereas they are mostly contained in the liver of white fish. Lipids in meat can be intramuscular (marbled), but are mainly extramuscular: this includes subcutaneous or external fat and visceral or internal fat. The overall biochemical composition, in particular the lipid content of the muscle, refers here to the composition of the skeletal muscle. Only intramuscular fat will be taken into account, as subcutaneous (often removed during slaughter) and visceral fat are generally not consumed.

2.1.1. *The structure and composition of meat and fish muscle*

2.1.1.1. *Tissue structure*

Meat

Meat and meat products vary significantly. This is demonstrated by the fact that there are almost 900 terms to define these products. Even if meat were restricted to only muscle, there would still be a wide diversity; more than 100 muscles with different structures and compositions are found on a carcass and each muscle is itself heterogeneous.

The carcass of terrestrial animals consists of several types of tissue: muscle, connective tissue, fat, blood, nerves and bones. Each contributes to the sensory quality of the meat: muscle and connective tissue to tenderness, blood tissue to color, fatty tissue to flavor and so forth. Bone tissue is generally not consumed, except in minced meat where a small proportion of bone is permitted by law.

Animals have two main types of muscles: striated and smooth muscle. They differ not only in their fibrous or non-fibrous appearance, clearly visible under an optical microscope, but also in their color: smooth muscle is white whereas striated muscle tends to be red. However, white striated muscle exists such as the pectoral muscle of chickens for example. Smooth muscle is mainly found in the organs (stomach, intestines, etc.) and is generally not consumed as meat. As a result, we will focus mainly on striated muscle.

Muscle tissue

Muscle tissue is highly differentiated and specialized to perform different tasks. It is composed of fibers with metabolic or contractile properties, which are held together by connective tissue. The main chemical component of connective tissue is collagen. The typical characteristics of fibers are based on the relative proportion of the constituent elements: myoglobin, interfibrous lipids, enzymes and so on. Thus, the chemical composition of a muscle depends on the relative proportion of each type of constituent fiber, but also the relative structure of the connective tissue (endomysium, perimysium, epimysium and tendon; Figure 2.1).

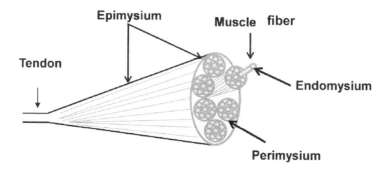

Figure 2.1. *Structure of muscle and connective tissue in meat*

Connective tissue

Connective tissue is present at different levels in the muscle. The first layer, known as the *endomysium*, consists of a thin sheath of connective tissue surrounding each individual muscle fiber. Groups of fibers form primary bundles that represent the "meat grain". Meat grain size varies with species and breed: Limousine cattle have fine-grained meat whereas Charolais cattle have coarser-grained meat. A second sheath of connective tissue, known as the *perimysium*, surrounds the primary bundles to form secondary bundles. Finally, the *epimysium*, or facia, surrounds the secondary bundles to form the muscle. This connective membrane, thick and white, is visible on certain muscles and is generally removed by the butcher.

The connective tissue at the end of the muscle forms tendons. These attach muscle to bone and transmit forces from muscular contraction. Tendons are removed when trimming meat.

Adipose tissue

Muscle connective tissue contains adipocytes in the *perimysium*; they form marbling, i.e. fat deposits visible to the eye. Apart from marbling, three types of adipose tissue can be found in a carcass: intramuscular fat, external fat that covers the carcass, and internal fat in the thoracic, abdominal and pelvic cavities (Figure 2.2).

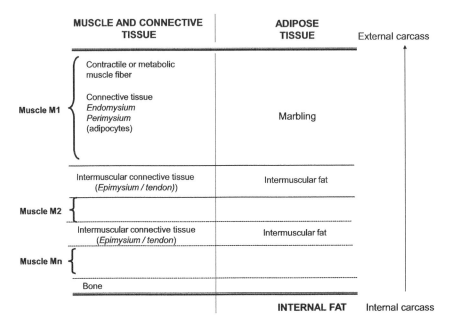

Figure 2.2. *Structure of muscle, connective and adipose tissues in meat*

Fish

Fish muscle is also composed of muscle and connective tissue, but the structure is different to that of meat muscle. Interpenetration of these two tissues is not as pronounced in fish as it is in meat. In addition, fish muscle has a metameric structure. Fish flesh is composed of long muscles divided into conical segments, the top of which point towards the head (muscle tissue). These segments, also known as *myotomes*, measuring 3 cm or less in length, are fitted into each other but remain separated by sheets of connective tissue known as *myocommata* (Figure 2.3). When the connective tissue is separated from the myotomes, the individual flakes fall apart; this is known as splitting of gaping.

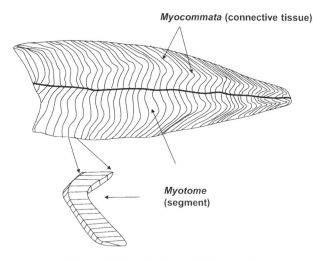

Figure 2.3. *Structure of fish muscle*

Fat is found in different locations in white and oily fish. In oily fish, fat serves as an energy reserve and is found in subcutaneous, visceral and connective tissues. In white fish, a small proportion of fat is located in the cellular membranes and therefore in the muscle tissue; the remainder is found in the viscera. Fat in muscle tissue is in the form of phospholipids. It does not serve as an energy reserve but can be used as such by white fish during periods of prolonged food shortage.

2.1.1.2. *Comparison of the biochemical composition of meat and fish muscle*

Overall biochemical composition

Despite having different compositions, it is possible to compare mammalian skeletal muscle and fish muscle since both are primarily composed of water, protein, fat, carbohydrates and minerals (Table 2.1).

The protein, sugar and mineral contents do not vary much according to the different physiological stages or diet of the animal (e.g. during the migration period or spawning season). However the lipid content fluctuates significantly: in the case of oily fish, for example mackerel, the fat content can vary from 4% in spring to almost 28% in autumn. The fat composition of meat also varies depending on the diet and age of the animal, but this variation occurs mostly in the adipose tissue and hardly in the skeletal muscle tissue.

	Fish muscle	*Mammalian skeletal muscle*
Water	70–80	65–72
Protein	16–22	20–23
Fat	1–10	4–15
Carbohydrates	0.3–1.0	0.5–1.0
Minerals	1.0–1.5	1.0–1.3

Table 2.1. *Comparison of the overall biochemical composition of fish and land animal muscle (in g per 100 g of muscle)*

These variations in the fat content mean that particular attention must be paid during the processing of these products. In the case of meat, composition control is easier, since processors can require that suppliers provide raw materials of homogeneous and well-defined quality. In the case of fish, however, manufacturers must adapt to "wild" products, the composition of which fluctuates substantially. The effectiveness of salting fish, for example, strongly depends on the fat content; a lean muscle can be salted faster since the salt transfer is not hindered by layers of fat. Manufacturers must constantly adapt the salting process to the raw materials they receive.

Meat and fish can be classified according to the fat content in the muscle (Tables 2.2 and 2.3), even if the fat content in the meat varies greatly from one muscle to another. In beef, a muscle such as the latissimus dorsi (lumbar area) contains 0.6% fat and the rectus femoris 1.5% fat; the same muscles in pork contain 3.4% and 1% fat, respectively.

	Rabbit	*Mutton*	*Pork*	*Beef*
Water (%)	77.0	77.0	76.7	76.8
Intramuscular fat (%)	2.0	7.9	2.9	3.4

Table 2.2. *Biochemical composition of the l. dorsi muscle of land animals (based on [LAW 98])*

Fish muscle is deemed white if its average fat content does not exceed 1%, oily if it is above 5%, and intermediate if between 1 and 5%. Table 2.3 shows the fat content of some fish, shellfish and crustaceans.

Content (%)	Fish, shellfish and crustacean (by English common name)			
Usually <1%	Cod	0.1–0.9	Coalfish	0.3–1.0
	Pollock	0.6–0.8	Whiting	0.2–0.6
	Ling	0.1–0.4	Scallop	0.3–0.9
Usually between 1 and 5%	Seabass	0.8–2.5	Plaice	1.1–3.6
	Shrimp	0.3–3.1	Seabream	0.8–3.3
	Halibut	0.7–5.2	Oyster	0.3–2.2
	Hake	0.4–2.7	Ray	0.1–1.6
	Dogfish	3.9–5.6	Sole	0.2–2.3
	Turbot	2.1–3.9		
Usually >5%	Anchovy	0.9–12	Eel	0.8–31
	Herring	0.8–25	Mackerel	0.7–23
	Mullet	0.2–14.8	Sardine	1–23
	Salmon	2.0–18	White tuna	0.7–18.2

Table 2.3. *Typical fat content (%) of some common fish, shellfish and crusteans (according to [PIC 87])*

Overall, the average fat content of fish muscle is less than that of meat muscle. Conversely, the water content of fish is slightly higher, making it more vulnerable to microbiological spoilage.

Polyunsaturated fatty acids in fish

The high level of polyunsaturated fatty acids in fish gives it valuable nutritional properties, which help to prevent cardiovascular disease. However, a large number of polyunsaturated bonds make fish products very vulnerable to oxidation. The high amount of polyunsaturated fatty acids limits the shelf life of oily fish, even when frozen, because once initiated, oxidation is difficult to inhibit by low temperatures or reduced water activity (a_w).

Polyunsaturated fatty acids in fish and meat represent up to 70% and 45% of total fat, respectively (Table 2.4).

Products	Saturated	Unsaturated	Monounsaturated	Polyunsaturated
Fish				
Cod	28.1	71.4	11.2	60.2
Whiting	20.4	77.5	41.3	36.2
Mackerel	30.3	69.7	43.9	25.8
Rainbow trout	20.4	71.4	32.3	40.1
Lemon sole	20.1	75	27.4	47.6
Albacore tuna	17.8	72.2	27.8	44.4
Bluefin tuna	34.1	63.8	36.2	27.6
Crustaceans				
Crab	16.8	80.2	27.4	52.8
Shrimp	21.8	75.4	29	46.4
Shellfish				
Oyster	30.4	67.4	15.7	51.7
California mussel	29.9	65.7	26.4	39.3
Grooved carpet shell	20.6–35.6	41.9–55.9	18.9–32.7	18.3–35.4
King scallop	31.7	65.8	9.1	55.7
Liver				
Dogfish	20.4	71.2	55	16.2
Cod	15–19	76–80	47–60	21–29
Oil				
Anchovy	30.5	69.7	32.6	37.1
Herring	26.1	73.5	55	16.2
Mackerel	27.5	72.5	48.9	23.6

Table 2.4. *Saturated and unsaturated fatty acid composition of some fish products (% of total fat) (according to [PIC 87])*

High level of non-protein nitrogen in fish

The main components of the non-protein nitrogen fraction of fish are ammonia, trimethylamine oxide (TMAO), creatine, free amino acids, nucleotides, purine bases and urea in the case of cartilaginous fish. They result in the formation of degradation compounds, which are often detrimental to quality.

TMAO is present in very large amounts in cartilaginous fish but absent in crustaceans, freshwater fish and mammals. It decomposes to trimethylamine by bacteria, thereby serving as a quality index for fish. In quantitative terms, creatine is also very important as, in its phosphorylated form, it provides energy for muscle contraction.

Free amino acids are present in significant amounts in fish (1.3 – 3.8% compared to 0.1 – 0.6% in meat). They are of relative importance that varies depending on the species: taurine, alanine, glycine and imidazole amino acids are prevalent in most fish. Histidine, a particularly abundant amino acid in clupeids and scombroids fish, has been the subject of numerous studies because its microbial decarboxylation results in histamine, a known source of food allergy. In addition, its heat resistance protects it from destruction during processing, which is why histamine is subject to very strict regulation.

2.1.2. *Muscle structure*

2.1.2.1. *Muscle cells*

Muscle tissue is composed of specialized cells called muscle cells or muscle fibers (Figure 2.4). These are large cells measuring 1 – 100 μm in diameter and 1 mm – 40 cm in length (3 cm maximum in fish). The cell membrane or *sarcolemma* is closely connected to a very fine layer of connective tissue called the *endomysium*, which ensheaths each muscle fiber.

The sarcoplasm (muscle cell cytoplasm) contains a number of organelles common to all cells:

– several 100 nuclei inside the sarcolemma membrane: the muscle cell is multinucleated;

– Golgi apparatus: a complex of flattened compartments bound by a double membrane. It plays a role in cell metabolism and the process of excretion;

– mitochondria, which play a crucial role in energy processes (electron transfer chain and oxidative phosphorylation);

– small lipid droplets.

However, it also contains organelles particular to muscle cells:

– glycogen granules: energy reserves for muscle contraction;

– lysosomes: small vacuoles containing many enzymes including acid proteases such as cathespins;

– a contractile system of protein filaments running parallel to the longitudinal axis of the fiber, giving it its striated appearance. Everything is held in place by a sort of internal scaffolding, the cytoskeleton.

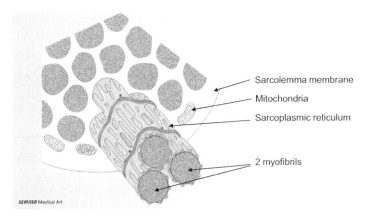

Figure 2.4. *Muscle cell structure. For a color version of this figure, see www.iste.co.uk/jeantet/foodscience.zip*

Myofibrils

Myofibrils are contractile elements that allow muscle cells to contract. They are aligned parallel to the longitudinal axis of the fiber and are responsible for the longitudinal striation of the muscle. They are the same length as the muscle fiber and have an average diameter of 0.1 µm. They have a heterogeneous structure.

Myofibrils repeat along the longitudinal axis of the fiber, approximately every 2.5 µm. In skeletal or striated muscles, this repetition forms a cross-striation pattern (Figure 2.5).

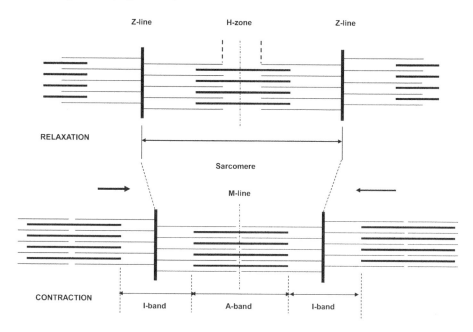

Figure 2.5. *Structure of myofibrils*

This striation is the result of alternating dense, dark or A-band (anisotropic) regions measuring 1.0 µm in length, and less dense, light or I-band (isotropic) regions measuring 1.6 µm in length. I-bands are separated by a "partition" known as the Z-line or Z-disc. Z-lines extend from one myofibril to the next, eventually connecting to the sarcolemma at the end of the muscle fiber. They form an internal skeleton for the cell and play an important support role. They also help to create synergy between the various contractile units. In the middle of the A-band is a less dense, lighter region called the H-zone. At the centre of the H-zone is a darker line called the M-line.

A complete structural unit between two Z-lines is called a sarcomere. It forms the basic contractile unit. In a relaxed muscle, a sarcomere measures 3–5 µm in length.

Myofilaments

Each myofibril consists of several parallel myofilaments. There are two types: thick and thin myofilaments. In I-bands (light regions), only thin filaments are present ($\varnothing$ = 6 nm). The dense regions of A-bands contain:

– thin filaments found in I-bands;

– thick filaments ($\varnothing$ = 15–17 nm) giving A-bands their characteristic birefringence.

Thick and thin filaments have a hexagonal arrangement, visible in a cross-section of muscle (Figure 2.6). A thick filament is surrounded by six thin filaments.

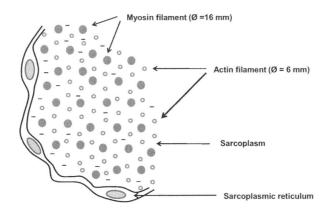

Figure 2.6. *Cross-section of a myofibril fragment*

Thick filaments extend from one end of the A-band to the other. However, thin filaments start at the Z-line and continue into the A-band as far as the edge of the H-zone.

Within the dense regions of the A-bands, there are bridges between the adjacent thick and thin filaments. The bridges are made of thick filaments and are the only existing connections between these filaments.

These various elements of the muscle cell are held in place in the sarcoplasm by longitudinal and vertical elements belonging to the sarcoplasmic reticulum and the cytoskeleton. The latter consists primarily of two insoluble proteins: connectin and desmin.

Change in the sarcomere during muscle contraction

During contraction and extension of the muscle, the size of the sarcomere varies considerably. During contraction, it can decrease in length by 50% and increase by up to 120% during extension.

Thick and thin filaments do not change in length, but rather the shortening of the muscle during contraction is due to the sliding of thick and thin filaments over each other. There is an interpenetration and overlapping of thick and thin filaments (Figure 2.5).

2.1.2.2. Red and white muscles

Red and white muscles are characterized by varying proportions of red and white fibers, which, apart from their color, can de distinguished by their contraction rate and metabolic type. Red fibers have the following characteristics:

– slow and prolonged contraction;

– strong vascularization and rich in myoglobin;

– significant respiratory processes and aerobic metabolism;

– rapid oxidation of unsaturated fats (palatable meat).

White fibers have the opposite characteristics, namely, fast and short contraction, low vascularization and poor in myoglobin.

Red and white fibers coexist in all muscles, thereby giving every muscle a different composition and metabolism. The relative proportion of both fibers determines the type of muscle, which also differs in morphology from other muscles.

2.1.2.3. Connective tissue

Connective tissue has a complex composition. It consists of:

– protein fibers (collagen, reticulin, elastin);

– the ground substance surrounding these fibers and cells. It plays a crucial role in exchanges between the blood and the muscle cell;

– characteristic cells, including fibroblasts that develop connective fibers, histiocytes (or macrophages) and fat cells;

– blood vessels;

– nerves – the finer the movement, the greater the number of nerves (e.g. hand muscles).

2.1.3. Proteins

There are three main types of proteins in the muscle: myofibrillar, sarcoplasmic and connective tissue protein. Table 2.5 shows the average protein composition of skeletal muscles.

	Fish muscle	Mammalian skeletal muscle
Sarcoplasmic proteins (soluble fraction)	20–35	30–35
Myofibrillar proteins (poorly soluble fraction)	60–75	50
Stromal proteins (extracellular proteins)	3–10	15–20

Table 2.5. *Protein composition of meat and fish muscles (in g per 100 g of total protein) (based on [LIN 94])*

2.1.3.1. Muscle tissue proteins

Soluble fraction: myoglobin

Myoglobin is an essential element in the color of meat, which affects consumer purchasing decisions. Meat color is related to the quantity and quality (chemical state) of myoglobin. The fatness of the animal and the pH of the meat also play a role. The fish muscle most closely resembling meat in terms of myoglobin content is tuna. Similarly, the quality of bluefin tuna is also closely linked to the quality of its myoglobin. However, white muscle is more prevalent in other species of fish.

Myoglobin is a heteroprotein with a molecular weight of 18.7 kDa. It is very compact and belongs to the group of respiratory pigments. The molecule is composed of two parts:

– a protein part (globin);

– a non-protein part (heme), consisting of a tetrapyrrole (or porphyrin) ring that coordinates ferrous iron (Fe^{2+}) in myogloblin or ferric iron (Fe^{3+}) in metmyogloblin. The iron in the heme is coordinated to the four nitrogen atoms of the pyrrole rings and also to a nitrogen atom from a histidine residue. The sixth position (coordination site) for the iron is occupied by water in myoglobin and oxygen in oxymyoglobin.

Figure 2.7 shows the two primary myoglobin reactions affecting meat color.

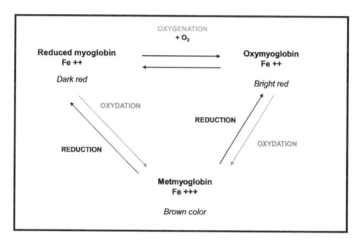

Figure 2.7. *Change in myoglobin and variation in meat color*

Three types of pigments (myoglobin, oxymyoglobin and metmyoglobin) coexist, but it is their relative proportion that gives meat its characteristic color. If, on the surface of the meat, more than 40% of the myoglobin is in the form of metmyoglobin, meat has a brown color and is therefore unsellable.

Many factors affect the chemical structure of heme and meat color. Some are linked to the characteristics of the animal:

– age: color tends to increase with age;

– physiological maturity, which can differ with breed;

– diet, fatness or iron deficiency tends to lighten meat color;

– type of muscle: each muscle has a specific color.

These differences are mainly due to the amounts of pigment in the muscle. The pH can also change the color of meat when it reaches abnormal values, especially in a stressful situation prior to slaughter. In addition, the storage and packaging of meat influence meat color. Vacuum packaging promotes dark red myoglobin, while modified atmosphere packaging is conducive to the formation of bright red oxymyoglobin. Brown metmyoglobin increases with storage.

Poorly soluble fraction

– Review of myofibril structure

Myofibrils are composed of protein myofilaments held in place by Z-lines that mark the boundaries between sarcomeres. An internal skeleton or cytoskeleton also contributes to the stability of the structure. All these structures are composed of poorly soluble molecules in addition to the regulatory proteins of muscle contraction.

– Myosin

Myosin forms the thick filaments. It has a molecular weight of 540 kDa and a pI close to 5.4. It is completely insoluble at the ionic strength of the sarcoplasm.

The myosin molecule is made up of two parts (Figure 2.8):

– a tail consisting of two polypeptide chains (two α-helices intertwined about a common axis to form a double super helix);

– a globular head composed of the two aforementioned chains randomly intertwined (heavy chains) to which two shorter polypeptide chains (light chains) are attached.

Myosin has two key properties:

– Constituent molecules spontaneously assemble into filaments under physiological conditions.

– Myosin behaves like a weak ATPase; it hydrolyses the terminal phosphate group of adenosin triphosphate (ATP) as well as guanosine triphosphate (GTP), inosine triphosphate (ITP) and cytidine triphosphate (CTP). This activity is stimulated by Ca^{2+} ions, inhibited by

Mg^{2+} ions, influenced by the concentration of KCl, and has two optimum pH for activity (6.0 and 9.5). The hydrolysis of ATP can be summarized as follows:

$$ATP + H_2O \rightarrow ADP + P_i + H^+ \qquad [2.1]$$

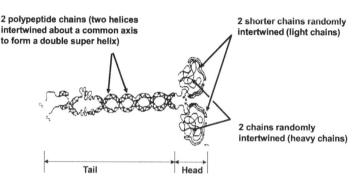

Figure 2.8. *Structure of myosin*

Actin

Actin exists in two forms: G-actin (globular monomeric actin) and F-actin (fibrous polymerized actin). At low ionic strength, actin is in monomeric form (G-actin). The polypeptide chain has a molecular weight of 422 kDa and is rich in N-methyl lysine, proline and cysteine (seven cysteines per molecule). G-actin is a powerful chelating agent of Ca^{2+} ions and can bind ATP and adenosine diphosphate (ADP). Under physiological conditions of ionic strength or in the presence of salts (KCl, $MgCl_2$), it can be polymerized as follows:

$$n \, (G\text{-actin} - ATP) \rightarrow (G\text{-actin} - ADP)_n + n \, P_i \qquad [2.2]$$

F-actin consists of two helically-wound strands of G-actin monomers. It forms the thin filaments (Figure 2.9).

Actomyosin

Myosin can combine with actin to form actomyosin. This is a crucial step in muscle contraction.

$$Myosin + Actin \rightarrow Actomyosin \qquad [2.3]$$

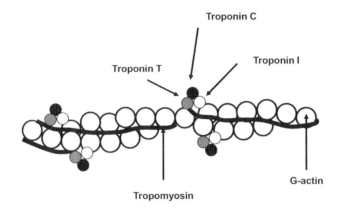

Figure 2.9. *Structure of actin*

A thin actin filament can bind to several myosin molecules. The head domain of the myosin molecule serves as the binding site. The characteristics of the actomyosin complex depend on pH and concentrations of protein, KCl and $MgCl_2$. The ATPase activity of actomyosin is 20 times greater than that of myosin only. It is activated by Ca^{2+} and Mg^{2+} ions. ATP and K^+ and Mg^{2+} ions can cause the dissociation of the actomyosin complex. In the relaxed muscle cell, the two proteins are in the dissociated state.

Regulatory proteins

Tropomyosin has a molecular weight of 70 kDa. It consists of two polypeptide chains in α-helical form arranged as a double super helix. It is rich in sulfur amino acids including several unpaired cysteines. It is found in the thin filaments of I-bands, in the groove between the two strands of the F-actin double helix (Figure 2.9). In relaxed muscle cells, it occupies the reactive site between actin and myosin, preventing any reaction between these two proteins. Through the intermediary of troponin I, calcium controls its position and thus regulates muscle contraction.

The troponin complex is found in thin filaments along the tropomyosin chain and is comprised of three subunits:

– troponin C that binds Ca^{2+} ions (reversible complex);

– troponin I that inhibits actin–myosin interaction. This action is neutralized when troponin C binds calcium;

– troponin T that binds to tropomyosin, interlocking them to form a troponin–tropomyosin complex.

α and β actinins are located in the Z-lines. Their main role is to accelerate the polymerization of globular actin to fibrous actin.

Cytoskeleton

The cytoskeleton is composed of many proteins such as filamin, desmin, vimentin and synemin found in Z-lines. Other proteins like titin, nebulin and connectin form a flexible network around the filaments ("gap filament") by linking thin filaments of adjacent sarcomeres and crossing Z-lines.

Ultrastructure of myofilaments

Thick filaments in A-bands consist of a cluster of myosin molecules arranged parallel to the longitudinal axis of each filament. Each thick filament of myosin is surrounded by six thin filaments of F-actin. The protruding heads of the myosin molecules are just long enough to reach the thin filaments.

During muscle contraction, the bonds between the myosin molecules and G-actin monomers form and break regularly. The filaments slide over each other and the fiber shortens (Figure 2.5).

Extracellular proteins

Extracellular proteins form the connective tissue. There is a strong correlation between the abundance of connective tissue and meat tenderness. Muscle hardness, for example, never occurs in fish, which has very little connective tissue. Connective tissue consists primarily of three proteins: collagen, reticulin and elastin. These are fibrous proteins that differ in their heat resistance.

Collagen

Collagen is the most abundant protein in mammals (15 – 35% of proteins), whereas its content is much lower in fish (3% in teleosts (ray-finned fish) and 10% in elasmobranchs (cartilaginous fish)). Its amino acid composition is characteristic and unique among proteins. The composition of mammalian collagen differs from that of fish collagen, especially in terms of the proline and hydroxyproline content, which is lower in fish; this difference is responsible for the different functional properties of gelatin. For example, fish collagen denatures and solubilizes at 35–40°C compared to 60–65°C for mammalian collagen. Collagen generally consists of:

– 33% glycine with a very regular distribution of this amino acid along the polypeptide chain; one glycine every three amino acids (-Gly-X-X-Gly-X-X-Gly-X-X-Gly-);

– 11% alanine;

– 12% proline (10.5% in fish collagen);

– 9% hydroxyproline (7% in fish collagen) and some hydroxylysine. These residues can be linked to a glucose and galactose disaccharide (o-glycosidic bond);

– no sulfur amino acids or tryptophan.

The presence of hydroxyproline is specific to collagen in animal proteins. It makes it possible to determine the quantity of protein in a mixture as well as detect potential fraud.

Depending on the amino acid composition, up to 16 different types of collagen can be distinguished, but types I to V are the most common. For example, collagen type I is found in skin, tendons, bones, and cornea, while type II collagen is found in cartilage, intervertebral discs, vitreous body and so on.

Collagen is a protein that is not easily hydrolyzed by digestive enzymes and therefore has a very low nutritional value. It is difficult to further process, except for the production of gelatin.

Structure of collagen

Collagen is a protein with a molecular weight close to 285 kDa. The molecule measures 1.4 nm in diameter and 300 nm in length, making it the longest known polypeptide (Figure 2.10).

The tropocollagen molecule consists of three helical polypeptide chains (α-helices), wound parallel along an axis to form a super triple helix. Its three-dimensional structure is stabilised by interchain bonds (2,000 hydrogen bonds per tropocollagen molecule) and by steric hindrance of the pyrrole rings of amino acids.

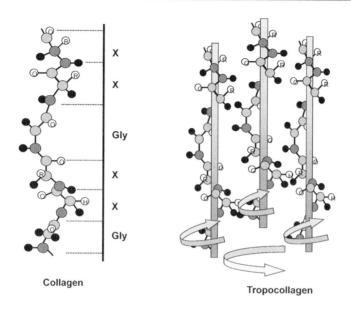

Figure 2.10. *Structure of collagen and tropocollagen. For a color version of this figure, see www.iste.co.uk/jeantet/foodscience.zip*

After synthesis in fibroblasts, collagen molecules combine to form fibrils. In these fibrils, each coiled tropocollagen monomer overlaps the next by one quarter of its length ('quarter stagger'); hence the clearly visible axial periodicity under an electron microscope. The bonds between the molecules, which are few in number in young animals, increase with age. The structure of the collagen therefore becomes more solid and rigid, in relation to meat tenderness. It has a quasi-crystalline structure.

Properties of collagen

Native collagen and meat tenderness

Collagen is a partially hydrated protein containing almost 50% dry matter. Unlike elastine, collagen is inelastic. The distribution of these two proteins in the walls of the blood vessels (collagen fibers are distributed along the longitudinal axis of the vessels and elastic proteins perpendicular to this axis) explains why they may increase in diameter but not in length. Collagen is very resistant to shear (18 $kg.cm^{-2}$), which means there is a relationship between meat tenderness and the amount of collagen. Collagen proportion can vary from one to four in muscles.

Tenderness is the most important sensory quality in meat, especially in red meat (beef and lamb). It is measured by shear force and/or compression. Measurements are challenging due to the complexity of the processes involved, the causes of which are difficult to analyze. There is an unclear correlation between tenderness and the amount of connective tissue, since it is not the only factor involved in tenderness. The quality of the connective tissue (the connective tissue of younger animals is more "tender" than that of older animals) and its distribution (for two muscles with the same amount of collagen, the shear force required to rupture the tissue is less if the structure is thinner) also play an important role.

The amount of connective tissue in the muscle varies with age. In beef, there is an increase in the amount of muscle connective tissue (and thus collagen) between 15 and 18 months, corresponding to puberty and the approximate time of slaughter of young bulls, whose meat is not particularly tender.

Genetic origin also explains variations in the collagen content of the muscle, as in the case of double-muscled animals. They are characterized by muscle hypertrophy, macroglossia, thinner skin than normal (<30% in weight) and less muscle connective tissue. This is particularly the case with forequarter muscles, which contain 25-30% less connective tissue than normal, and are therefore more tender.

Another characteristic of the collagen molecule is its slow turnover rate, which decreases with age.

Thermal stability of collagen

When collagen or tropocollagen is heated, significant changes in the physical properties occur at certain temperatures (drop in viscosity, loss of helical structure). The temperature at which half the helical structure of tropocollagen is lost is known as the melting temperature (T_m). For intact collagen fibers, an equivalent index is the shrinkage temperature (T_s), the temperature at which fibers shorten considerably. The T_m and T_s of collagen depend on its amino acid composition; they also correlate with the body temperature of the species.

Hydrothermal shrinkage of collagen

Native collagen fibers are practically inextensible. Heated in an aqueous medium, they shrink suddenly at around 60°C. In their shrunken state, collagen fibers have high elasticity. This shrinkage, which can be up to 75% of the length of the fiber, is accompanied by the disappearance of an X-ray diffraction pattern (loss of the crystalline structure). The hydrothermal shrinkage of collagen is the result of denaturation involving the loss of hydrogen bonds as well as an increase in hydrophobic interactions and hydration. Collagen is one of the few proteins whose affinity for water increases with denaturation: from 200 g of water per 100 g of collagen to 1,000 g of water per 100 g of denatured collagen, which partly explains improved tenderness.

However, if the temperature exceeds 60°C, the heat generally ruptures all bonds. The collagen fiber relaxes, releases water and loses its elasticity. The molecule is then completely denatured.

All of this has consequences for cooking meat: above 55°C, sarcoplasmic proteins denature and release water, which is partly recovered around 60°C during the initial phase of collagen denaturation. For optimal juiciness in red meat, the cooking temperature in the center should not exceed 60°C (rare meat). In the case of marinades, the acidic pH causes a swelling of collagen fibers, an increase in hydration and consequently a decrease in the shear force required to rupture the fibers. The meat becomes more tender but can no longer be grilled but rather needs to be stewed or boiled, like stewed steak.

Salt also increases the thermal resistance of collagen (boiled salted meat is generally hard and stringy).

Hydrothermal solubility of collagen

When collagen is heated in aqueous media, the fiber structure is damaged above 60°C. Some of the collagen solubilizes, forming gelatin. If, during cooling, the gelatin concentration is sufficient, a gel is obtained. The hydrothermal solubility of collagen depends on the age of the animal.

Enzymatic hydrolysis of collagen

Proteolytic enzymes have a poor capacity to hydrolyze collagen. Only a few specific proteases can degrade the molecule, such as cathepsin B_1, which attacks collagen at the end of the chains, and tissue collagenase, which acts synergistically with cathepsin B_1 to cut the collagen molecule into two pieces (3/4 and 1/4), and *Clostridium histolyticum* collagenase (gas gangrene), which cuts the molecule at more than 200 sites.

During the maturation of meat, the hydrolysis of collagen is very small since the lysozomal enzymes (cathepsins) are not in contact with their substrate (collagen is outside the cells).

In summary, there are two approaches to improving meat tenderness in terms of its connective tissue: a biological and a technological approach. The biological approach involves genetic selection (animals with a small amount of collagen in the muscle), castration (avoiding the peak of puberty), the type of farming (animals that exercise produce less collagen) and the age of the animal. The technological approach includes tenderizing (difficult, with some associated hygiene risks), mincing, marinating, cooking or else specific enzymatic treatments such as:

– the Proten® process from the USA: injection of papain before slaughter;

– use of microbial collagenases (*Clostridium histolyticum*);

– use of tissue collagenases.

2.1.3.3. *Other proteins in the connective tissue*

The other proteins in the connective tissue are mainly reticulin and elastin. Reticulin is very similar to collagen. Elastin is even more resistant than collagen to heat treatment and enzymatic attack. It is found in artery walls and certain elastic ligaments, but hardly in muscles. It contains two atypical amino acids, desmosine and isodesmosine, also found in the constitutive proteins of egg shell membranes.

2.1.4. Carbohydrates

Meat and fish are low in sugar; these carbohydrates, serving primarily as energy reserves, are mostly found in the form of glycogen granules but also, after the death of the animal, in the form of ribose from the degradation of ATP. The glycogen content of the muscle depends on species, muscle type, the physiological state of the animal, and is generally lower in fish muscle (1% in meat and less than 1% in fish).

2.1.5. Vitamins and minerals

The vitamin content of mammalian and fish muscle is very similar except for vitamins A and D, present in large quantities in the muscle of oily fish and the liver of white fish (cod, halibut), but present only in trace amounts in meat. Fish is a good source of minerals, especially calcium, phosphorus, iron and copper, as well as iodine in sea fish. Meat, and in particular beef, is rich in iron.

2.2. Biological and physicochemical changes in muscle

2.2.1. *Muscle contraction*

2.2.1.1. *Excitation and contraction coupling*

The stimulus for muscle contraction is the transmission of an impulse from the motor nerve to the motor end plate of the sarcolemma, which occurs very rapidly. Normally, the potential difference (pd) between the inside and outside of the muscle cell is +60 mV. During the transmission of the nerve impulse, pd is zero; this is known as depolarization. It is caused by changes in cell membrane permeability to K^+, Na^+ and Ca^{2+} ions.

For a long time, it was not understood how all the myofibrils in the muscle fiber could contract simultaneously. The diffusion of a chemical mediator from the sarcolemma to the myofibrils did not explain the rapidity of the response (<1 s). Electron microscopy provided the solution. Sarcolemmal invaginations run along the Z-line across the muscle cell so that these tubules are in close contact with all the myofibrils in the cell. This system of transverse tubules is known as the T (triad) system. When a nerve impulse reaches the sarcolemma and depolarization occurs, the entire T system is also stimulated, thus communicating nerve impulses to all the sarcomeres in the muscle fiber.

Electron microscopy was also used to explain the conversion of electrical signals in the T system to chemical signals in the myofibrils. Surrounding each set of adjacent sarcomeres are double-membrane vesicles, which are arranged longitudinally and contain many perforations. Each vesicle extends from one A-I junction to the next (Figure 2.5). These vesicles form the sarcoplasmic reticulum. The internal compartments of the vesicles or cisternae are linked to each other by sac-like channels called terminal cisternae. Parallel pairs of terminal cisternae run along the myofibrils in close contact with the T system.

When the sarcolemma is excited and the T system is depolarized, the permeability of the membranes next to the sarcoplasmic reticulum increases. As a result, Ca^{2+} ions are released from the cisternae of the reticulum into the sarcoplasm, where they stimulate myosin ATPase activity, thereby triggering muscle contraction. In the rest of the cell, the lack of calcium prevents the hydrolysis of ATP by myosin. The concentration of Ca^{2+} in the sarcoplasm of relaxed muscles is estimated to be less than 1 µM. The minimum concentration of Ca^{2+} required to trigger muscle contraction is in the range of 10 µM.

2.2.1.2. Relaxation

When the nerve impulse has crossed the sarcoplasmic reticulum causing Ca^{2+} ions to be released into the sarcoplasm, the sarcolemma and the sarcoplasmic reticulum return to their initial state of polarization (+60 mV between the inside and outside of the cell). Calcium is therefore retained in the reticulum cisternae; the muscle is ready to receive further excitation.

The accumulation of calcium in the sarcoplasmic reticulum is due to the action of an ATP-dependent calcium pump located in the membranes of the vesicles. This transfer occurs against a concentration gradient via active transport. The energy required for this transport comes from ATP hydrolysis by sarcoplasmic ATPases in the reticulum membranes. Vesicles are able to pump calcium from the surrounding medium up to a Ca^{2+} ion concentration of 1 µM. The Ca^{2+} ions in the vesicles are then released into the sarcoplasm when a new impulse arrives.

2.2.1.3. Sources of energy for muscle contraction

During muscle contraction, lactic acid is formed by glycogenolysis and the glycogen content of the muscle decreases. It was thought, therefore, that glycogenolysis provided the energy needed for muscle contraction. However, if this metabolic pathway is inhibited, contraction still occurs. So it was then assumed that ATP was the energy source, which also proved to be wrong for two reasons:

– the ATP concentration of the muscle is insufficient to supply the energy needed for contraction. For every minute of muscle contraction, 10^{-3} mol of ATP per g of muscle is required. However, it only contains 5×10^{-6} mol.g^{-1}, which corresponds to 0.5 second of activity;

– before and after muscle contraction, concentrations of ATP and ADP hardly vary in the muscle.

The energy required for muscle contraction is generated mainly by phosphocreatine, a compound capable of storing energy and found in all muscles of vertebrates, at a concentration four to five times higher than that of ATP.

The phosphate group of phosphocreatine can be quickly transferred to ADP by the action of creatine kinase:

$$\text{Creatine-P} + \text{ADP} \leftrightarrow \text{Creatine} + \text{ATP} \qquad [2.4]$$

At the pH of the sarcoplasm (pH 7), the equilibrium is shifted towards the formation of ATP; therefore the concentration of ATP does not drop during muscle contraction. If glycolysis and respiration are blocked, phosphocreatine may be deficient and the ATP concentration drops.

Rephosphorylation of ATP and phosphocreatine varies with muscle type: for very active or red muscles containing a lot of myoglobin and respiratory pigments (cytochromes), respiration is the main supplier of energy through oxidative phosphorylation like, for example, in the flight muscles of birds, the leg muscles of mammalian runners and the muscles of pelagic fish. For less active or white muscles containing a small amount of myoglobin and respiratory pigments (pectoral muscles of birds or benthic flatfish), glycolysis is the main source of energy.

Some ADP can also be converted to ATP using adenylate kinase as follows:

$$2\,\text{ADP} \leftrightarrow \text{ATP} + \text{AMP} \qquad [2.5]$$

2.2.2. Changes in muscle after death

2.2.2.1. Transport of animals

Animals are not slaughtered at their production sites, apart from fish, which are slaughtered and eviscerated in aquaculture centers. Legislation requires that animals be killed by specialized and approved slaughterhouses.

Animals are transported from the production site to the slaughterhouse using trucks adapted to each type of animal: cattle, pigs, poultry, or fish. They are designed to avoid stress as much as possible, as stress causes a depletion of glycogen. Once at the slaughterhouse, animals must have time to "de-stress" in order to replenish their glycogen stores and provide quality meat.

While it is possible with farmed fish to create the least stressful conditions possible by avoiding excessive numbers of fish in pools or by cooling the storage water of fish before capture, the same is not possible for wild fish that often undergo real stress prior to death and may struggle for hours in a trawling net, which is detrimental to their quality. Thus, line-caught fish are of better quality than those captured in trawling nets.

2.2.2.2. Stunning and death

Prior to death by exsanguination (bleeding to death), animals are anesthetized to prevent any possible suffering. The anesthesia is achieved by destroying vital nerve centers using a captive bolt gun. This method is used primarily for cattle and horses. For pigs, sheep and poultry, electronarcosis is used, which involves the passage of an electric current through electrodes placed on the head or immersion in a water bath electrified for poultry. Carbon dioxide anesthesia can also be used for pigs. In the case of fish, a combination of carbon dioxide and cooled water, or electronarcosis, is used.

Exsanguination is performed by opening the carotid artery and jugular vein in cattle and sheep or the anterior vena cava in pigs. This method does not yield blood of high microbiological quality. Only by inserting a trocar into the vena cava of the pig and allowing the blood to flow into a refrigerated tank via a tube can high quality blood be obtained. Since the emergence of bovine spongiform encephalopathy (BSE or mad cow disease), bovine blood is generally removed and burned in cement kilns.

2.2.2.3. Pre-rigor phase

Overview

In the hours after slaughter, before the onset of rigor mortis (see section 2.2.2.4), muscle fibers can asynchronously shorten. The muscles undergo spontaneous, sometimes violent, fibrillary contractions, the frequency and intensity of which decrease with time. The biochemical mechanism of these contractions is identical to those observed *in vivo*, but the stimulating agent is no longer the nervous system. Stimulants may be thermal, mechanical or chemical. The length of this phase depends on temperature: it is maximal at 10°C (3 h for beef carcasses in good conditions).

During this phase, skinning (removal of leather), evisceration (removal of internal organs) and finishing (removal of external fat) are performed and the carcass is split in two. Fish muscle, which is naturally more tender because it contains less connective tissue proteins, can be handled either during the pre-rigor or rigor mortis phase. However, it is best to fillet fish after the rigor mortis phase to avoid fillet shrinkage and liquid loss during freezing.

Immediately after slaughter, the temperature of the carcass increases slightly (few degrees). This temperature increase, which can be explained by the activation of different exothermic reactions due to slaughter stress, can have a negative impact on meat quality. In the case of fish, the duration of this first phase can vary depending on species, mode of capture and storage temperature after death (Table 2.6): 30 min for cod caught in a trawling net and stored at 30°C, and up to 22 h for a redfish caught in a trawling net and stored at 0°C.

Species	Condition	Temperature (°C)	Time between death of the fish and onset of rigor mortis (h)	Time between death of the fish and end of rigor mortis (h)
Cod	Trawl fishing	0	2–8	20–65
		10–12	1	20–30
		30	0.5	1–2
	At rest (without stress or effort)	0	14–15	72–96
Tilapia	At rest	0–2	2–9	26.5
Grenadier	Trawl fishing	0	<1	36–55
Plaice	Trawl fishing	0	7–11	54–55
Coalfish	Trawl fishing	0	18	110

Table 2.6. *Onset of rigor mortis in different species of fish (based on [HUS 88])*

Carcass classification

Carcass classification depends on the type of animal. With cattle, classification is based on the SEUROP system, which assesses meat yield and quality, especially tenderness. The letter S is assigned to high quality carcasses. Fat classes range from one to five, with one corresponding to low fat. The fat class of pigs is assessed by measuring the thickness of the back fat using an automatic infrared device. The color of veal is assessed by comparison with a fixed color scale ranging from pale pink to red.

2.2.2.4. Rigor mortis

Physical changes

Rigor mortis gradually sets in as the previous phase (pre-rigor phase) comes to an end. The muscles become stiff and inextensible, causing the limbs to become difficult to move. The solidification of fat due to a temperature drop in the carcass also helps to increase the firmness of the meat. The abundance of polyunsaturated fatty acids in fish oil preserves its fluidity.

Rigor mortis develops in a certain order in the carcass, starting with the head and neck, and spreading to the forelimbs, dorsal region and hind limbs.

Biochemical changes

In vivo, blood flow via respiratory pigments (hemoglobin and myoglobin) constantly supplies muscles with oxygen. This gas maintains the conditions necessary for aerobic glycolysis, and in particular the Krebs cycle, the electron transport chain and oxidative phosphorylation. Under aerobic conditions, free glucose or glucose from glycogen is a rich source of ATP. Degradation of a glucose molecule produces 36 ATP molecules. The end products of degradation are CO_2 and water.

After slaughter, blood circulation stops and oxygen no longer reaches muscle cells: the anaerobic stage quickly takes over. In this case, the Krebs cycle, the electron transport chain and oxidative phosphorylation can no longer continue. ATP is produced by anaerobic glycogenolysis, which is a metabolic pathway providing much less energy since the degradation of a glucose molecule produces no more than two ATP molecules. The final product obtained after the death of the animal is lactic acid, which accumulates in the muscles due to the cessation of blood circulation, and contributes to lowering muscle pH, which is the most significant outcome of rigor mortis.

Mechanisms of rigor mortis

Immediately after slaughter, the muscle has a sufficient supply of ATP; it maintains the dissociation of actin and myosin, and consequently muscle elasticity. Hydrolyzed ATP is replaced by new molecules from anaerobic glycolysis. It can also be regenerated from phosphocreatine and ADP, according to reaction [2.4]. New ATP molecules can also be synthesized directly from ADP (reaction [2.5]).

Together with the production of ATP, anaerobic glycolysis produces lactic acid that accumulates in the muscle. This results in a reduction in pH, which inhibits sarcoplasmic ATPase allowing Ca^{2+} ions to pass outside the reticulum vesicles. When the calcium concentration in the sarcoplasm exceeds 10 µM, ATPase activity of myosin begins. ATP is hydrolyzed, actin binds to myosin (formation of actomyosin complex) and the muscle fiber stiffens. Reserves of ATP, phosphocreatine and glycogen in the muscle are gradually depleted and the steady accumulation of lactic acid in the muscle inhibits the enzymes involved in anaerobic glycogenolysis. Therefore, less ATP is synthesized, and more actin binds to myosin. The "shortening" of the muscle cell increases and finally a state of rigor mortis is reached. At this point, the pH is about 5.8 (final pH). To drop from 7.3 to 5.8, about 100 µM of lactic acid per g^{-1} of muscle is required, corresponding to the consumption of 50 µM of glucose.

The time needed to reach the final pH depends on various factors: species, breed, muscle type, temperature and so on. It can vary from 10 min for exudative pork to 48 h for a large muscle mass of beef.

In general, the final pH of fish after the onset of rigor mortis ranges between 6.2 and 6.5; the pH is higher than in meat because the glycogen content is lower in fish. There are exceptions, as in the case of tuna, which has a final pH of less than 6, and flatfish, where the pH can drop to around 5.5.

Factors influencing the kinetics of rigor mortis

Initial muscle glycogen content

Lowering the pH from neutral to 5.8 during the development of rigor mortis is an important factor in obtaining good quality meat. Achieving this final pH depends on the glycogen level in the muscle during slaughter. If it is insufficient, which is the case with many animals having suffered major stress before slaughter, the final pH of the muscle will not be low enough after rigor mortis sets in. The result would therefore be bad quality meat such as dark firm and dry (DFD) meat.

Initial levels of ATP and phosphocreatine

The initial levels of ATP and phosphocreatine play a major role in the rate of development of rigor mortis. The greater the reserves of ATP and phosphocreatine at the time of slaughter, the slower rigor mortis develops. Rigor mortis begins when the ATP level drops below 90% of its initial value. If the animal is numb before slaughter or if it bleeds out under anesthetic, this removes all sources of ATP consumption by muscle excitation and rigor mortis develops very slowly. In contrast if the muscles are excited (e.g. by electrical current) before slaughter, rigor mortis develops quickly.

Temperature

As the temperature drops from 37 to 5°C, rigor mortis is slower to develop since the cold slows down all exothermic biochemical processes in the muscle. However, below 5°C, the onset of rigor mortis is as rapid as at 15°C. This phenomenon is due to an increase in the ATPase activity of actomyosin by the cold (300-fold rate increase), which is accompanied by muscle contractions and consequent shortening depending on the pH: this is known as cold shortening or cold shock. The ATPase activity of actomyosin is increased in the presence of Ca^{2+} ions. Under cold conditions, these ions cannot be maintained within the vesicles of the sarcoplasmic reticulum because the sarcoplasmic ATPases of the calcium pump have a very high Q_{10} and are therefore inhibited by low temperatures. In the case of fish, and especially those living in cold water, refrigeration temperature is recommended to slow down the rigor mortis phase. However, with some tropical species, refrigeration temperatures are not shown to extend the pre-rigor phase and a temperature of close to 20°C is preferred.

If the carcass is cooled too quickly after slaughter, cold shortening can occur, with rigor mortis developing in the contracted state. This can have a negative impact on meat quality, and in particular tenderness.

The ideal cooling temperature for mammalian muscle is 10°C for 10 h after slaughter. A compromise should be found to avoid cold shortening and excessive bacterial growth.

Other changes in the muscle after slaughter

Between slaughter and the onset of rigor mortis, there is a drop in pH from 7 to 5.8 for meat and 6.2 for fish, a decrease in redox potential due to the lack of oxygen supply (after death, the muscle acquires reducing properties), an increase in conductivity, an overall loss of extensibility and a reduction in the water retention capacity. The latter is due to the drop in pH (the final pH of meat is close to the average pI of muscle protein) and the release of Ca^{2+} ions, which shields protein charges resulting in a closer myofibrillar network. This is very important because it determines the succulence of the meat and influences, to varying degrees, weight loss during the bleeding of the carcass and cooking.

Abnormal rigor mortis and consequences

Dark firm dry meat

DFD meat is abnormal, dark red in color, bland, unappetizing and does not keep well (1–2% of cattle and 5–10% of young bulls). Its pH is abnormally high, with glycogen reserves having been exhausted before the death of the animal. DFD meat occurs in animals that have been subject to stress prior to slaughter, and more frequently in younger animals. Stress triggers a discharge of catecholamines, which accelerates glycogenolysis. For this reason, it is advised to transport animals a few days before slaughter. The administration of tranquilizers before transport is also effective, but prohibited by law. The final high pH promotes a high level of hydration (sticky texture), dark color (low reflectance) and uncontrolled microbial growth. The causes of this abnormal meat may be:

– excitation prior to slaughter;

– breed: Limousine is more sensitive than Holstein;

– type of farming: animals reared outdoors produce more glycogen and DFD meat is less frequent.

Pale, soft and exudative (PSE) pork

Exudative pork is flaccid, pale and has a poor water retention capacity. After slaughter, the muscle temperature is abnormally high and the final pH is reached very quickly (sometimes in 10 min). The processing yield of PSE pork can be 10% less than normal pork.

– *Causes*: In "normal" pork, many muscles are rich in white fibers with anaerobic metabolism producing lactic acid. In exudative pork, this is exacerbated and a large amount of lactic acid is produced. These animals also have a hormonal imbalance due to genetic selection, which favors anabolic (energy consuming) rather than catabolic (energy generating) hormones. Finally, they are hypersensitive to stress due to a lack of calcium permeability in the sarcoplasmic reticulum and mitochrondrial membranes. In addition, the reticulum is abnormally sensitive to high temperatures, causing the release of a high amount of calcium as well as violent muscle contractions.

High stress levels in living animals can therefore have serious consequences in the form of violent muscle contractions and elevated temperatures of up to 43–44°C, causing the death of the animal: this is known as malignant hyperthermia syndrome, and is not the result of a heart condition in the animal. This disorder is relatively common in older double-muscled breeds with poor blood circulation and a high potential for producing lactic acid. It is less common in cattle because of their more efficient ATPase regulatory systems. In pigs, this defect is controlled by an autosomal recessive gene. Only male or female homozygous animals have this defect, which also depends on breed: it is common among Belgian Landraces and Pietrains, less common among French Landraces and non-existent in Large Whites. The method of slaughter also has an impact on the PSE defect (slaughter performed under poor conditions, stressed animals). Stunning animals by electronarcosis is preferable to using CO_2, which increases the harmful effects of stress.

– *detection of PSE defect in vivo*: This defect can be detected in living animals using the halothane (or fluothane) test. This gas interferes with the movement of calcium in PSE pork, causing a release of Ca^{2+} ions and strong muscle contractions. The animal becomes stiff, cyanotic and its temperature increases. The animal recuperates by no longer breathing halothane. It is also possible to measure creatine phosphokinase, which exists in small quantities in PSE pork;

– *impact on meat quality*: At the time of slaughter (high stress), anaerobic glycogenolysis produces large amounts of lactic acid. Due to the deficiency of vasodilatory adrenal steroids, lactic acid is inadequately removed and quickly accumulates in the muscle. The drop in pH after the death of the animal is rapid and, due to a high body temperature, results in a severe denaturation of muscle proteins that lose their ability to retain water. To compensate for the loss in processing yield, PSE pork should be cooled as soon as possible after slaughter. Cooling the core is difficult; it can be done by injecting very cold brine (-10°C).

2.2.2.5. *Post rigor phase: maturation*

The maturation of meat results in a significant improvement in tenderness. Two protein fractions play a key role in determining tenderness: connective tissue protein (collagen) and myofibrillar proteins. No distinct change in the connective tissue has ever been shown during maturation. The increase in tenderness would therefore appear to be mainly due to the change in myofibrillar proteins. The following is observed during maturation:

– a partial destruction of bonds between actin and myosin during rigor mortis;

– a change in myofibrillar proteins by intracellular proteases;

– a partial destruction of bonds between thin actin filaments and Z-lines.

2.2.2.5.1. Physical changes

From the onset of rigor mortis, several histological changes occur in the muscle fibers: folds, ripples, or knots. During maturation, the sarcoplasm may shrink, allowing water and soluble substances to pass through the sarcolemma. Some fibers may lose their structure (loss of striation) and take on a homogeneous appearance. These major changes only affect a limited number of fibers, however.

During maturation, the shear force and work of the muscle as well as the resistance to compressive force decrease with increasing tenderness. The shear index α (dimensionless), related to meat hardness, is expressed as:

$$\alpha = \frac{W}{F_M \, l_0} \qquad [2.6]$$

where W is the shear work (J), F_M is the maximum force necessary to ensure shear (N) and l_0 is the initial thickness (m). During rigor mortis, α is always greater than 0.5. After maturation, it ranges between 0.1 and 0.2.

2.2.2.5.2. Biochemical changes

The biochemical mechanisms that lead to the maturation of meat are still poorly understood. They result mostly from the release of enzymes maintained *in vivo* in lysosomes. While they are significant and essential in beef, they are less so in pork, which can be eaten fresh or after processing (charcuterie or deli meat products). Maturation is carried out under cold conditions to limit microbial growth. In fish, the maturation phase is quickly followed by the autolytic phase and therefore putrefaction. Enzymatic activity in fish is more intense than in meat; this characteristic is exploited, in particular in the manufacture of sauces like nam pla (Thai fish sauce) or fermented pastes. The cathepsin activity of a fish fillet is almost ten times greater than that of pork muscle.

Changes in nitrogen fractions

The muscle cell contains specific membrane-bound organelles of different sizes called lysosomes. Like mitochondria, they are surrounded by a triple-layered lipoprotein membrane, and contain a number of enzymes, the main ones being cathepsins B, D, L and H. These are acid proteases, active within a pH range of 4 – 6. Two other intracellular enzyme systems are also involved in meat maturation: calpains and proteasome. After the onset of rigor mortis and the final pH of about 5.8 has been reached, approximately 15% of lysosomal enzymes are released and become active. They are largely responsible for the changes in muscles during maturation, in particular the increase in meat tenderness. In fish, this is less important since the final pH is much higher.

During maturation, the non-protein nitrogen content of meat does not increase much; after 50 days of experimental maturation under cold conditions, it only increased by 6%. The sarcoplasm contains calcium-activated factor and cathepsin inhibitors. Proteolytic activity in the muscle is therefore quite low. Actin–myosin complexes formed during rigor mortis are hardly altered and the muscle remains inextensible. In addition, at the end of maturation, the amount of actomyosin extracted by a high ionic strength solution remains high and often greater than that extracted a few hours after slaughter (at the peak of rigor mortis, it drops to 75%). This is due to a disintegration of Z-lines (solubilization of α-actinin), allowing the release of actomyosin complexes and an increase in intracellular osmotic pressure up to 500–600 mOsmol (twice the normal physiological value). An important consequence is the increase in meat tenderness. The rate of maturation depends on many factors, the most important being the breed of animal.

Finally, as previously mentioned, connective tissue proteins (collagen, reticulin, elastin) hardly change during maturation. Only cathepsin B1 has limited activity on collagen. As already mentioned, the rate of development of rigor mortis depends on the initial level of ATP in the muscle. Immediately after death, ATP can be regenerated from anaerobic glycogenolysis, phosphocreatine (reaction [2.4]) and ADP (reaction [2.5]). When the level of phosphocreatine is less than 30% of its initial level, ATP is degraded as follows (where AMP is adenosine monophosphate and IMP is inosine monophosphate):

ATP → ADP + P

ADP → AMP +P with AMP

AMP → IMP + NH_3 with IMP [2.7]

IMP → Inosine + P

Inosine → Hypoxanthine + Ribose

Ammonia, which slightly raises the pH of meat during maturation, and hypoxanthine are flavor enhancers that give meat its characteristic flavor. Ribose, a reducing sugar, is involved in the Maillard reactions that take place during the cooking of meat.

Myoglobin also undergoes changes during maturation. The decrease in the redox potential favors the formation of myoglobin in its reduced form (purple-red color). However, the acidic pH promotes the oxidation of myoglobin to metmyoglobin (brown color). Meat color depends on the proportion of the three pigments in the muscle (myoglobin, oxymyoglobin and metmyoglobin) and its surface condition (DFD meat). Its intensity is measured at 525 nm, the isosbestic point of the three pigments.

The non-protein nitrogen content is particularly high in fish. These components are easily degraded from the onset of the maturation phase and increase during autolysis or putrefaction. These degradation reactions are described in section 2.2.2.6.

Change in carbohydrates

After rigor mortis sets in and the final pH is reached, residual glycogen undergoes enzymatic degradation to glucose:

Glycogen → Dextrin → Maltose → Glucose [2.8]

In addition, the catabolism of mucopolysaccharides is closely linked to the activity of lysosomal enzymes (β -glucuronidases in particular).

Change in lipids

During the maturation of meat, there is a low level of lipolysis characterized by the release of fatty acids from triglycerides, mainly under the action of microbial lipases. The increase in free fatty acids in the muscle can cause flavor defects, especially if the released fatty acids undergo oxidation (rancidity). The change in lipids only poses problems for frozen meats, especially pork (a meat rich in fat and unsaturated fatty acids) since the oxidation of fatty acids is autocatalytic and occurs at low temperatures. Lipid oxidation also limits the storage of frozen oily fish, high in unsaturated fatty acids (mackerel, sardines, sprat, etc.).

2.2.2.6. Autolysis or putrefaction

Meat and fish are no longer consumable during the autolytic phase. In the case of fish, this phase develops quickly during maturation. The muscle components, and in particular non-protein nitrogenous matter, are degraded. Reactions include the degradation of nucleotides (reaction [2.7]), the deamination and decarboxylation of free amino acids (arginine to putrescine or lysine to cadavarine by decarboxylation), the degradation of trimethylamine oxide (TMAO) to trimethylamine or dimethylamine and formaldehyde, and the degradation of urea to ammonia and CO_2. These degradation products are responsible for the foul odor of putrefying or rotting fish. Their concentration is often used as an index of freshness or spoilage.

2.3. Meat and fish processing technology

2.3.1. *Meat processing technology*

Charcuterie products are extremely diverse, so only an example of a cooked whole product (ham) and a processed raw product (dried sausage) are described in this section. Other products classified according to the "French code of practice for charcuterie", cured meats and canned meats are shown in Table 2.7. They are classified based on their raw materials and process technology.

Category	Product type
1, 2, 3	Pieces and cuts 1) Uncooked, cured, pickled, steamed and/or smoked pieces 2) Matured/dried uncooked pieces 3) Cooked pieces and cuts of meat
4, 5, 6	Sausages and salamis 4) Sausage meat, sausages and uncooked or cooked salamis 5) Dry sausages and salamis 6) Cooked sausages and salamis
7	Pâtés, mousses, terrines, creams, *"galantines"*, *"ballottines"*, *"confits de foie"*
8	*"Rillettes"* (terrine of cooked shredded meat), *"frittons"*, *"grattons"*
9	Head or tongue products
10	*"Andouille"* (coarse-grained smoked pork sausage), *"andouillette"* (coarse-grained chitterling sausage)
11	Tripe, *"tripoux"* (stuffed sheep stomach or haggis), feet
12	Black pudding
13	White pudding, *"quenelles"* (fish or meat dumplings)
14	Canned beef products
15	*"Foie gras"* and *"foie gras"* products
16	Other products

Table 2.7. *Classification of charcuterie and canned products [FRE 16]*

2.3.1.1. Manufacture of ham

The manufacture of ham is one of the most common processes in the food industry. Some hams are uncooked and cured with dry salt, known as dry-cured ham (e.g. Parma ham, Serrano ham or Bayonne ham), while others are cooked such as bone-in ham (cooked with the bone), York ham (matured for over a week before being cooked with the bone) or baked ham. These hams are said to be *superior* (or premium) when they have no additives to increase water retention during cooking (polyphosphates, sugar, carrageenan, etc.) or "*de choix*" when polyphosphates are added. For example, the manufacturing process of a premium cooked ham is based on the following steps:

– *Choice of raw material*: Before processing, it is important to ensure the freshness of the product, the maintenance of the cold chain during transport, the absence of defects (hematoma, torn rind, etc.) and low pH meat (no PSE meat; pH <5.7). The latter is essential, since pH strongly influences the quality of the ham and processing yield.

– *Preparation of the meat*: The preparation of the meat involves separating the hock, completely skinning and removing external fat, deboning, internal trimming (i.e. separating the bones) and carefully peeling them.

– *Injecting brine/tenderizing*: brines contain a number of ingredients/additives such as:

- salt (NaCl), which slows bacterial growth by lowering a_w, promotes solubilization thereby increasing the techno-functional properties of muscle proteins (emulsifying capacity, binding capacity etc.), gives a salty taste and enhances flavor,

- nitrite salts (NO_2^-, NO_3^-), which play a role in color (reaction of nitrogen oxide [NO] on metmyoglobin), flavor and microbial growth (they inhibit the growth of *Clostridium botulinum* in particular),

- sugars (dextrose, sucrose etc.), which enhance the reducing power of the medium, and in particular serve as a nutrient medium for the bacteria responsible for reducing nitrate to nitrite; they also contribute to the flavor of cured products,

- sodium ascorbate, a reduction catalyst for NO_2^- to NO, which protects myoglobin from oxidation,

- various flavorings (onion, bay leaf, thyme, etc.),

- water, which hydrates the product and gives a soft texture after cooking.

Brine is injected into the muscles using a multiple needle device. During this phase, the meat is also tenderized (by a system of small blades that pierce the connective tissue) to reduce collagen shrinkage during cooking, and promote the extraction of cellular contents after massaging:

– *Massaging*: This operation involves bringing sufficient quantities of soluble proteins to the surface of the muscle, where they coagulate during cooking and bind the muscle (producing a surface glue (*"limon"*) contributing to better ham texture.

– *Moulding/maturation/cooking:* Individual pieces (1 unit = 1 ham) or combined pieces (several hams in the same mold) are vacuum molded. After molding, maturation is often necessary to improve the sensory properties (maturation at 4–5°C for 24–48 or 72 h). The ham may be cooked in water or steam at a constant temperature (68–70°C) or at an incrementally increasing temperature (first stage at 60–62°C for 5–7 h, coagulation of proteins on the outer layer limiting the shrinkage of collagen; second stage at 68–70°C in order to reach 65–66°C in the center).

– *Cooling*: This is carried out in two stages. In the first stage, at the end of cooking, the ham is sprayed with cold water to allow it to cool to room temperature (1–2 h to reach 55°C). In the second, longer stage, the ham is stored at 2–3°C for 48–72 h. This phase allows the coagulated protein gel to set (mix of intracellular protein coagulum and collagen gel), improves the integrity of the slice of ham and processing yield, and limits exudation after slicing. It also helps to prevent microbial growth.

– *Unmolding/unpacking/repacking*: Unmolding and unpacking are essential, as cooking juices or water should be removed before final packaging. This stage must be carried out under good hygienic conditions (in a clean room) to avoid contaminating the product that is ready for consumption.

2.3.1.2. Production of dried sausage

Dried sausage is an uncooked, cured, chopped, fermented and dried product. It is composed of approximately 2/3 lean meat and 1/3 fat. The lean meat may be from different types of animals (usually beef and pork). The fat, generally pork, should be firm and non-oily (back fat, fat covering ham, etc.).

The production process begins with pre-curing the lean meat (with salt or nitrate) and/or the fat (with salt only). The lean meat and fat are then minced at slightly above 0°C and slightly below 0°C, respectively. Additives are then mixed in (salt, sugar, and nitrate or nitrite salts). The mixture is allowed to stand for a number of hours at 0–5°C (additives act on the proteins) before being stuffed into natural or artificial casings. An external flora (surface bloom) is applied and allowed to grow during the parboiling stage (1–5 days, H_R 85–90%, temperature 20–28°C), at which point dehydration begins. The product is then dried for 4–6 weeks in a dryer (12–16°C, H_R 75–85%).

Good product development depends on a series of successive or simultaneous physical, chemical and microbiological changes.

Changes in the physical appearance of dried sausage

The physical appearance of dried sausage changes during curing and drying. The color changes from brown, due to the oxidation of the meat by salt, to dark red through the formation of nitrosometmyoglobulin. It becomes firmer due to the coagulation of proteins by acidification and dehydration. The casing shrinks, causing grains of fat to protrude.

Change in water activity

The a_w of a product changes over the various stages of the production process. Just after salting, the salt dissolves on the surface of grains and causes a local reduction in a_w from 0.98 to 0.93; a_w then increases to 0.96 after complete dissolution of the salt. During curing and after drying, a_w drops to 0.85.

Change in pH

The change in pH is an important feature in the production of dried sausage. During the curing stage, acidifying microorganisms (such as lactobacilli, staphylococci, streptococci, etc.) from the sugars in the meat and the sugars added during seasoning, produce organic acids (especially lactic acid), causing a drop in pH. This acidification is important because it selects an acidophilus flora used in the production of dried sausage, promotes the coagulation of dissolved proteins (binding fat grains with lean meat), and reduces the carbohydrate content and the water retention capacity of proteins, thereby facilitating drying. However, it can cause the development of acidic flavors, inhibit certain microorganisms and enzymatic reactions responsible for aroma, and induce excessive dehydration.

Change in redox potential

When producing the mixture, the sausage meat is highly oxygenated and the redox potential is high. During the maturation stage microbes multiply, depleting the oxygen, and the mixture becomes anaerobic. The redox potential decreases, promoting the growth of useful bacteria (lactobacilli, staphylococci, etc.).

Change in proteins

Proteins undergo several changes during the production process: salting causes their dissolution, acidification leads to their precipitation (aiding cohesion of the sausage meat), and enzymatic proteolysis results in the production of amino acids, peptides, ammonia and amines during the maturation stage, which is responsible for the sensory qualities of the sausage.

Change in lipids

First, lipolysis causes the release of free fatty acids from glycerides, at which point oxidative rancidity may occur. The reaction is catalyzed by light, oxygen, temperature and the presence of oxidants (salts, nitrates, metals). Alcohols, ketones and aldehydes responsible for rancid flavor are then produced.

Change in internal bacterial flora

From the beginning, the meat mixture is naturally contaminated with various microorganisms. This contamination should, however, remain low (maximum 10^6 microbes per gram). The microorganisms originate from the various processes from slaughtering to cutting of the meat; the flora is therefore very complex and varies from one site to the next. During the production and maturation of sausages these microorganisms grow, depending on their tolerance to the environmental conditions (temperature, salt content, pH and redox potential).

Dried sausage is a fermented product, taking its sensory characteristics from fermentation, but which should be stable at room temperature. It is essential to destroy unwanted bacteria while promoting the growth of bacteria necessary for preserving the product and developing sensory qualities. The flora of dried sausages can be divided into four general categories (Table 2.8):

– *useful flora* contributes to the acidification of the product and inhibits the growth of harmful bacteria. It promotes a reduction in nitrates, which by reacting with metmyoglobulin, gives charcuterie meats their pink color. It is also involved in the lipolysis and proteolysis necessary for the release of aromatic compounds;

– *flora harmful to health* corresponds to pathogenic flora capable of surviving processing and causing food poisoning;

– *spoilage flora* often causes manufacturing defects (production of unpleasant odors and gases);

– *neutral flora*, although present in significant amounts, does not play any known role in the production of dried sausage.

Type of flora	Microorganisms	Function or effect
Useful flora	Lactic acid bacteria (*Lactobacillus, Pediococcus, Leuconostoc*)	Fermentation (sugars)
	Non-pathogenic staphylococci and micrococci	Fermentation (glucose, lactose); reduction (nitrate to nitrite); lipolysis
	Yeasts	Lipolysis; formation of aromatic compounds
Pathogenic flora	Fecal coliforms, anaerobic sulfite-reducing, *Staphylococci aureus, Salmonella*	–
Spoilage flora	*Leuconostoc, Pseudomonas, Enterobacter, Clostridium*	Formation of gas
Neutral flora	Fecal streptococci (group D), *Bacillus, Corynebacterium, Microbacterium*	–

Table 2.8. *Composition of the different types of flora present in the production of dried sausage*

Change in external bacterial flora

After stuffing, the sausages are covered in a flora called "surface bloom". This includes fungi from the *Penicillium* genus *(nalgiovensis, chrysogenum, caseicolum, candidum,* etc.*)* and yeasts (*hansenii, prisca,* etc.), which inhibit the growth of different-colored natural blooms (green, grey, brown, yellow, blue).

In conclusion, all the physical, chemical and microbiological changes that occur during the production process give dried sausages their many sensory properties while also enabling them to be stored at room temperature. The principle of preservation is very well explained and summarized by the principle of hurdle technology [LEI 85], showing complementarity between the main preservation factors of dried sausage (Figure 2.11).

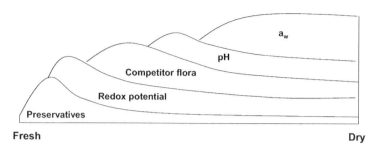

Figure 2.11. *Hurdle technology [LEI 85]*

2.3.2. *Fish processing technology*

Most fish is sold as whole fish or fillets, either refrigerated or frozen. Shellfish and crustaceans are also seldom processed, since they are mostly sold alive. If processed, fish products can be salted, dried, smoked, pickled or made into spreads like seafood charcuterie, butter, cream or taramasalata (fish roe paste). Fish can also be processed into surimi. As with meat, there are many different types of fish processing. Only pickled fish and surimi will be dealt with in this section.

2.3.2.1. *Production of pickles*

Pickling is a method of stabilizing fish while at the same time giving it specific and desired sensory qualities. It involves immersing the whole fish, fillet or shellfish meat into a pickling marinade, that is a solution containing acid and salt. Usually vinegar is used, but any organic acid permitted in the food industry may also be used (e.g. acetic acid). Other ingredients like sugar, spices and condiments are often added to the marinade also.

During pickling, part of the water content of the product is replaced by the marinade. There are several different types:

– *cold marinades* (rollmops, pickled herring fillets wrapped around a pickle or onion);

– *cooked marinades* (shellfish): shells are removed by cooking and the meat is then placed in the marinade;

– *fried marinades* (whiting marinade): the fish is fried in oil before being packed in a marinade;

– *jelly marinades* (sardines): after soaking in the marinade, the sardines are rinsed and then preserved in a jelly of potassium carrageenan.

In all marinades, the combined effects of salt and acid are necessary for the preservation and stability of the product. In the case of hot marinades, it is also possible to add a heat treatment by pasteurizing after marinating. The products obtained can be stored for several months, away from light, at below 15°C. In the case of cold marinades, products must be stored below 3°C for 2 – 4 weeks.

Marinating is carried out in two stages:

– immersing the product in a marinade of vinegar (5 – 10% acetic acid) and salt (10 – 15%) for several days. The fish acquires its characteristic texture and flavor;

– after washing, the product is immersed in a milder marinade of vinegar (1 – 2% acetic acid), salt (2 – 4%), spices and sometimes sugar.

During marinating, two phenomena occur:

– a characteristic softening of the fish meat, mainly due to proteolysis by autolytic enzymes in the fish tissue;

– the removal of water and coagulation of tissue proteins due to the acid and salt content.

Effect of salt

The rate of salt penetration in fish meat is expressed according to Fick's law (see Chapter 9 of Volume I). It depends on the salt concentration of the brine, the diffusion coefficient of the salt, the temperature at which salting is performed and the conductivity of the muscle. It is more difficult for the salt to migrate through oily fish; however in more mature fish, it penetrates the inner layers faster (better conductivity of the muscle).

At the beginning of salting when the salt concentration is between 2 and 4%, the amount of water bound to proteins increases (swelling); this is known as turgor pressure. Increasing the ionic strength generates negative charges on the surfaces of the proteins, which leads to an increase in repulsive forces within and between the polypeptide chains and an increase in hydration. As the NaCl concentration increases (between 3 and 12%), myofibrillar proteins solubilize, and above 12% some proteins precipitate (e.g. albumin). At salt concentrations close to saturation, almost all the proteins precipitate (salting-out effect).

Effect of acid

Acid penetrates fish meat faster than salt. It inhibits the growth of certain microorganisms by lowering the pH of the meat. The pH of marinades is usually 4.5, which inhibits the growth of most bacteria responsible for food poisoning. However, some yeasts and moulds can tolerate a lower pH (1.5–2.5). As a result, the preservative effect of marinades is based on the combined effects of salt and acid.

During marinating, proteins are partially hydrolyzed by tissue enzymes (proteolysis is however limited by the presence of salt) with the release of peptides and amino acids, which gives marinades their characteristic flavor. Some of the released amino acids may migrate to the marinating liquid.

Finally, the pickled product is preserved in sealed acid-resistant containers. These tend to be glass jars with metal lids capable of withstanding corrosive substances, plastic capsules, plastic buckets with lids or plastic containers with heat-sealed lids.

2.3.2.2. Production of surimi

Surimi, a Japanese word literally meaning "ground meat", is a fish-based product that has been pulverized and washed. It is made from whole fish or fillets caught exclusively for surimi production. Surimi made from fish fillets produces a higher quality protein base. After the addition of salt and heat treatment of the surimi base, a protein gel is formed called a *kamaboko*, which is mainly used for crab sticks. Other surimi-based products include imitation lobster tail and scallops.

Production of the surimi base

The surimi base can be made from white or oily fish, although white fish (Alaska pollock, blue whiting, pouting, etc.) is more widely used because of its light color (premium quality). Surimi can also be made from low-value oily fish due to its characteristic sensory qualities (e.g. *Sardinella*). Using oily fish is more difficult because, despite washing the fish meat several times, especially in an aqueous ozone solution, the result is a darker-colored surimi, making it less suitable for the production of crab sticks for example. The second aspect of the surimi base is its techno-functional properties, in particular gelling, emulsifying and foaming qualities. For this, it is essential to make surimi from very fresh raw materials. As a result, it is often made directly on board factory ships.

The production process involves repeatedly washing the fish flesh with water or low ionic strength solutions so as to remove sarcoplasmic proteins (enzymes, pigments, blood and heme compounds). The washed pulp contains myofibrillar proteins (actin and myosin) and connective tissue proteins (mainly collagen, as well as reticulin and elastin). It is drained and cryoprotectants are added such as sugar or polyols (sorbitol), and polyphosphates. The surimi is then frozen in the form of 20 kg blocks at -20°C (Figure 2.12).

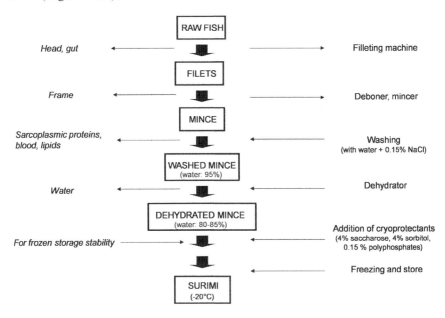

Figure 2.12. *Production process of the surimi base*

The surimi base has a high nutritional value since it is rich in protein (~16%) and low in fat (<0.2%); however, the washing process reduces the vitamin (B_{12}) and mineral (K^+) content.

Production of kamaboko products

Kamaboko is obtained after the heat-induced gelation of proteins in the surimi base. The production process is based on the solubilization of proteins (mainly myofibrillar such as actin and myosin, but also tropomyosin and troponin) using salt and the denaturation of these proteins by heat treatment. The increase in temperature induces the formation of an ordered protein network by aggregation of denatured proteins. The quality of the gel (gel strength) depends on:

– maintaining the proteins in their native state prior to heat treatment, which requires the use of raw materials that are as fresh as possible;

– the heating conditions during the gelling stage.

The different stages in the production of kamaboko can be summarized as follows:

– dissociation of the F-actin–tropomyosin–troponin complex;

– unfolding of the α-helix of F-actin;

– dissociation of myosin into heavy and light chains;

– partial unfolding of the α-helix in the heavy chain of myosin;

– formation of several intermolecular associations resulting from hydrophobic interactions by exposing hydrophobic residues;

– dissociation of the F-actin–myosin complex;

– the aggregation of heavy myosin chains by hydrophobic interactions and disulfide bond formation.

It is best to carry out protein gelation at two different temperatures (Figure 2.13).

The first phase of gelation is carried out at around 45°C, which is known as the *suwari* stage: the interactions between the tail portions of myosin facilitate the formation of a three-dimensional network containing water molecules. The second stage, called *ashi*, results in kamaboko. It involves strengthening the protein network by promoting intermolecular hydrophobic interactions at high temperatures (>80°C) and forming disulfide bonds within the head portion of myosin. The resulting gel is firm and elastic.

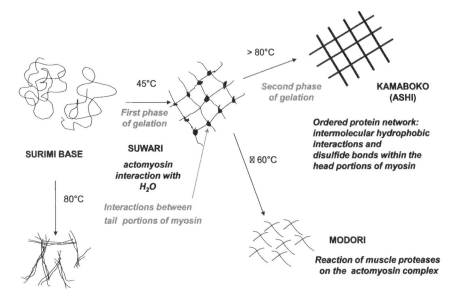

Figure 2.13. *Formation of kamaboko gel*

Heating the fish paste to 80°C in a single step does not allow a satisfactory gel network to form because the network is fixed without sufficient time for structural rearrangements. Moreover, gelation at 60°C results in a more fragile network, and therefore a less firm gel. This is mainly due to the reaction of muscle proteases on the actomyosin complex. The addition of egg white, rich in anti-protease activity, to the surimi paste limits this gelation defect.

Kamaboko gel is a protein base that is very easy to color and flavor, which is why is it used for a wide variety of products of different shapes, colors and flavors. Japan produces a much larger range of such products compared to Europe.

3

From Eggs to Egg Products

Chicken eggs are a multifunctional ingredient because, apart from their nutritional value, they can simultaneously serve several technological functions in a formulated food product. They are a universal ingredient in the home and food industry alike due to their emulsifying, foaming, gelling, thickening, coloring and aromatic properties (Table 3.1). In some preparations, more specific properties are required, whereby the egg white and yolk are used separately: egg white is known for its foaming properties while egg yolk is an excellent emulsifier.

	Whole egg	*Egg white*	*Egg yolk*
Biscuits *Pastries* *Custards, etc.*	Colouring Binding - Gelling Foaming	Foaming Gelling	Emulsifying Colouring
Confectionary		Control of crystallisation Foaming	
Ice cream	Emulsifying - Binding		Emulsifying
Charcuterie (pâtés, *dumplings), surimi*	Emulsifying Binding	Gelling	
Pasta	Colouring – Binding Gelling		
Mayonnaise *Sauces*			Emulsifying Thickening
All industries	Nutritional value and aromatic properties		

Table 3.1. *Main techno-functional properties of eggs and their parts in food applications*

Chapter written by Marc ANTON, Valérie LECHEVALIER and Françoise NAU.

The exceptional foaming properties of egg white form the basis of many traditional recipes, with meringue being the most well-known one. The simplicity of the formulation (egg white and sugar, flavorings are optional) facilitates the optimal foaming of egg white. Foamed egg white is added to many other products, whether with fat (ladyfingers, sponge cake, soufflés) or without (angel food cake). These foaming properties are also used in the preparation of a variety of products where foaming is carried out after all the ingredients, including fats, are mixed together. The foaming properties of whole egg are also used, such as in whipped omelettes, boudoir biscuits, meat, vegetable or fish mousses, and so forth.

In the food industry, egg yolk is incorporated into several products because of its exceptional emulsifying properties and the fact that it gives food a desirable color and flavor. Egg yolk is thus an essential ingredient in the production of hot (e.g. Béarnaise and hollandaise sauces) and cold emulsions (e.g. mayonnaise and salad dressings). It contributes to the formation and stabilization of emulsions, on the one hand by reducing the interfacial tension between oil and water, and on the other hand by forming a protective barrier around oil droplets preventing breakup.

The gelling properties of egg, mainly whole egg or egg white, are used in several food applications (e.g. pastry, cakes, biscuits, *charcuterie*, pasta, etc.). In all cases, gelation is achieved by heating (cooking stage).

With the exception of the coloring and flavoring properties of egg, linked respectively to pigments and aroma compounds in the egg yolk, all of the techno-functional properties of egg and its parts are textural or structural, primarily involving proteins as well as some fat molecules (emulsifying properties of lipoproteins).

Numerous studies have been carried out and are still underway to understand the physicochemical phenomena behind the various techno-functional properties of egg, in terms of the role of each major constituent. The difficulties encountered in modeling these phenomena are due to a poor understanding of the composition and complex structure, of the egg yolk for example. The synergies between the constituents are also not easily understood. In addition, in many food preparations and formulations, the complexity and the juxtaposition of phenomena involved further complicate the understanding of the mechanisms: this is the case with certain *charcuterie* products where foaming, emulsification and gelation take place simultaneously or successively. One of the major challenges in controlling the techno-functional properties of egg can be overcome by understanding the behavior of the different components, even though extrapolation from model systems (especially simple protein solutions) to "real" food systems is not always possible.

3.1. Chicken egg – raw material in the egg industry

3.1.1. *Structure and composition*

Due to intensive breeding for many decades and the genetic homogeneity of hen breeds, the average weight of a hen's egg varies little, ranging from 55 to 65 g. The egg consists of three main parts: the shell (about 10% of the total weight), the egg white (60%) and the yolk (30%) (Figure 3.1).

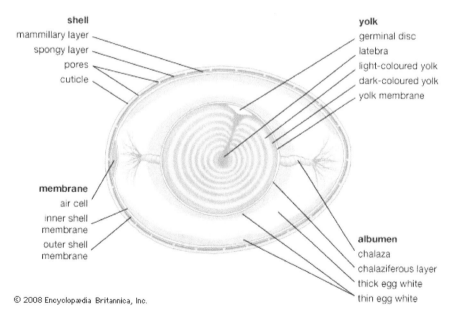

Figure 3.1. *Internal structure of a hen's egg. For a color version of this figure, see www.iste.co.uk/jeantet/foodscience.zip*

The shell and its membranes, although not edible, are essential for the quality of the egg contents due to their role as a physical barrier against microbiological contamination. The shell itself is inorganic (95% (w/w) minerals, of which 93.5% is calcium carbonate), while the cuticle surrounding it as well as the two shell membranes separating it from the egg white are organic (composed of protein). The latter, formed by the superposition of layers of interwoven protein fibers, act as an effective barrier against bacteria and mold, which could pass through the shell pores if the cuticle were no longer intact.

The egg white is primarily an aqueous solution of proteins, carbohydrates and minerals (Table 3.2). Despite this simple composition, it is a heterogeneous medium made up of four distinct layers in fresh egg (Figure 3.1):

– *chalaziferous layer*, a very firm thick albumen layer surrounding the vitelline membrane, which extends to opposite ends of the egg by the chalazae. It represents 3% (w/w) of the total egg white:

– *outer thin albumen* (23% of egg white), next to the shell membranes;

– *thick albumen* (57% of egg white), attached to both ends of the egg;

– *inner thin albumen* (17% of egg white), surrounding the yolk.

These four parts differ in water content (ranging from 84 to 89% from the inner to the outer layers) and protein concentrations. Thick albumen contains four times more ovomucin than thin albumen, giving it a gel-like structure and much higher viscosity. In addition, physicochemical changes during storage result in an increase in pH due to the loss of CO_2 (from 7.5 at the time of laying to 9.5 after a few days), liquefaction of the thick albumen and a transformation of ovalbumin to more heat-stable S-ovalbumin.

The egg yolk has a solid content of close to 50% (Table 3.2), comprising mainly lipids (around 65% solids) and proteins (33%). It also contains a large quantity of minerals including calcium, iron and phosphorous. The yolk is a dispersion of particles (profiles and granules), maintained in equilibrium in an aqueous protein solution (plasma). The profiles, which are low-density lipoproteins (LDLs) measure 12–48 nm in diameter. The granules are in the shape of more or less flattened spheres measuring 0.3–2 µm.

	Whole egg	Egg white	Yolk
Water	76	88	50
Proteins	12.5	10.6	16
Lipids	10.5	-	33
Carbohydrates	0.5	0.8	0.5
Minerals	0.5	0.6	0.5

Table 3.2. *Average composition of whole chicken egg, egg white and yolk in % of total weight (according to [THA 94])*

The yolk can be separated into two parts by dilution in NaCl (0.17 M) and centrifugation (10,000 g):

– *plasma* (orange supernatant) constitutes 75–80% of yolk solids and consists of LDL (85%) and soluble proteins called livetins (15%). It accounts for approximately 55% of the proteins and 85% of the phospholipids in the yolk;

– *granules* (whitish pellet) constitute 20–25% of yolk solids and contain high-density lipoproteins or HDL (70%), phosvitin (16%) and residual LDL or LDL_g (12%) (Table 3.3). Granules account for around 47% of the proteins and 15% of the phospholipids in the yolk. HDL and phosvitin, the main components of granules, contain a large proportion of phosphorylated serine residues that allow their association by divalent calcium ions. The large number of calcium phosphate bonds produces a compact low-moisture structure that is poorly accessible to enzymes and protects proteins from denaturation and thermal gelation.

Whole egg is made up of 76% water, with solids distributed in almost equal proportions of proteins and lipids (Table 3.2). This composition makes the egg a low-energy food (6.25 J g^{-1}). It does however have a high nutritional value due to its proteins (no limiting amino acid for adult humans), lipids (very good digestibility with a naturally high level of unsaturated fatty acids) and abundance of phosphorus, iron and many vitamins. This universally-consumed food is however low in calcium, completely deficient in vitamin C and fiber, and is known to be a major food allergen, especially among young children.

		% yolk solids	% yolk lipids	% yolk proteins	Composition (%)	
					lipids	proteins
Yolk		100	100	100	64	32
Plasma		78	93	53	73	25
	LDL	66	61	22	88	10
	Livetins	10	-	30	-	96
	Others	2	-	1	-	90
Granules		22	7	47	31	64
	HDL	16	6	35	24	75
	Phosvitin	4	-	11	-	95
	LDL_g	2	1	1	88	10

Table 3.3. *Distribution of the constituents of chicken egg yolk [POW 86]*

3.1.2. *Biochemical and physicochemical properties of the protein and lipid fractions of egg*

3.1.2.1. *Egg white proteins*

Proteins constitute the main part of egg white solids (more than 90% of the dry matter). The total number of egg white proteins is not precisely known. Until recently, only the main egg white proteins were identified. The recent development of powerful separation and analytical techniques has made it possible to find many minor proteins, some of which have already been identified. Table 3.4 shows the main properties of the currently-known proteins. 0

Egg white proteins are mostly globular proteins with an acidic isoelectric point apart from lysozyme and avidin. They are all glycoproteins, except for lysozyme and cystatin, and are rich in sulfur amino acids. Many of these proteins have biological properties that contribute to the protective role of the egg white during embryonic development, in particular with regard to certain pathogens. Some are very heat sensitive and/or highly susceptible to surface denaturation, thus contributing to the remarkable techno-functional qualities of egg white.

Protein	%	MW (kDa)	pI	Biological properties
Ovalbumin	54	45	5.0	Immunogenic phosphoprotein
Ovalbumin Y	5.0	44	5.2	ND (not determined)
Ovalbumin X	0.5	56	6.5	ND
Ovotransferrin	13	76	6.7	Binds iron, bacteriostatic activity
Ovomucoid	11	28	4.8	Trypsin inhibitor
Ovomucin	1.5–3.5	230–8,300	4.5–5	Highly glycosylated, inhibits viral haemagglutination
Lysozyme	3.5	14.4	10.7	Lyses walls of Gram+ bacteria
Ovoinhibitor	0.1–1.5	49	5.1	Serine protease inhibitor
Ovoglycoprotein	0.5–1	24.4	3.9	ND
Flavoprotein	0.8	32	4.0	Binds riboflavin (vitamin B2)
Ovostatin	0.5	760–900	4.6	Serine protease inhibitor, highly allergenic
Cystatin	0.05	12.7	5.1	Cysteine protease inhibitor
Avidin	0.05	68.3	10	Binds biotin
Ex-FABP	ND	18	5.5	Lipocalin family
Cal gamma	ND	20.8	6.0	Lipocalin family
TENP	ND	47.4	5.6	BPI family (bactericidal permeability-increasing protein)
Hep 21	ND	18	6.4	uPAR/Ly6/snake neurotoxin family

Table 3.4. *Protein composition of egg white (based on [LIC 89, STE 91, GUE 06]*

Ovalbumin is the main protein component of egg white, accounting for more than half of the total protein content. It is a globular, phosphorylated (usually two phosphates per molecule) protein with a molecular weight of around 45 kDa that belongs to the serpin family, despite lacking protease inhibitory activity. Of the 385 amino acid residues comprising the protein, half are hydrophobic and one third are charged, mostly negatively at physiological pH. Ovalbumin also has six Cys residues buried inside the protein, two of which are involved in a disulphide bond (Cys^{73}-Cys^{120}). Thus, it is the only egg white protein with free thiol groups capable of rearrangements depending on storage, pH and surface denaturation conditions. During storage or moderate heat treatment in alkaline conditions, ovalbumin acquires a more heat-stable conformation called S-ovalbumin through the isomerization of specific amino acid residues (Ser^{164}, Ser^{236} and Ser^{320}, which take on a D configuration).

Ovotransferrin, also called conalbumin, belongs to the transferrin family. It is a polypeptide chain of 686 amino acid residues (~77.7 kDa) arranged in two lobes, each with a binding site for iron (or Cu^{2+}, Zn^{2+}, Al^{3+}). Ovotransferrin is the most heat-sensitive protein in egg white: its denaturation temperature at pH 7 is approximately 63°C; it makes this protein a limiting factor when heating treatments are applied. Iron (or aluminum) binding causes a conformational change in the protein and an increase in its thermal stability.

Ovomucoid is an acidic protein (pI 4.1) of about 28 kDa, which can contain up to 25% carbohydrates (w/w). It is the most allergenic egg white protein. At pH 7, its denaturation temperature is approximately 77°C. However, it is very heat-resistant at acidic pH. In these conditions, ovomucoid retains its antitrypsin activity after heat treatment of several minutes at 100°C.

Ovomucin is also a highly glycosylated protein of high molecular weight (up to around 10^4 kDa). It is insoluble in diluted egg white (lower ionic strength), and after precipitation, its resolubilization is very difficult. Ovomucin can associate with other egg white proteins (ovalbumin, ovotransferrin and especially lysozyme) by electrostatic interactions. At the natural pH of egg white (between 7.5 and 9.5), sialic acid carboxyl groups of ovomucin can interact with the amine groups of lysine residues in lysozyme, forming a lysozyme–ovomucin complex, which is insoluble in water and is responsible for the gel structure of egg white, especially thick albumen.

Lysozyme is an enzyme responsible for the lysis of Gram-positive bacterial cell walls by the hydrolysis of β-1,4-bonds between N-acetylmuramic acid and N-acetylglucosamine of mucopolysaccharides in the cell wall of the bacteria. It is a globular protein consisting of 129 amino acid residues, 40% of which are hydrophobic and one-third are charged; the majority of charged residues are basic, giving protein its particularly high pI (10.7). Lysozyme has a rigid structure stabilized by four disulfide bonds (Cys^6-Cys^{127}, Cys^{30}-Cys^{115}, Cys^{64}-Cys^{80} and Cys^{76}-Cys^{94}).

3.1.2.2. Protein constituents of egg yolk

LDL are major constituents of egg yolk: they represent 66% of its dry matter and 22% of its proteins. They consist of 83–89% lipids and 11–17% proteins. Lipids are divided into 74% neutral lipids (triglycerides and cholesterol) and 26% phospholipids. There are two types of LDL: LDL_1 (20% of LDL) with a molecular weight of 10,300 kDa and LDL_2 (80% of LDL) with a molecular weight of 3,300 kDa.

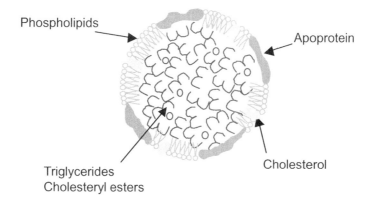

Figure 3.2. *Schematic representation of egg yolk low-density lipoprotein. For a color version of this figure, see www.iste.co.uk/jeantet/foodscience.zip*

LDLs have a typical lipoprotein structure (Figure 3.2) consisting of a neutral lipid core (triglycerides and cholesterol esters) surrounded by a monolayer of phospholipids and proteins in contact with the aqueous phase. They are spherical in shape with a diameter ranging from 17 to 60 nm. Egg yolk LDLs derive from lipoproteins synthesized in the chicken's liver, known as very-low-density lipoproteins (VLDL). VLDLs are transported by the bloodstream to the oocyte or immature egg cell. Specific receptors in the oocyte membrane allow binding of VLDLs which are then secreted into the egg yolk by endocytosis and changed to LDLs.

LDL proteins are particularly difficult to study because of their low solubility due to the large proportion of hydrophobic amino acid residues (40%), making them some of the most hydrophobic proteins identified. They are glycosylated with a pI ranging from 6.5 to 7.3.

Livetins are lipid-free globular proteins found in plasma. They represent 11% of the dry matter and 30% of the protein in the egg yolk. They are blood serum proteins deposited in the yolk, with a pI of between 4.3 and 5.5.

HDLs represent around 16% of egg yolk dry matter and 36% of its proteins. They consist of two subunits, α and β, which have a very similar amino acid composition, but differ in their degree of phosphorylation and sialic acid content. These subunits are both dimers with a molecular weight of 400 kDa. They contain 80% proteins and 20% lipids, divided into 65% phospholipids and 35% neutral lipids.

Phosvitin represents 11% of yolk proteins. Its molecular weight is between 36 and 40 kDa and its pI is 4.0. It is glycosylated and contains approximately 10% phosphorus (60% of the protein phosphorus of the yolk). More than half (54%) of phosvitin amino acid residues are serine, exclusively present in the form of phosphoric acid esters, giving it cation-binding properties: mainly iron and calcium. In addition, the polypeptide sequence of phosvitin contains hardly any or no cysteine, cystine, methionine, tryptophan and tyrosine, and only 10% hydrophobic amino acids. Phosvitin is therefore particularly hydrophilic compared to most proteins and has a negative net charge.

3.1.2.3. *Egg yolk lipids*

Lipids, the main components of egg yolk (60%), are exclusively distributed in lipoproteins (LDL and HDL). They are composed of triglycerides (65%), phospholipids (29%), cholesterol (5%), free fatty acids (<1%), and other lipids including carotenoids (<0.1%) that give yolk their yellow colour. Yolk phospholipids are very rich in phosphatidylcholine (76% of phospholipids) and phosphatidylethanolamine (22%).

The fatty acid composition of lipids, based on a standard hen diet, is about 30–35% saturated fatty acids, 40–45% monounsaturated fatty acids and 20–25% polyunsaturated fatty acids. The main fatty acids are oleic acid (C18:1, 40–45%), palmitic acid (C16:0, 20–25%) and linoleic acid (C18:2,

15–20%). However, this composition is subject to wide variation, especially depending on the type of fatty acids ingested by the hen.

3.2. Physicochemical properties of the different egg fractions

3.2.1. *Interfacial properties*

3.2.1.1. *Foaming properties of egg white*

Egg white is an excellent foaming agent; compared to other plant or animal protein ingredients, it possesses the best foaming properties. These properties are due to the very good interfacial properties of egg white proteins and their ability to maintain this foamed structure, especially after heat treatment. The foamability of globular proteins such as egg white proteins depends on the successive stages at or close to the surface of the gas bubbles:

– the diffusion of proteins to the air–water interface;

– the conformational changes and shrinkage of proteins adsorbed at the interface;

– the irreversible rearrangement of the protein film.

Therefore, the ability of egg white to form a foam depends on intrinsic factors, such as the structure and conformation of proteins, themselves dependent on environmental factors such as pH, ionic strength, protein–protein and protein–water interactions.

On the whole, egg white can be compared to a solution of effective surfactants. Its proteins are amphiphilic, and have a relatively high level of surface hydrophobicity, thus quickly diffusing to the air–water interface where they can efficiently adsorb. Their molecular flexibility allows conformational rearrangements (expansion) at the interface causing a significant decrease in surface tension. The ability to form a continuous intermolecular network, especially when a certain level of denaturation has first been reached, allows them to form a viscoelastic interfacial film ensuring foam stability. However, egg white proteins are heterogeneous and do not contribute to the same extent to the foaming properties. Many studies have attempted to attribute certain properties to the different protein fractions. Globulins have long been associated with foamability, ovomucin with foam stability at room temperature, and ovalbumin with foam structure after cooking. However, it is now known that the various roles of each egg white protein cannot be separated from each other due to the complexity and synergy of the phenomena. Thus, ovalbumin, lysozyme and ovotransferrin exhibit different behaviors at the air–water interface depending on whether they are on their own in solution or in a mixture [LEC 03].

Even though it is not possible to model the interfacial behavior of proteins according to their physicochemical properties, parameters such as surface hydrophobicity (which determines the efficiency of interface adsorption), the number of disulfide bonds (which determines protein flexibility), and the number of free thiol groups (which determines "reactivity") appear to be crucial with regard to the structural changes that take place at the air–water interface. However, the difficulty of establishing physicochemical rules linking the structure of each protein to its interfacial properties is compounded by the difficulty of extrapolation to the complex protein mixture of egg white. Competition for the interface and possible exchanges between proteins at the interface can occur over time. In a mixture of ovalbumin, globulins, ovotransferrin, lysozyme and ovomucoid in the same proportions as in egg white, [DAM 98] have shown that only ovalbumin and globulins adsorb at the interface. The exclusion of ovotransferrin, lysozyme and ovomucoid could be linked to their lower diffusion rates: since they reach the interface after the other two proteins, they would not be able to move them. But in a simpler solution of ovalbumin, ovotransferrin and lysozyme in the same proportions as in egg white, [LEC 05] demonstrated the involvement of the three proteins in the formation of the interfacial film, in which electrostatic interactions and covalent bonds were observed. In addition, analysis of the structural changes in each of these three proteins upon contact with the air–water interface revealed a phenomenon called "denaturation synergy": when ovalbumin, ovotransferrin or lysozyme are mixed, the structural changes following contact with the interface are greater than or equal to the changes observed when only one of these proteins is present in a solution.

Nevertheless, it is clear from a number of studies that the foaming properties of isolated egg white proteins are always much lower than those of egg white, which tends to confirm the existence and role of protein interactions in egg white. It is generally accepted that the natural coexistence of basic proteins (lysozyme) and acidic proteins (ovalbumin) in egg white is responsible for the electrostatic interactions, which partly explains the high level of stability of egg white foams.

3.2.1.2. *Emulsifying properties of egg yolk*

Food emulsions containing egg yolk are oil-in-water emulsions, i.e. consisting of oil droplets dispersed in an aqueous phase, such as mayonnaise, hollandaise sauce, custard and so on. An emulsion is thermodynamically unstable due to the high level of interfacial tension that develops at the interface between the two immiscible liquids: the greater the interfacial tension, the quicker the emulsion tends to destabilize. The main forms of destabilization are creaming, flocculation and coalescence (see Volume 2). All the aggregation states of droplets such as flocculation and creaming or the use of large-volume fractions of oil promote coalescence.

Nevertheless, it is possible to form visually-stable emulsions by slowing aggregation or droplet migration through the addition of emulsifiers and/or thickeners prior to homogenization (see Volume 2). To understand the main contributing factor to the emulsifying properties of egg yolk, several authors have separated the yolk into its main fractions: plasma and granules. By comparing the stability and particle size of yolk, plasma and granule emulsions, [DYE 93, NNA 93, ANT 97] demonstrated the similarities between yolk and plasma emulsions, whereas granule emulsions exhibited distinctly different properties (Figure 3.3) depending on their solubility [ANT 00]. Under emulsification conditions where the energy input creates oil droplets of less than 1 µm, the size of droplets in emulsions made of insoluble granules is greater due to a higher level of coalescence. In contrast, when granules are soluble (pH 7.0 and an ionic strength of 0.55 M), coalescence does not occur. Under these conditions, oil droplets are similar in size to those obtained with egg yolk. However, at equal protein concentrations, granules are less effective than plasma. This suggests that the molecules responsible for the emulsifying properties of egg yolk are located in the plasma.

Many authors have shown that yolk LDLs are better emulsifiers than bovine serum albumin (BSA) [MIZ 85] and β casein [SHE 79]. These authors and others [ALU 98, MIN 00] demonstrated that LDLs quickly replace caseins at the interface, thus proving their excellent surface properties. Soluble yolk proteins do not have the penetrative power of lipoproteins at the oil–water interface, especially under high surface pressure. However, at equivalent dry matter, they form finer and more stable emulsions in terms of creaming.

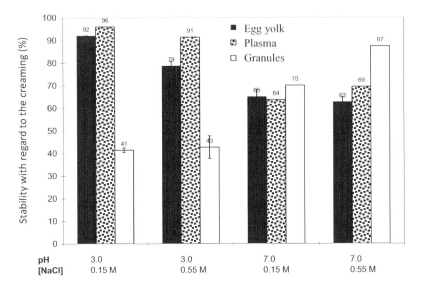

Figure 3.3. *Stability with regard to the creaming of oil-in-water emulsions containing yolk, plasma and granules; 30/70 O/W emulsions, protein concentration in the aqueous phase: 25 mg ml^{-1}*

Conversely, livetins are unable to adsorb when emulsified in a mixture containing caseins or egg yolk LDL. These results demonstrate the crucial role of LDL among plasma constituents and, consequently, in egg yolk.

[MAR 02] compared the emulsifying properties of LDLs and HDLs in egg yolk. It appears that LDLs produce emulsions with smaller oil droplets than those obtained with HDLs, regardless of the pH and ionic strength conditions. Combining all these results underlines the major implication of LDLs in the emulsifying properties of egg yolk.

We have seen that LDLs have a particular structure with a neutral lipid core surrounded by a film of phospholipids and proteins. The integrity of the structure of LDLs seems to be essential for their interfacial properties; but is it maintained when adsorption at the interface occurs? It is usually assumed that LDL micelles break up close to the interface through the weakening of protein–protein interactions. Then, the core lipids coalesce with the oil phase, and the proteins and phospholipids spread out at the interface [SHE 79, KIO 89]. The direct adsorption of proteins and phospholipids contained in the outer layer of LDLs is not easy due to their low solubility in water (high hydrophobicity). Thus, the interactions between proteins and phospholipids, which contribute to the formation of LDL structure, are essential for transporting soluble surfactants close to the interface before releasing them.

A comparison of the compression isotherm of LDLs (using a Langmuir film balance) with those of their individual components (proteins, neutral lipids and phospholipids) has provided a better understanding of the adsorption mechanism of LDLs at interfaces [MAR 03]. Three phase transitions at 20, 41 and 54 mN m^{-1} are visible in the compression isotherm of LDL (Figure 3.4). The single transition observed in the isotherm of neutral lipids corresponds exactly to the transition visible on the isotherm of LDL at 20 mN m^{-1}. The first transition of the LDL film is therefore attributed to neutral lipids that are definitely present in the interfacial layer created by the LDL. Neutral lipids form the core of lipoproteins and their presence in the LDL film requires LDL dissociation at the air–water interface. LDL should therefore be dissociated upon contact with the interface to allow the release and spreading of neutral lipids. The transition observed on the isotherm of phospholipids (54 mN m^{-1}) corresponds to the last transition of the LDL isotherm. The isotherm of total lipids also shows a transition at 54 mN m^{-1}, which is consequently attributed to the interfacial restructuring of phospholipids. However, given their insolubility in water, LDL proteins cannot be spread out at the interface. And the transition observed at 41 mN m^{-1} on the LDL isotherm can therefore not be assigned, even though proteins are highly suspected.

Given these results, it is therefore evident that LDLs must break up upon contact with the interface and spread out at the surface, thus enabling the adsorption of their constituents. LDLs thus serve as transport vehicles for emulsifiers (proteins and phospholipids) that are insoluble in water and adsorb once they are released near the interface.

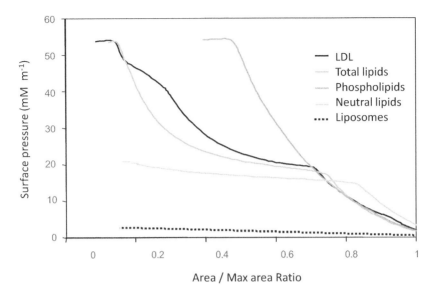

Figure 3.4. *Compression isotherm at the air–water interface of LDL, liposomes and the different lipid components of LDL. For a color version of this figure, see www.iste.co.uk/jeantet/foodscience.zip*

[MIZ 85] have shown that the increasing hydrolysis of LDLs by proteases (trypsin and papain) leads to a decrease in the properties involved in emulsion formation and stabilization. It was therefore suggested that only a small proportion of LDL phospholipids participate in adsorption at the oil–water interface and that the protein fraction may play a major role. This hypothesis was confirmed by measuring the interfacial concentration of proteins and phospholipids in emulsions containing egg yolk, plasma and granules [LED 00]; the interfacial protein concentration is correlated with emulsion stability and particle size, while the interfacial phospholipid concentration is not correlated to any indicator of emulsion stability, suggesting the important role of the protein fraction. Compression isotherms of phospholipids in solution or in the form of vesicles (liposomes) were compared with those of LDLs to understand the roles of proteins and phospholipids in the adsorption of LDLs. Phospholipids in the form of liposomes showed low surface pressure along the isotherm (Figure 3.4). This means that liposomes are not able to break down and spread at the interface. Conversely, inserting proteins at the surface of liposomes results in similar isotherms to those of LDLs. Apart from the existence of a core of triglycerides in LDLs, the main difference between liposomes and LDLs is the absence of protein on the liposome surface (phospholipid bilayer). It could therefore be suggested that the proteins located on the surface of LDLs play an initial role in LDL anchoring at the interface. This anchoring induces protein denaturation, resulting in the destabilization of the outer layer of LDLs and an overall spreading of LDLs at the interface.

Using atomic force microscopy, it was observed that the second transition of the compression isotherm of LDLs was due to the disassociation of triglyceride–protein complexes, while proteins and phospholipids appear to be miscible all along the isotherm. This clarifies the mechanism of interaction that occurs between the molecular species released during the adsorption of LDLs at an interface (Figure 3.5). For practical reasons, these experiments were carried out at the air–water interface. At the oil–water interface (in emulsions), triglycerides probably associate with the oil phase and the interface in this case is comprised solely of protein–phospholipid complexes.

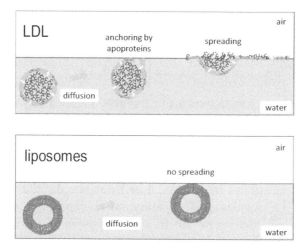

Figure 3.5. *Schematic description of the adsorption mechanism of egg yolk LDL and liposomes at the air–water interface. For a color version of this figure, see www.iste.co.uk/jeantet/foodscience.zip*

3.2.2. *Gelling properties*

3.2.2.1. *Egg white*

The gelling properties of egg white are used in many food applications (*charcuterie*, fish or vegetable terrine, pastries, etc.). The physicochemical process is the thermogelation of proteins; egg white gels are known as thermotropic protein gels.

Egg white loses its fluidity at around 60°C, but maximum firmness is reached above 70°C. With the exception of ovomucin and ovomucoid, all egg white proteins coagulate with heat. However, they do not all react in the same way to heat treatment; for example in egg white at pH 7, the denaturation temperatures of ovalbumin, lysozyme and ovotransferrin are 85, 74 and 63°C, respectively [DON 75]. The latter is thus the most heat-sensitive protein in egg white, which is why it is often regarded as the "initiator" protein in gelation. The coagulation temperature of egg white can be significantly increased and the rheological properties of the gels improved by selectively removing ovotransferrin or binding metal ions (Al^{3+}, Fe^{3+}) on this protein, thereby increasing its thermal stability.

The thermogelation of egg white proteins, resulting from a change in balance between the attraction and repulsion forces between proteins, corresponds to the long-established model of heat-induced gelation of globular proteins. It involves two stages: heat denaturation of proteins and aggregation of denatured proteins (see section 8.2.4).

The macroscopic characteristics of the gels obtained (firmness, breaking strength, elasticity, water retention capacity, etc.) depend on the number and type (energy) of interactions created. In thermotropic egg white gels, the bonds are mainly low-energy bonds (hydrophobic and electrostatic interactions), but high-energy covalent bonds (disulfide bonds) also exist. Regardless of the type of interaction, their formation depends on the protein conformation, i.e. the level of denaturation reached after the first stage, which in turn will determine the level of exposure of reactive groups or portions of the protein sequence. The formation of interactions also depends on the medium, which can be promoting or limiting. This will either accelerate or slow down the aggregation stage and thus restrict or increase the level of denaturation at the end of the first stage (reaction rate ratio of the two stages).

These mechanisms have been studied in the case of ovalbumin thermogelation and the impact of ionic strength on the structure and properties of gels. Thermal denaturation results in an overall increase in the surface hydrophobicity of ovalbumin. When heat is applied at high ionic strength, the surface charges carried by the protein are screened, causing a decrease in electrostatic repulsion and favoring the formation of hydrophobic interactions. Under these conditions, random aggregates of partly denatured proteins quickly appear, resulting in opaque gels, and lower retention capacity, elasticity and rigidity. Conversely, at low ionic strength, the electrostatic repulsions delay aggregation, thus promoting a greater level of denaturation (first stage). Moreover, this aggregation can only occur at certain contact points (hydrophobic patches), resulting in the formation of linear aggregates (Figure 3.6). Further replaced at higher ionic strength, these linear aggregates can interact to form translucent gels.

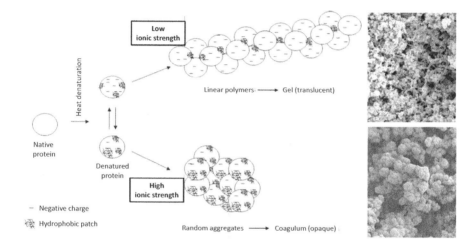

Figure 3.6. *Effect of ionic strength on the thermogelation of ovalbumin: mechanistic model (based on [DOI 93]) and electron microscopy images of corresponding egg white gels [CRO 02])*

Since ovalbumin plays a major role in the gelling properties of egg white, the impact of ionic strength as just described can be transposed to egg white. A two-stage heating process was proposed to prepare transparent, very firm and elastic egg white gels with exceptional water retention capacity [KIT 88]. The first stage consists of heating demineralized (by dialysis) egg white to form linear polymers of denatured proteins in solution. The second stage involves heat treatment, this time in the presence of salts, which induces the interactions between the linear polymers of denatured proteins. Similarly to ionic strength, which impacts egg white protein gelation by modifying the apparent net charge of proteins, pH is also a major factor influencing the gelation of egg white. When close to their pI, proteins tend to form random aggregates, similar to what has been described for high ionic strength. This results in low rheological properties for egg white at around pH 5. In contrast, egg whites exhibit the best gelling properties at extreme pH values, especially at basic pH (around 9). Under these conditions, strong electrostatic repulsions delay aggregation, increase denaturation and promote the formation of linear aggregates. But the higher reactivity of thiol groups in this pH region may also be the cause of the improved gelling properties by facilitating the formation of disulfide bonds.

3.2.2.2. Egg yolk

Egg yolk can gel when subjected to heat treatment or a freeze–thaw process. With a dry matter content of 24%, a NaCl concentration of 0.17 M and a pH of 6.1, egg yolk and plasma can gel after a heat treatment of 72°C for 2 min 30 s, whereas the granules do not gel [LED 99]. The components responsible for the gelation of egg yolk are plasma constituents, especially LDLs. The other yolk constituents either do not play a role or have little effect.

The gelation of plasma solutions (and therefore egg yolk) is favored at neutral pH and/or at high NaCl concentration. These conditions correspond to a significant neutralization of LDL protein charges, which limits the intensity of short-range electrostatic repulsions, and thus promotes protein aggregation. Gelation times are very short irrespective of pH and ionic strength. This is unique to the coagulation process. Gels are opaque, indicating light diffraction by large particles (protein aggregates).

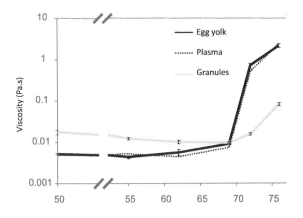

Figure 3.7. *Effect of temperature on the viscosity of solutions of egg yolk, plasma and granules (pH 6.1, 0.17 M NaCl, 24% solids)*

The apparent viscosities of egg yolk and plasma after heat treatment strongly correlate to the loss of protein solubility in the samples (Figure 3.7). The amount of soluble protein decreases when gelation occurs. However, the correlation coefficient between apparent viscosity and protein solubility is low for granules.

At an ionic strength of 0.17 M and pH of 6.1, granules seem to withstand greater heat treatment than egg yolk and plasma without undergoing gelation. The structure of the granules is preserved under these conditions:

– only 24% of granule proteins are soluble;

– calcium phosphate bridges are maintained in the granules;

– granule constituents, in particular phosvitin and HDL, are maintained in the aggregates.

Although the granular structure does not prevent the denaturation of some of the proteins (HDL-α and LDL), the resistance of phosvitin and HDL-β allows granules to retain their structure, limit their aggregation and restrict the formation of a gel network.

In contrast, under conditions where granule constituents are dissolved (0.5 M NaCl), gelation occurs during heat treatment. The gels formed are opaque, owing to their particulate structure. Due to the very small proportion of apo-LDL (3% of granule proteins) and the resistance of phosvitin to heat treatment, it can be deduced that HDL is responsible for the gelation of granule solutions.

Solutions containing purified LDL (4% w/v) begin to denature at 70°C and form gels above 75°C. The same solutions heated to 80°C for 5 min form more stable gels than those obtained with BSA. Unlike BSA, LDL can gel upon heating over a wide pH range. Between pH 6 and 9, LDL forms opaque gels, whereas at extreme pH values (pH 4 – 6 and pH 8 – 9), the gels are translucent.

LDL gelation upon heating is governed by protein denaturation. The first step is the breakdown of LDLs promoted by protein denaturation. The second step involves an increase in protein–protein interactions, especially hydrophobic interactions.

As regards the gelation of LDL upon freezing, it appears that the dehydration caused by the formation of ice crystals promotes protein denaturation and subsequent protein–protein interactions, resulting in gel formation, similar to that in heat-induced gels.

3.3. The egg industry: technology and products

Egg products refer to products obtained from eggs, their components or mixtures thereof, excluding the shell and membranes, which are intended for human consumption; they may be partially supplemented by other foodstuffs or additives; they can be liquid, concentrated, dried, crystallized, frozen or coagulated. Based on this definition, egg products can be classified into two general categories: egg products after primary processing, corresponding to the whole egg, egg yolk or egg white presented in different forms for use as techno-functional ingredients; and egg products after secondary processing, corresponding to the industrial-scale preparation of eggs based on classic recipes, which include cooked or precooked products such as peeled boiled eggs, poached eggs, fried eggs, omelets, stiffly beaten egg whites and so forth.

The egg industry is relatively new (1970s) and has grown considerably over the years. As with any industry, this sector is subject to a number of regulatory constraints, with the main objective being hygiene control. To this end, requirements have been put in place for the organization of industrial sites and the quality of raw materials: only non-incubated eggs fit for human consumption with fully-developed intact shells may be used in the manufacture of egg products. The content of an egg from a healthy hen is in most cases sterile and, if the shell is intact, will remain so for a long period of time without any special storage conditions. If the egg is cooled (12–15°C) between laying and processing, the risk of microbial growth in damaged eggs (e.g. those with a crack in the shell) is greatly reduced. However, once the egg's natural protection (i.e. the shell) has been removed, its content will inevitably become contaminated; sterile breaking is impossible on an industrial scale and storage poses several difficulties. While egg white is not a favorable medium for microbial growth, since it lacks many nutrients necessary to bacteria and since it contains many antimicrobial factors (lysozyme, ovotransferrin, ovomucoid, avidin, etc.), egg yolk or whole egg are ideal culture media for microbial growth.

In order to control the microbiological quality of egg products, it is necessary to limit their initial contamination, remove all or part of the contamination flora, and limit or even prevent its development by using different stabilization methods. The difficulty facing manufacturers is the fragility of egg proteins with regard to stabilization processes as well as any resulting loss in techno-functional properties, which is particularly harmful in the case of the primary processing of eggs. The egg industry therefore offers a range of products in order to respond more closely to the techno-functional applications of egg products (Figure 3.8).

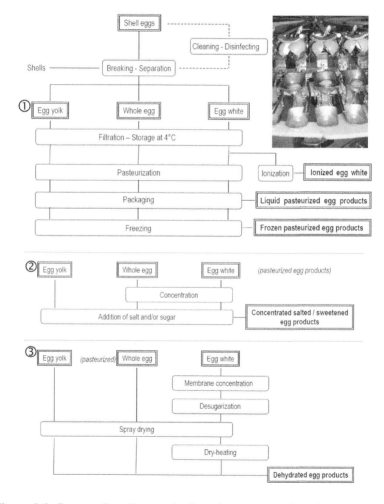

Figure 3.8. *Process flow diagram for the primary processing of egg products*

3.3.1. Decontamination of shells

Apart from following good hygiene practices (cleaning and disinfecting equipment, personal hygiene, "workflow from dirty to clean areas"), one way of limiting the contamination of egg products is to clean and disinfect the shells before breaking. Removing the surface flora by brushing, washing, disinfecting and drying the eggs can considerably reduce the microbial population of the egg product, provided however that these operations are carried out very carefully; the quality of drying in particular is a determining factor in the final result. The disinfection of shells, carried out on farms in some countries, requires compliance with the cold chain and limits the shelf life of eggs since the cuticle, removed during these operations, can no longer act as a barrier. In several countries, the washing of eggs is prohibited in the production and sale of fresh eggs but permitted just before breaking. In practice, very few European egg manufacturers wash eggs before breaking.

3.3.2. Breaking and separation of the egg white and yolk

Eggs must be individually broken, either mechanically or manually. Specific equipment (known as a "breaker") has been developed, which can to more than 200,000 eggs an hour with instant separation or only a very slight delay between breaking and separation of the white and yolk. The products are then filtered to remove shell fragments (target is less than 100 mg of shell per kg of product), and cooled to 4°C before being sent to storage tanks. Here, adjustments can be made to the dry matter, and salt, sugar or permitted additives (e.g. guar or xanthan to egg white) may be added.

After breaking and separating white and yolks, the basic products obtained are egg white, with a dry matter content of 10–11% and a pH of 8.5–9.5, egg yolk with a dry matter content of 42–48% and a pH of 6.5, and whole egg with a dry matter content of 20–24% and a pH of 7–7.5. Variations in the dry matter content of yolk and whole egg are mainly linked to equipment performance (quality of white–yolk separation), which itself depends on the type of device and production rate. Shells removed at this stage are a by-product with a relatively high moisture content (almost 30%) due to traces of egg white still attached to the shells. This residual moisture creates a significant risk of microbial growth in the by-product that must therefore be controlled. The shells are crushed and "spun dry" by centrifugation, which reduces the moisture content by about 50%. They can then be dried in hot air tunnels before being used as a liming material for agriculture. A fraction of these shells is also intended for human consumption (calcium supplements) after having undergone heat treatment to ensure the complete destruction of microorganisms.

3.3.3. *Primary processing of egg products – decontamination and stabilization*

3.3.3.1. *Heat treatment*

Heat sensitivity is one of the properties of egg proteins which has the greatest impact on processing. The most sensitive egg white proteins begin to denature at 57°C, which means it is not possible to sterilize egg products or apply pasteurization conditions like those in the dairy industry, for example. Despite this, the most effective and common method of decontaminating egg products remains heat treatment, which includes pasteurization.

The treatment used (time–temperature combination) depends on several factors:

– desired shelf life: ranging from several days at less than 4°C for liquid egg products in bulk intended for the food industry, to several weeks (up to 60 days) for smaller packagings (1–2 kg) intended for cooks or pastry chefs, restaurants, small businesses; use-by dates of several months at room temperature can also be obtained for products concentrated and supplemented with high concentrations of sugar or salt;

– desired functionality (foaming or gelling properties of egg white in particular): the negative effect of heat treatment on the functional properties is proportional to its intensity.

In all cases, the aim is the complete destruction of pathogens, in particular *Salmonella*, and the safe preservation of the product up to the use-by date.

Conventional tubular or plate heat exchangers are used to pasteurize egg products. The pasteurization conditions for whole egg and yolk are usually 65°C for 2–6 minutes and 57°C for 2–6 minutes for egg white. There are also some specific devices, such as ohmic heating devices or concentric tube heat exchangers, which enable the eggs to be heated to a higher temperature for a shorter time (e.g. 70°C for a 100 seconds). This type of treatment, followed by ultra-clean packaging, produces egg products with a long shelf life.

Since liquid egg white cannot be subjected to high temperatures, industrial operators have developed a specific heat treatment for this product in a dry state. The dry-heating of egg white has become a routine decontamination process, which has the advantage of requiring little investment. The powder is kept in a hot chamber (around 65°C for about 10 days) to destroy pathogens and a large portion of the harmless flora. More intense dry-heating conditions (75 – 80°C for about 10 days) improve the functional properties of egg white (Figure 3.9). The improvement in gelling and foaming properties can be explained by the structural changes caused by this type of treatment. Increased protein flexibility, exposure of reactive groups including an increase in surface hydrophobicity as well as a decrease in the rate of protein aggregation and an expansion of the polymers formed are responsible for the improved techno-functional performance of dry-heated egg white [KAT 89, KAT 90a, KAT 90b].

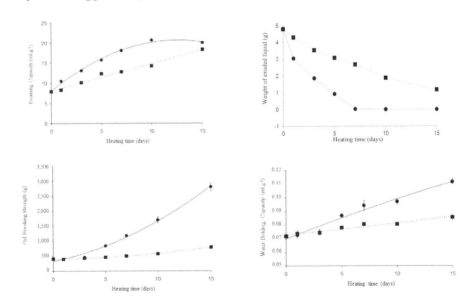

Figure 3.9. *Change in foaming and gelling properties during dry-heating of egg white at 67°C and 75°C [BAR 03]*

3.3.3.2. Ionizing radiation

The ionization of frozen egg white or egg white powder is an effective technique with few side effects, as egg white contains no lipids. However, the required machines (irradiators) are few in number, treatment is expensive and users are generally very hesitant with regard to this type of technology. Moreover, this process is unsuitable for egg yolk or whole egg since it causes lipid oxidation, with detrimental consequences for sensory properties (off-flavor).

3.3.3.3. Decrease in a_w

Different stabilization techniques, conventionally used in other food sectors, are applied to egg products to prevent microbial growth. They are all based on the principle of lowering water activity below thresholds that allow microbial growth.

Freezing can thus extend the shelf life of products by up to 24 months. It significantly affects the egg yolk and whole egg, however, causing irreversible gelation and a change in functional properties (considerable increase in the viscosity of the product after thawing). It has less impact on the egg white. Nevertheless, this stabilization method is hardly used in practice because the defrosting times are relatively long with a high risk of microbial growth during this stage. In addition it can only be used for small or medium packs (i.e. below 20 kg).

The concentration of whole egg or egg white by ultrafiltration or reverse osmosis, followed by the addition of high concentrations of sugar and/or salt can result in products that are stable for several months at room temperature. The concentration of whole egg by a factor of 2 (around 48% dry matter) combined with the addition of 50% sugar or 9% salt gives an a_w of 0.80 and 0.85, respectively. In the case of egg yolk, which is naturally highly concentrated (about 50% dry matter), the simple addition of 12% salt or 50% sugar yields the same a_w values. In the case of egg white, membrane concentration is usually only performed up to a final dry matter content of 33% (volume concentration ratio of 3), which does not yield such low a_w values, even after the addition of sugar and/or salt; nevertheless, a significant decrease in a_w (to 0.88) can still be obtained using this method.

Drying produces powdered egg products, which is the easiest, safest, most flexible and long-lasting form of preservation. The numerous benefits explain why egg powder production has developed across the world, with certain countries using egg products only in this form. The technology used is spray drying in horizontal or vertical dryers. In the case of egg white, prior "desugarization" is necessary to prevent the Maillard reaction occurring during drying, even more during the dry-heating stage. This desugarization consists of removing glucose (around 0.5 g l^{-1} in liquid egg white). This is done by controlled fermentation (using yeast or bacteria) or by enzymatic means using a glucose oxidaze–catalaze enzyme system.

3.3.4. *Secondary processing of egg products*

Processed egg products, intended almost entirely for the catering industry, represent about 10% of total egg products, depending on the country, and of those, peeled hard-boiled eggs represent the main part of the market. The industrial process is similar to that carried out in the home. The cooking stage consists of boiling in water at 98–100°C or steaming. After lowering the temperature by immersion in an ice-water bath, the shells are cracked up and then removed. This is probably one of the most difficult steps to control on an industrial level, which explains the large number of methods and pieces of equipment currently available. Shells are generally broken by mechanical shock. Machines for removing egg shells are equipped with a system of wheels, bars and rubber "fingers" as well as high-pressure water jets to facilitate the operation. In all cases, a visual inspection and a manual removal of residual shell fragments is necessary after the machine operation. The peeled hard-boiled eggs are then rinsed, drained and packed in brine or wrapped in plastic, under a modified atmosphere if necessary. The two key stages that must be controlled by manufacturers during the production process are:

– the degree of cooking, which must result in the yolk being fully cooked without being over-cooked, as this can lead to the appearance of a green/black ring around the yolk. This defect occurs as a result of the formation of iron sulfide through a chemical reaction between sulphur, abundant in the egg white, and iron in the yolk, the kinetics of which is temperature-dependent;

– the quality of shelling, which depends on the ease with which the shell fragments can be separated from the cooked egg white. Ease of peeling is correlated to the pH of the egg white before cooking, which should be above

8.5; this criterion therefore excludes the use of very fresh eggs for this type of production.

Several methods have been developed to produce a large range of products on an industrial scale, such as poached, fried and scrambled eggs, omelets, whipped egg whites, etc. Despite the apparent simplicity of these products, several technical challenges exist in the various production processes.

3.3.5. *Egg extracts*

It is possible to extract a large number of constituents from eggs, especially proteins, on a laboratory or pilot plant scale. While these products are mostly available from manufacturers of biochemical products, lysozyme is the only one used in both the agri-food and pharmaceutical industries for its antibacterial properties, and is therefore produced in large quantities worldwide. Lysozyme is extracted by cation exchange chromatography (see section 10.1.2).

PART 2

Food from Plant Sources

4

From Wheat to Bread and Pasta

Wheat is one of the most common cereal grains; it is rich in starch and can be ground into flour and semolina. Cereals have been regarded for thousands of years as an important nutritional component of the human diet due to their energy value, despite being deficient in some amino acids and vitamins as well as having antinutritional factors (decalcifying effects caused by phytic acid in the bran). Separation and fermentation processes can improve the nutritional value [AUB 85]; the natural fermentation of cereals by yeasts and lactic acid bacteria leads to an increase in vitamin and amino acid content, improved digestibility and better microbiological quality.

Unlike berries and fruit that can be consumed immediately in their natural state, grains can be stored but must be processed before being fit for human consumption; Figure 4.1 shows:

– the mechanical dehulling, milling and sieving processes used in the production of flour;

– fermentation processes (spontaneous or controlled fermentation), which can improve the processing, nutritional and sensory quality of the grains;

– heat treatments, which have a significant effect on sensory and nutritional quality; two types of cooking can be carried out:

- *High temperature baking*, in an oven, on coals or on heated stones or plates (Figure 4.2); this significantly improves flavor because of the Maillard reaction and caramelization. Changes in texture make it possible to directly

---

Chapter written by Hubert CHIRON and Philippe ROUSSEL.

consume certain types of liquid or semi-liquid mixtures, like pancake batter for example. Thick mixtures like dough only become palatable if they have been allowed to rise prior to baking,

– *Cooking in water* improves the sensory quality, texture and digestibility of dough even if the flavor is not greatly improved.

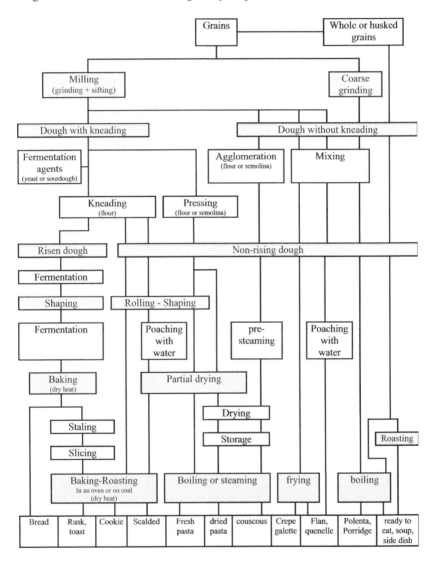

Figure 4.1. *Traditional methods of processing wheat*

Figure 4.2. *Baking in a vertical tandoor oven in Pakistan*

Over the centuries, an improvement in palatability has led to the creation and diversification of textures and flavors. Consumers have always sought properties such as tenderness, friability and softness in food, which has resulted in an ongoing effort to improve the quality of fresh produce and increase shelf life [MAU 32].

Other ingredients can be used in the processing of cereals, and the term "cereal products" applies only if the proportion of cereal grains remains high. In the cereal sector, distinctions are made between:

– products from the primary processing of grains, such as flour, semolina, hulled grains, flakes or malt;

– products from the secondary processing of grains, such as bread, pastries, cakes, biscuits, pancakes, pasta, starch or beer;

– the animal feed industry is also included in the cereal sector.

4.1. Biochemistry and physical chemistry of wheat

4.1.1. *Overall composition*

Cereals belong to a range of plant families that include:

– Gramineae (true grasses) or Poaceae: wheat (common wheat, durum wheat, Einkorn wheat and spelt or dinkel wheat), rye, triticale, rice, barley, oats, corn, millet and sorghum;

– Polygonaceae: buckwheat;

– Chenopodiaceae: quinoa.

These grains have similar compositions (Table 4.1). They are classified as starch products due to their high starch content, and have a medium protein content and a low fat content.

Species	Water	Starch and simple carbohydrates	Protein	Fat	Fiber	Minerals (ash content)
Oats	13–15	50–54	12–13	5.0–6.0	14–15	2.5–3.0
Common wheat	13–15	64–68	10–12	1.7–1.9	5.0–5.5	1.7–1.9
Durum wheat	13–15	62–66	13–14	1.8–2.0	5.0–5.5	1.8–2.0
Corn	13–15	58–62	9–11	5.0–5.5	10–11	1.0–1.1
Barley	13–15	57–63	10–11	2.0–2.5	10–11	2.5–2.7
Rice (cargo)	13–15	70–72	7–8	1.8–2.4	2–3	1.0–1.5
Buckwheat	13–15	57–63	10–11	2.0–2.5	11–12	1.9–2.1
Rye	13–15	62–66	9–11	1.7–1.8	7–8	1.9–2.1
Triticale	13–15	61–65	12–13	1.7–1.8	6–7	1.9–2.1
Quinoa	13–15	56–60	12–14	5.0–7.0	8–10	2.2–2.5

Table 4.1. *Chemical composition of cereal grains (%)*

The two main wheat species are:

– common wheat, which contains 42 chromosomes (21 pairs). Its kernel (endosperm) is relatively friable or "floury" making it very suitable for the production of flour. It is valued for its use in the production of bread, pastries, cakes, biscuits, pancakes, sauces, etc.;

– durum wheat, which contains 28 chromosomes (14 pairs). Its hard kernel makes it suitable for the production of semolina, used in pastas and couscous. The production of flour is nevertheless possible even though the grain size is larger or rounder (Table 4.2). It can also be used in bread making (common in Mediterranean countries).

Characteristics	Durumwheat	Common wheat
Species	Triticumdurum	Triticumaes
Specific weight (kg hl^{-1}) Thousand-kernel weight	75–85 (often>80) 25–60 g	70–80 35–50 g
Appearance(Figure4.3)Length Width Thickness	Elongated grain, open crease, white bran, yellow endosperm; bearded 6–9 mm 2.5–4.0mm 2.2–3.2mm	Round short grain, very shallow crease, redbran, slightly bearded 5–8 mm 3–4 mm 2.5–3.5 mm
Physical characteristics of kernel	Vitreous, hard	Floury, vitreous, soft
Products after milling	Semolina: 70–75% Middlings: 18–22%Durum groats (flour):	Flour: 75–80% Bran:12–15% Sharps: 5–7%
Ash content of kernel(% ash)	~ 0.60–0.70%	0.30–0.35%

Table 4.2. *Main physical characteristics of wheat grains*

Both species contain proteins that form gluten when the flour or semolina is hydrated and mixed. The gluten of common wheat gives a better bread texture. Each species includes varieties with varying compositions, especially protein, making them more or less suitable for bread (common wheat) or pasta (durum wheat); (Figure 4.3).

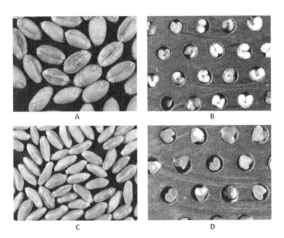

Figure 4.3. *Different aspects of wheat grains: a) common wheat; b) common wheat (section); c) durum wheat; d) durum wheat (section)*

4.1.1.1. Histological characteristics of wheat grain

Wheat grain is a caryopsis, which is a botanical term referring to a dry one-seeded fruit (achene) where the outer layer of the fruit (pericarp) is fused to the outer layer of the seed (seed coat or testa). It is an indehiscent fruit, which means it does not open at maturity.

4.1.1.2. Structure of the starchy endosperm (kernel)

The endosperm (82% of the grain) is in the form of longitudinal cells (Figure 4.4) containing starch granules surrounded by protein matrices or filaments with a thickness in the order or microns. A distinction is made between large A-starch granules (20–40 µm) and small-B starch granules (<10 µm).

4.1.1.3. Structure of the bran

Bran (13–15% of the wheat grain) comprises the hard outer layers of cereal grain. It consists of the pericarp (fruit wall), the seed coat and the aleurone layer (the outermost portion of the endosperm) (Figure 4.6). These layers are fused together to form a flexible membrane that is difficult to grind. It is not possible to peel the outer layer from the kernel as can be done with oranges for example. The outer layer penetrates the grain at the crease, which makes abrasion (in the case of rice) unsuitable for completely removing the bran.

Figure 4.4. *Longitudinal cells of starchy endosperm (×200 magnification)*

Figure 4.5. *A and B starch in the cells of starchy endosperm (scale: 20 μm)*

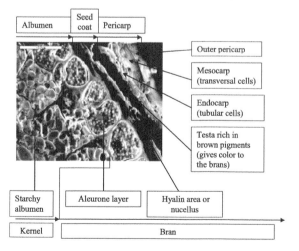

Figure 4.6. *Outer layers of the grain*

The pericarp (4% of the grain) and seed coat (2% of the grain) consist of a high proportion of cellulose and minerals; the aleurone layer (7–9% of the grain) is rich in protein, fat, vitamins and minerals. This outer shell of the grain, which is rich in nutrients, is rarely included in white flour, but instead forms the "middlings".

4.1.1.4. Structure of the germ

The germ constitutes around 3% of the wheat grain, contains a high proportion of protein, fat, vitamins and minerals, and exhibits high enzymatic activity. It is composed of two main parts: the embryo and the scutellum located next to the starchy endosperm.

The germ is removed from most flours by the roller milling system and forms the middlings (bran and sharps).

4.1.2. Structure and properties of the constituents

4.1.2.1. Carbohydrates

Starch

Wheat starch is made up of granules measuring 10–40 µm in diameter (Figure 4.7). These granules store sugar (glucose) created during photosynthesis.

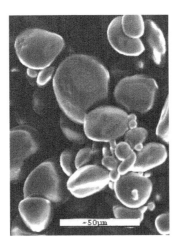

Figure 4.7. *Native starch granules of common wheat*

Long carbohydrate chains called starch chains are arranged in such a way on the inside that the starch looks like a granule on the outside. These chains consist of glucose molecules linked by α–1, 4 bonds, and can include hundreds of glucose units. They can be linear, such as amylose, comprising 100 – 300 glucose units, representing around 27% of total starch, or branched, such as amylopectin, comprising 1,000 – 5,000 glucose units, representing 73% of total starch. Branching occurs between carbon 6 of a glucose molecule and carbon 1 of another molecule (α 1–6 bonds).

In the granule, amylose chains are arranged in highly-ordered crystalline structures, thereby decreasing their affinity for water. In the case of amylopectin, branched regions appear to be amorphous because they are non-crystalline and accessible to water and enzymes, whereas linear regions are crystalline and can be identified under X-ray. This shows that the outer layer of starch is crystalline (Figure 4.8) making it difficult to hydrate.

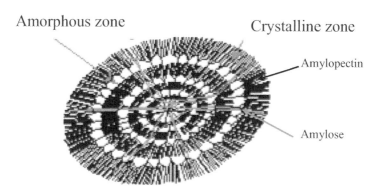

Figure 4.8. *Diagram of starch chains in a granule. For a color version of this figure, see www.iste.co.uk/jeantet/foodscience.zip*

As a result, at cold and lukewarm temperatures water only partially penetrates the starch granules. When a mixture is left to stand without stirring, like a pancake batter for example, the starch granules sink to the bottom as they are denser than water.

Above 55–60°C, the strong bonds between the starch chains in the crystalline structure are broken and the chains separate; water can then penetrate, the granules swell and open (Figure 4.9) and the contents thicken:

this is known as starch gelatinization (Figure 4.10). This mechanism is well known in the preparation of sauces using wheat flour or corn starch.

During baking (bread, pastry or pancakes), starch gelatinization also occurs. For example, pancake batter thickens and the starch chains in the different granules bind to give a smooth continuous structure, replacing the granular form of starch.

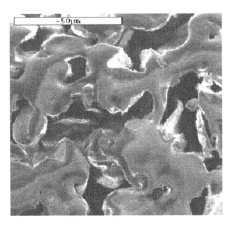

Figure 4.9. *Wheat starch granules at the start of gelatinization at 60°C*

After gelatinization and during cooling, the starch becomes firm and rigid leading to the formation of a continuous gel.

Amylose and some regions of amylopectin tend to recrystallize. This retrogradation, which results in a partially crystalline state, is accompanied by the expulsion of water. Despite partial recrystallization, amylopectin is still able to maintain its water retention capacity. The hardening that results from retrogradation is one of the causes of staling of dough-based products. At a given temperature, it varies depending on the composition and interaction between the components in the material. This rigidity of the non-crystalline amorphous material, which looks like a gel, changes from a liquid to a glassy state via an elastic rubbery state. The transition from a flexible rubbery state to a hard and brittle state is known as glass transition. The glass transition temperature is the temperature at which the transition between these two states occurs. Due to its low value, processing is possible at room temperature.

The increase in the amylase activity of flour by the hydrolysis of starch to sugar in a hydrated medium and during baking reduces such hardening; the baked product therefore appears softer.

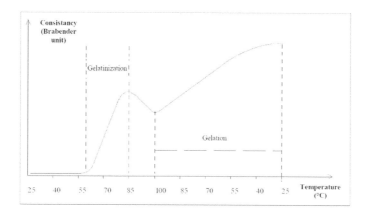

Figure 4.10. *Gelatinization and gelation curve from a Brabender amylograph*

Fiber

Fiber can be divided into two categories: water-insoluble crystalline fiber and water-soluble non-crystalline fiber. Their level of solubilization varies depending on the chain configuration.

Cereal fibers consist of carbohydrate chains (polysaccharides) such as cellulose (soluble and insoluble pentosan and β-glucan). These fibers (Table 4.3) are all indigestible by humans.

Tissues	Product equivalent	% of tissue	Polysaccharide	Solubility in water	
Endosperm	Flour	2–7%	Cellulose	2–4%	Insoluble
			Arabinoxylan	64–70%	25–30%
			Arabinogalactan	2–3%	Soluble
			β-glucan	20–30%	25–35%
Pericarp	Bran	70–80%	Heteroxylan	60–65%	Insoluble
			Cellulose	25–30%	Insoluble
			Lignin	10–15%	Insoluble

Table 4.3. *Types of polysaccharides in wheat as a percentage of total cell wall polysaccharides [PLA 97]*

In wheat flour type 55 (standard white flour for baking), the total percentage of fiber is generally 2–3%, or 1–2% of total pentosans (carbohydrate chains).

Cellulose is formed from a chain of β D-glucose molecules (up to 10,000 units) linked by β–1, 4 bonds. Cellulose fibers join to form insoluble crystalline structures with a low affinity for water. These structures are resistant to acid, enzymatic hydrolysis and physical deformation.

β-glucans are polysaccharides consisting of linear chains of glucose (β D-glucopyranose) and are linked by 1, 3 or 1, 4 bonds [GRO 96].

Pentosans are mainly comprised of C5 sugars (pentose). Arabinose–xylose bonds (arabinoxylans) (Figure 4.11) and arabinose–galactose (arabinogalactans) are the most common pentosans in cereals and wheat (Table 4.3). Hemicellulose contains a mix of pentose and hexose sugars.

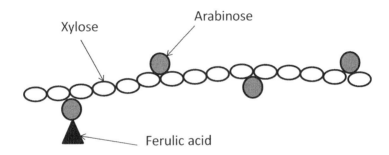

Figure 4.11. *Simplified illustration of an arabinoxylan chain*

A distinction is made between water-soluble pentosans, often referred to as gums or mucilage, and water-insoluble pentosans or hemicellulose. Many food scientists agree that the soluble fraction improves baking results. Pentosans in the grain kernel, mainly from endosperm cells, are more soluble than the pentosans in the pericarp or bran. Regardless of the degree of solubility, non-crystalline pentosans easily absorb water at hot and cold temperatures, unlike cellulose. Their hydration capacity varies between six and ten times their weight in water depending on the desired viscosity. They therefore have thickening properties and also help to form relatively strong gels. A low proportion of these pentosans in cereal flours (2–6%) can account for 25–30% of the total hydration of wheat flour dough and more with rye flour dough. The viscous characteristics of these gels affect the behavior, extensibility and expansion of dough by stabilizing the gas cells.

Hemicellulose, mainly arabinoxylan, binds with other carbohydrates, proteins or ferulic acid; the latter is involved in oxidation reactions between carbohydrate chains [PET 96] and proteins, and increases the stability and consistency of dough by gelation. According to [REN 98], pentosans form a biochemical fraction, which plays a significant role in the alveograph parameters P and G (Figure 4.15), more by their structure (level of arabinose/xylose branching) than by their content (soluble or total). The composition varies with wheat variety and climatic conditions [ROU 96]. The percentage of ferulic acid and the molecular weight of pentosans are variables that influence viscosity.

Simple sugars

Simple sugars include hexoses, pentoses and oligosaccharides. Unlike starch and fiber, these sugars are soluble in alcohol. Flour contains glucose, fructose, sucrose, maltose and short chains of pentose or glucose. They represent around 1.5–2% of the dry matter content of flour. These constituents are similar to fermentable sugars, which are quickly used by yeast at the beginning of fermentation, before starch hydrolysis by amylase. They have a low structuring capacity; their presence in large quantities, in particular when wheat has begun to germinate before harvest, yields sticky doughs that are difficult to handle.

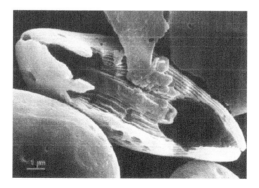

Figure 4.12. *Wheat starch partly hydrolyzed by amylase (enzymes). This enzymatic attack cuts the chains, allowing the separation of glucose molecules*

4.1.2.2. *Proteins*

The amino acid composition and structure of wheat proteins gives wheat a variety of functional properties that differentiate them from other cereals. Their foaming capacity is activated during the kneading of dough, whereby proteins facilitate the formation of very rigid three-dimensional structures capable of retaining gas, thus allowing the production of bread. Proteins not only contribute to the cohesion of dough but also the formation and stabilization of gas bubbles in the dough. The protein structure of other cereals is much more fragile and less resistant.

Characterization of wheat proteins

Wheat flour proteins are complex and abundant. Some are water-insoluble (e.g. gliadins and glutenins) and when hydrated can associate to form gluten. Protein agglomeration gives the product viscous and elastic properties; it can then stretch to form a membrane capable of holding fermentation gases during bread making. This feature makes wheat proteins suitable for bread making and gives the crumb its alveolar structure (or honeycomb-like structure due to the network of air bubbles).

Traditionally, proteins were classified according to their solubility properties. In 1907, Osborne proposed a classification of wheat proteins based on four types:

– *Albumins* (5–10% of total wheat proteins) are globular and water-soluble. They are mainly concentrated in the periphery of the grain and the germ;

– *Globulins* (5–10% of total wheat proteins) are globular and soluble in dilute salt solutions. Like albumins, they are found mainly in the peripheral parts of the grain;

– *Gliadins* (40–50% of total wheat proteins) are soluble in alcohol solutions. They are mainly concentrated in the kernel or endosperm of the wheat grain. They are also found in gluten, giving it viscous properties (fluidity, extensibility). Their molecular weight varies from 3.5×10^4 to 9×10^4 Da;

– *Glutenins* (30–40% of total wheat proteins) are soluble in acid or alkaline solutions. They give gluten its cohesion and elastic characteristics as well as its resistance to deformation. Their molecular weight can vary from

10^5 to 3×10^6 Da. Like gliadins, they are primarily found in the endosperm of the grain.

Wheat proteins are characterized qualitatively and quantitatively by chromatography and electrophoresis. Chromatographic methods (gel permeation) separate high-molecular-weight protein aggregates (glutenins) from low-molecular-weight aggregates (gliadins), and electrophoretic methods separate gluten into 22 distinct bands (Figure 4.13).

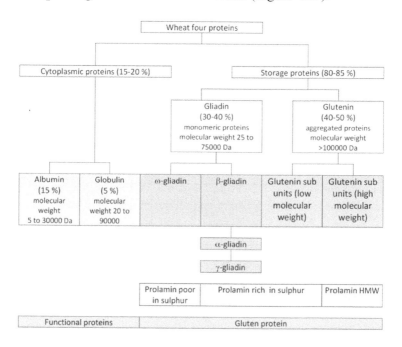

Figure 4.13. *Qualitative classification of wheat proteins [FEI 00]*

This classification also includes a family of sulfur-rich proteins belonging to the albumin–globulin fraction, which includes friabilins (present in soft wheat) and puroindoline a and b. These proteins have surfactant properties that are crucial to the alveolar structure of the dough.

4.1.2.3. Lipids

Lipids mainly occur in the form of triglycerides; they are present in small quantities in cereals, and do not play a major functional role. However, interactions between endogenous lipids and proteins modify the functional properties of gluten and contribute to the regularity of the alveolar structure.

Storage lipids (mono-, di- and triglycerides, free fatty acids) represent the largest fraction of lipids (Table 4.4). These lipids are composed of mostly unsaturated long-chain fatty acids, which are valuable from a nutritional perspective, but are sensitive to oxidation. Structural lipids (mainly glycolipids and phospholipids) contribute to lipid–protein–carbohydrate interactions and ensure the formation of lipoprotein complexes. Due to their surfactant properties, they play an important role in the stability of air inclusion and gas retention.

	Neutral lipids (nonpolar)	Glycolipids	Phospholipids	Unsaponifiables
Lipid fractions	70	19.5	10	5.5
Fatty acids (% of total neutral lipids)	C16		17–24	
	C16:1		1–2	
	C18		1–2	
	C18:1		8–21	
	C18:2		55–60	
	C18:3		3–5	

Table 4.4. *Proportion of lipid fractions of wheat [DRA 79]*

The lipid content is greater in the germ than in the aleurone layer, the endosperm and pericarp; the lipid level increases with the wheat flour extraction yield (Table 4.5). The amount of lipids is therefore an indicator of the extraction yield, but also of the improper storage of flour (based on the level of oxidized lipids).

	Flour T45	Flour T55	Flour T65	Sharps flour	Wheat germ	Whole wheat flour
Fat	1.2 – 1.4	1.4 – 1.7	1.8 – 2.0	3 – 5	10 – 15	2.2 – 2.8

Table 4.5. *Proportion of fat (%) in different milling products*

During storage, some glycerides are hydrolyzed by lipases releasing fatty acids (hydrolytic rancidity) that are prone to oxidation. However, the hydrolysis of lipids partly bound to proteins in gluten makes the latter more brittle and elastic, and results in lower gas retention during fermentation of the dough. The risk of hydrolysis increases if:

– the lipase activity is higher (greater in the germ and outer parts of the grain);

– the temperature increases (accelerated enzymatic activity);

– the water content is higher (substrate of the reaction).

The amount of free fatty acids is important in assessing the state of preservation of flour or "flour aging".

4.2. Biological and physicochemical factors of wheat processing

The main objective of processing wheat is to make it more suitable from a practical, nutritional, food safety, human consumption and storage point of view. Processing methods are constantly being adjusted to improve the sensory properties of cereal products. As previously mentioned, the whole grain, which is hard and compact, is difficult to consume; by cooking it in water, the texture softens and the starch gelatinizes.

Flour or semolina must then be formed into dough to give cohesion and shape. Generally thick doughs with viscous properties (as opposed to liquid doughs or batters) are preferred, making it possible to form different shapes.

Baking these doughs in an oven makes them firmer by drying. Only cooking in hot water (pasta and couscous) reduces firmness without changing the flavor. To achieve good texture (soft and friable) and flavor (development of appealing aroma compounds during baking), it is necessary to reduce the firmness of cooked/baked products. A number of different methods cab be used (Figure 4.14):

– make the dough thinner (e.g. pancakes, cornflakes.);

– break up the continuous structure by adding fat (e.g. biscuits, puff pastry);

– introduce air (bread and pastry) into the continuous structure.

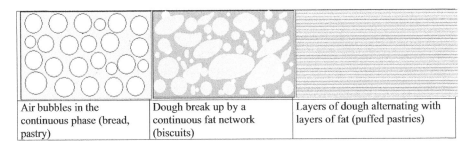

| Air bubbles in the continuous phase (bread, pastry) | Dough break up by a continuous fat network (biscuits) | Layers of dough alternating with layers of fat (puffed pastries) |

Figure 4.14. *Different structures to reduce the firmness of flour dough. For a color version of this figure, see www.iste.co.uk/jeantet/foodscience.zip*

The two products mentioned in this chapter are bread (leavened dough) and pasta (unleavened dough). Bread is obtained by baking dough in the oven after it has been kneaded, fermented and shaped. It is mostly made from flour (wheat or rye), water, salt and a fermentation or leavening agent (e.g. yeast or sourdough). Pasta is made from a kneaded dough of semolina flour and water, which is shaped with a press and stabilized by drying (dried pasta); unlike bread, it does not rise, except minimally during cooking when the starch gelatinizes.

4.2.1. *Development of texture*

4.2.1.1. *Dough structure*

The structure of dough primarily depends on the desired quality and texture of the final product (Figure 4.15). Light, crispy and crunchy textures are characterized by a continuous network of baked dough containing air pockets. The ability of dough to deform, due to the thickness of the cell walls and the flow of the material, gives a soft texture. Inability to deform leads to friability, giving a crispy or crunchy texture.

The presence of water or fat gives a pliable, elastic texture; conversely, insufficient water yields harder or brittle products (e.g. rusks, zwieback).

Kneading (Figure 4.16) stretches and reassembles gluten proteins; its effects depend on the type of mechanical stress (compression, shear, extension) and the flow pattern of the dough.

From Wheat to Bread and Pasta 165

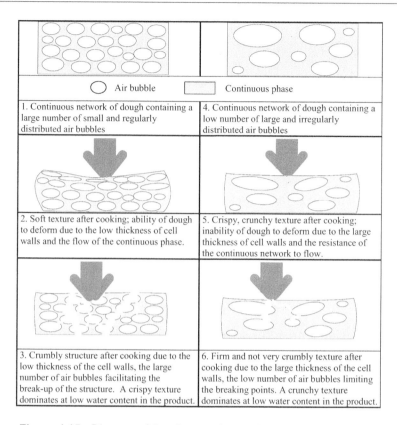

Figure 4.15. *Diagram of the changes in two dough structures before (1, 4) and after (2, 3, 5, 6) cooking. For a color version of this figure, see www.iste.co.uk/jeantet/foodscience.zip*

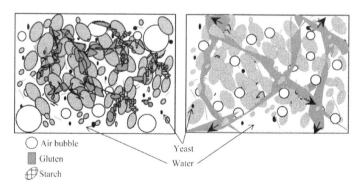

Figure 4.16. *Diagram of dough at the beginning of and during mixing. For a color version of this figure, see www.iste.co.uk/jeantet/foodscience.zip*

Water, in contact with flour, diffuses between the particles and binds them; air gradually escapes but bubbles are trapped in the dough during its formation. The diffusion of water throughout the particles breaks up the starch granules and causes protein aggregates to swell, forming gluten. Gluten, which has elastic and viscous properties, is gradually developed by the mechanical effects of kneading. Free water dissolves soluble elements (sugars, soluble proteins and fibers, enzymes, salt), disperses insoluble components and forms a medium of varying viscosity, which determines the workability of the dough.

During kneading, the action of folding and stretching also incorporates additional air (through air bubbles) and stabilizes the protein film and the viscous or liquid phase formed mainly from fibers, starch and water. This structure looks like a foam. The more the dough is mixed, the more gluten is developed and the more capable the structure is of holding the fermentation gases. At the same time, the number and regularity of air bubbles increases while their size decreases, giving rise to the crumb structure.

Intense mixing produces a crumb with a large number of small, regular air pockets with thin walls giving a soft texture, whereas insufficient kneading results in poorly developed gluten yielding a more irregular alveolar structure with fewer air bubbles and a lower gas retention capacity; the bread is therefore flat and the crumb is more firm.

Developing both the gluten network and aeration facilitates oxidation reactions causing changes in color, odor and the stability of the protein network. Heavily kneaded dough becomes whiter, with a noticeable change in aroma; in addition, protein oxidation (formation of disulfide bonds) increases significantly, thereby strengthening the gluten structure.

4.2.1.2. *Structure of pasta*

The smallest possible quantities of water should be used to prepare pasta dough to limit the amount of water that evaporates during drying, given its energy cost. The process of homogenously dispersing a small quantity of water to prevent the irregular formation of gluten is difficult to achieve. Under these conditions, the semolina mixture must be compacted using a screw press.

The first step of mixing is carried out in a vacuum mixer, which avoids the formation of air bubbles in the dough and thus obtains (Figure 4.17):

– a more regular, smooth, shiny and translucent structure;

– a less marbled or cracked appearance;

– better homogeneity, giving a more yellow color;

– a reduction in oxidation reactions;

– a less porous structure, which affects the penetration rate of water into the dough and therefore the cooking time.

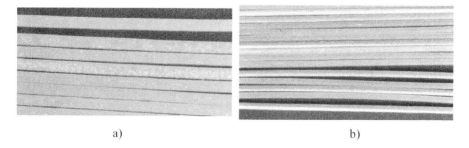

Figure 4.17. *a) Pressing without vacuum (the presence of too many air bubbles); b) Insufficient pressure in the screw press*

In the screw press, the water in the dough hydrates proteins and helps form a continuous gluten network containing starch. This structure makes durum wheat pasta qualitatively superior to pasta made from other cereals; the gluten network, which is both flexible and elastic, ensures the stability of the pasta constituents during cooking in water and allows them to swell while remaining stable without sticking together.

4.2.1.3. *Shaping bread*

Bread dough is shaped before it is baked. The final shape of the product depends on its intended use but also the ability of the dough to maintain its form after shaping, during fermentation (leavened dough) and baking.

In general, it is easier to shape dough if the consistency is soft. However, it should not flow under its own weight; shaping is not possible in this case. With very soft dough, the final shape is obtained either by using a baking pan or mould, or by spreading on a hot plate or pan (pancakes).

The elasticity of the dough also hinders good workability, since stretching it can involve tearing. If over-stretched, the dough retracts and must gradually be subjected to alternating periods of work and rest until it loosens or relaxes. Elasticity decreases by limiting gluten formation during mixing (shorter kneading time), controlling protein oxidation or adding fat to the dough mixture.

The effects of compression during these operations cause degassing (reducing the amount of gas in the dough) and a distribution of air bubbles (Figure 4.18).

| Unshaped dough after division | Dough shaped manually | Dough shaped mechanically |

Figure 4.18. *Effects of compression at the shaping stage*

In the case of pasta, the structure results from the combined action of the high pressure of the screw press in the production of the pasta and its extrusion through dies (perforated plates) giving it its shape. The cutter attached to the draw plate cuts the pasta at the desired length. Pasta can also be shaped by rolling strips of dough from a press and cutting (lasagne sheets) or pinching ("bow tie pasta").

4.2.1.4. Expansion

A simple dough made of cereal flour and water always has a density greater than 1 (starch 1.5; protein 1.3), which is incompatible with the formation of light textures. As with some pastry doughs, beating does not create sufficient aeration to significantly reduce the density of the dough. Nevertheless, the creation of air bubbles is a favorable factor in the expansion of the dough. The fermentation capacity of yeast cannot create the initial nucleation which leads to air bubble network [BAK 41]. The additional gas is provided by carbon dioxide produced by yeast fermentation and retained by its impervious structure. The dough will develop if the pressure of these gases increases, provided that the gluten structure is capable of holding the gas bubbles; deformability depends on the viscoelastic properties of the dough.

Most of the gas is obtained during the fermentation of the dough; in the oven these gases expand and vaporize, causing significant deformation in the first few minutes of baking. The level of development of the dough is determined by its gas retention capacity and deformability, which in turn affect the level of expansion at the start of baking (oven spring).

4.2.1.5. Stabilization

The texture of dough is stabilized by drying. In the case of pasta, drying (moisture content <12.5%):

– stabilizes the extruded pasta shape;

– ensures homogeneous color and appearance;

– ensures rigidity and elasticity;

– ensures a long shelf life.

The technique should be well controlled to avoid internal stresses in the pasta, which could lead to cracking (Figure 4.19). These cracks appear on the surface and inside the pasta either immediately during drying (internal ruptures of the structure) or later during storage, causing a change in color, brightness and strength.

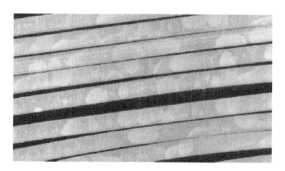

Figure 4.19. *Cracked pasta*

In the case of bakery products, dehydration should not occur before the stabilization of the alveolar structure, which is ensured by the starch and protein in the dough during its maximum expansion.

After being placed in the oven, the dough gradually increases in temperature, its viscosity decreases but paradoxically its stability increases. The thermal expansion of gases results in an increase in the internal pressure of the dough, which in turn causes the dough to rise. The starch begins to gelatinize under the effect of heat (between approximately 60 and 85°C) causing the dough to thicken, which is against the natural flow phenomenon associated with a temperature increase. Above the maximum temperature of gelatinization, viscosity decreases without the dough collapsing since protein coagulation is sufficiently developed to permanently stabilize the structure. At this stage, the set dough has become bread by passing from a visco-elastic liquid to a visco-elastic solid. This limited stability still allows bread to be sold as "pre-baked" using modern techniques such as freezing or modified atmosphere packaging.

4.2.2. *Development of color and flavor*

4.2.2.1. *Oxidation reactions*

Unsaturated fatty acids (linoleic and linolenic) are very prone to oxidation; the most sensitive are non-esterified or free fatty acids, which develop after the hydrolysis of fats. As already mentioned in Volume 1 [JEA 16a], oxidation causes several chain reactions that successively lead to the formation of peroxides and hydroperoxides. Changes in hydroperoxides result in the formation of volatile aromatic compounds (aldehydes, ketones, etc.), which are responsible for the rancid flavor associated with the acidification of fat after the hydrolysis of triglycerides. This oxidation can be autocatalytic or catalyzed by lipoxygenase. The hydroperoxides formed induce oxidation reactions among proteins or carotenoid pigments, causing the color to change from cream to white tones. Bread with a honeycomb-like structure and soft texture requires intense kneading, which is automatically accompanied by a whitening of the dough resulting from pigment oxidation. However, if a specific aroma is desired, oxidation in the dough must be reduced as much as possible, thereby limiting the development of gluten and resulting in a denser bread with a more creamy colored crumb.

Pasta retains its yellow color, largely due to the addition of egg ingredients, if pressed under reduced oxygen conditions. Reduced lipoxygenase and polyphenol oxidase activity is necessary to maintain a high yolk index close to the color of the grain kernel and to limit browning.

The current selection of wheat is based on genetic aspects that induce color factors and oxidative enzymatic activity.

4.2.2.2. Fermentation

Fermentation involves the aerobic or anaerobic activity of microorganisms added to or already present in the dough as well as enzymatic activity in the dough. These biological changes generate molecules with aromatic properties, flavor or aroma and flavor precursors. Enzymatic activity mainly involves hydrolases (amylases, proteases, lipases) and oxidases.

Under anaerobic conditions, yeast mainly produces ethanol and carbon dioxide (95%) from glucose, maltose and sucrose; secondary fermentation produces aromatic compounds (higher alcohols, carbonyl compounds, organic acid esters).

The presence of bacteria in flour has different consequences on the formation of flavor and aroma compounds. In spontaneous fermentation (creating a sourdough starter) lactic acid bacteria dominate, with their number exceeding that of yeast cells. Two families exist: homofermentative lactic acid bacteria forming lactic acid; and heterofermentative lactic acid bacteria forming lactic acid, acetic acid, carbon dioxide and secondary aromatic compounds that contribute to the diversity of flavor. When yeast dominates the fermentation process, these bacteria become ineffective and their influence on flavor and taste is very limited.

4.2.2.3. Maillard reactions and caramelization

Browning reactions such as caramelization and Maillard reactions (non-enzymatic browning; see Chapter 5, Volume 1 [JEA 16a]) play a key role in the formation of color and aroma.

Simple sugars on the outside of the dough melt from 130–140°C and react to give colored transformation products of high molecular weight at around 180°C through the process of caramelization.

The presence of proteins facilitates the degradation of sugar through the Maillard reaction. This reaction occurs between free NH_2 groups of proteins (mainly lysine) and C=O groups of reducing sugars, leading to the formation of water and glycosylamine; successive degradations of this molecule result in melanoidins, which are responsible for changes in aroma and color, and volatile substances.

Maillard reaction kinetics depend on temperature, water activity and pH; optimum conditions are an a_w of 0.7 and near-neutral pH: sodium or potassium hydroxide is used as a neutralizing agent in the production of pretzels to color the crust.

4.3. The technology of milling, bread making and pasta making

4.3.1. *Processing of wheat into flour and semolina*

Milling includes all milling operations from the arrival of wheat at the mill to its departure as flour. In the strictest sense, it refers to the reduction of wheat to flour. The most common milling machines include multiple pairs of cast-iron cylinders; however, some stone grinding mills are still in operation.

4.3.1.1. *Processing of wheat to flour*

According to the compendium of uses of bread in France [REC 77], "the designation 'wheat flour' or 'flour' without further qualification refers exclusively to the powdered product obtained from a batch of wheat, of the species *Triticum aestivum* subsp. *vulgare,* which is healthy, marketable, suitable for milling and industrially pure".

Cleaning and tempering wheat

Wheat should be cleaned and prepared before the milling stage. Impurities are removed based on their physical characteristics:

– *size*: sieving (grain cleaner/separator);

– *density*: aspiration, settling (stone remover), centrifugation (cyclone);

– *shape*: sorting into round or long grains;

– *magnetic properties*: magnetic separation.

In addition to these cleaning operations, wheat is prepared by tempering or wetting and leaving it to rest, usually for 24 hours, to facilitate the separation of the outer layers and the reduction of the kernel during milling. The aim is to make the outer layers more flexible and elastic so as to avoid large quantities of small broken grains in the white flour and facilitate the separation of bran. Wetting the kernel also helps to grind it to semolina flour. Tempered wheat contains 16–17% water.

Principle of milling using millstones and roller mills

Milling involves the reduction and separation of wheat and its fractions as well as the separation of the endosperm from the germ and bran:

– *Millstone system*: a pair of millstones consists of two superimposed stones, the upper turning stone (runner stone) and the bottom stationary stone (bedstone). The grain is fed into the center of the runner stone and channeled to the outer edges before being released. The grain is crushed by the "scissoring" or grinding action of the stones.

– *Roller milling system*: this system uses corrugated cylinders (break rolls) and smooth cylinders (sizing and reduction rolls). The fine grooves and the high speed differential (the speed ratio between slow and fast rolls is 1:2.5) result in the greater shearing action of roller mills (Figure 4.20(a)). The angle of the grooves or flutes can be varied in order to produce different results: a small angle has a dull edge used for crushing whereas a large angle has a sharp edge used for cutting. Shearing or crushing processes therefore depend on the position and orientation of the grooves. Sieving is carried out after each break passage (Figure 4.20 (b)).

a) b)

Figure 4.20. *Roller milling system: a) roller mills in a modern mill; b) gyrating sifter (sieves)*

The various break rolls (numbered B1, B2, B3, B4, and B5) are distinguished by:

– the spacing between the rolls;

– the number of grooves;

– the position of the grooves (Figure 4.21).

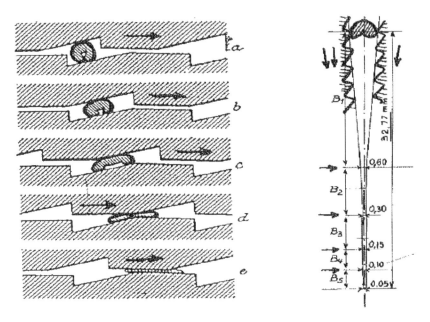

Figure 4.21. *Comparison of grain reduction between millstones and roller mills [WIL 90]*

The outer layers of the grain contain increasingly less kernel from the first break roll (B1) to the last (B5); the outer layers are gradually scrapped away to expose the kernel. It is composed of large particles (semolina) and small particles of <200 μm (flour). Flour is extracted whereas semolina is sent to the smooth rolls (the speed ratio between the slow and fast rolls is 1:1.25 to 1:1.4) suitable for crushing (processing into flour). Coarse semolina is directed to the sizing rolls and fine semolina to the reduction rolls. The number of passages on these smooth rolls is generally ten (Figure 4.22).

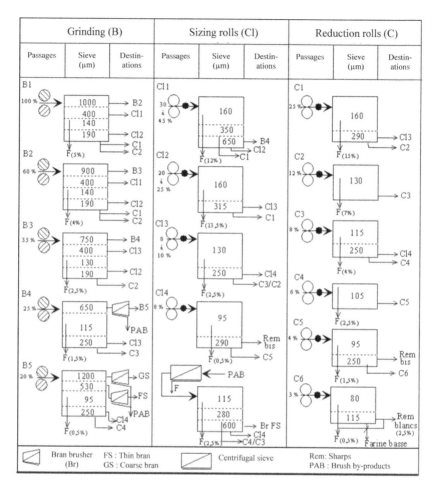

Figure 4.22. *Block diagram of the milling of wheat*

Unlike millstone milling, roller milling produces a large variety of flours, in particular white flour. Standard flour, such as white flour type 55, obtained on roller mills contains very little aleurone and few germ components, the latter having been removed; the nutritional composition is therefore modified (Tables 4.6. and 4.7). Millstone flour obtained by crushing the whole grain contains some of these components. To produce a less refined flour using a roller mill, the finely ground outer layers of wheat (sharps) are subsequently added; since these fractions are relatively large, the flour is therefore coarser than that produced from millstone milling.

Flour type	Nutritional value			
	Water (g)	Protein (g)	Fat (g)	Carbohydrates (g)
150	14–15	10.0–11.5	1.5–2.0	68–73
80	14–15	9.5–11.0	1.2–1.6	69–74
55	14–15	9.0–10.5	1.0–1.2	70–75

Table 4.6. *Nutritional value per 100 g flour*

Type	Fiber (g)		Minerals (mg)						Vitamins (mg)			
	Cellulose	Hemi-cellulose	Na	K	P(*)	Mg	Ca	S	B_1	B_2	B_3	C
150	1.5-2.5	6-8	10	300-500	200-400	100-200	25-80	180	0.35-0.55	0.1-0.2	4.0-8.0	0
80			5	230-250	200-300		22-25	100	0.25	0.15	3.0	0
55			3	100-150	110-150		15-20	60	0.06-0.1	0.08	0.6-0.9	0

Table 4.7. *Average micronutrient composition per 100 g flour*

If millstone milling is carried out correctly, further reduction produces finer flours with a higher fat content, which favors yeast activity, but does not keep as long as equivalent roller milled flour. While millstone flour may not be superior in terms of its baking properties compared to roller milled flour, there are tiny differences in bread quality, in particular with regard to "dough touch" and crumb color.

Milling quality

The aims of milling are:

– to separate the wheat kernel from the bran and germ in order to achieve the best flour yield (Table 4.8) for a specific type of flour (Table 4.9);

– to reduce kernel fragments into sufficiently fine particles to make flour.

In France, the classification of flour is based on the ash or mineral content. The type ranges from 45 to 150, starting with the whitest flour (low extraction rate flour) to wholemeal flour, containing a high level of bran (high extraction rate flour). This differentiation is based primarily on the concept of purity or whiteness, and does not correspond to the processing value even if it is easier to work with white flour compared to wholemeal flour.

Flour type	Average extraction rate (%) Roller milling
45	70-75
55	75-80
65	78-83
80	82-86
110	87-90
150	90-98

Table 4.8. *Relationship between flour types and extraction rate*

Flour type	Ash or mineral content (% relative to dry matter)	Appearance	Uses
45	<0.50%	White	All-purpose flour
55	0.50–0.60%	White	Bread, pastry, Danish pastries
65	0.62–0.75%	White	Biscuits
80	0.75–0.90%	Brown	Light to medium brown or wholemeal bread
110	1.00–1.20%	Brown	Light to medium brown or wholemeal bread
150	>1.40%	Wholegrain	Wholemeal or whole-wheat bread

Table 4.9. *Commercial classification of French flours*

The art of milling lies in the ability to obtain the maximum extraction of flour from wheat with the least amount of outer layers (rich in minerals; Table 4.10).

Constituents	Percentage of minerals relative to dry matter
Pericarp	2–4
Seed coat	12–18
Aleurone layer	6–15
Germ	5–6
Endosperm	0.35–0.60
Whole wheat	1.6–2.1

Table 4.10. *Distribution of minerals in wheat*

Some types of wheat are more suitable for achieving these aims than others. Wheat and wheat varieties are differentiated by their different milling properties:

– the *extraction rate* depends on the fiber content, the bran/endosperm ratio, the mineral content, ease of sifting, friability and kernel resistance.

– the *hardness and vitreousness index* is used to classify wheat varieties into categories (hard, medium-hard, soft). It is linked to the proportion of starch broken down during the grinding of the endosperm. Hardness is a genetic characteristic: a major gene (*Ha*) linked to hardness has been identified. The most common theory suggests the presence of puroindoline A and B, where the A/B ratio gives different hardness levels, and the presence of a protein called friabilin; it is possible that certain lipids are also involved. Vitreousness is a visual characteristic related to the degree of grain compaction. It increases proportional to the protein content, which results in a transition from a floury to a glassy appearance. Hardness and vitreousness affect the mechanical properties of grain during reduction operations.

– *Friability, component cohesion, particle size.* Friability decreases as the hardness index increases (hard wheat) and therefore particle size increases. In contrast, there is a lower cohesion of kernel components (starch granules, protein, fiber) in soft wheat: in this case, fragmentation is easier resulting in finer flour (Figure 4.23). The easy separation of components makes soft wheat suitable for air classification (i.e. the separation of particles based on their size, shape and density using air currents).

4.3.1.2. *Processing of durum wheat into semolina*

The milling of durum wheat is different to that of common wheat since the objective is to produce pure semolina and not flour.

The preparation of durum wheat is the same as that of common wheat; tempering to raise the moisture level to around 17% is needed to soften the husk and separate it from the kernel. However, water should not penetrate too deeply into the endosperm during this stage to avoid reducing its resistance; this would lead to the production of flour and cause the wheat to lose its shine and yellow color. Tempering is therefore carried out in three stages with short periods of rest in between (3–6 h for the first two rest periods and 30 min for the last rest period).

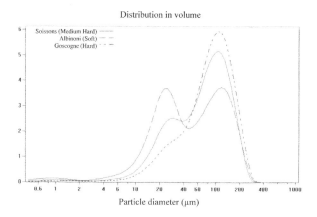

Figure 4.23. *Particle size of hard and soft wheat. For a color version of this figure, see www.iste.co.uk/jeantet/foodscience.zip*

The grinding phase is a gradual process; it requires up to seven passages with the corrugated rolls set in a sharp-to-sharp configuration. Semolina is divided into pure fractions and denser fractions containing bran fragments known as coarse semolina, both of which have the same particle size. They are separated in a purifier according to density and particle size, and then placed in a drum detacher to remove any parts that might affect color. The objective of durum milling is not to reduce the endosperm to flour particle size, but instead to convert it to semolina. Sieving depends on the final production objectives (balanced distribution of roller mill feeding) and the desired particle size. (Table 4.11).

	Coarse semolina				*Average semolina*		*Fine semolina*			*Flour*
Sieve					42	60	80	100	120	
Opening (μm)	1,150 1,000	900	700–800		530	350	250	187	161	140
Classes	CC	MC	SSSC	SSSSE-SSSS	SSSE		→	←		D groats
	Unsifted		Sifted				SSSF	→		(animal
Uses	Soup		Couscous	Soup	Pasta					feed)

Table 4.11. *Example of the classification of semolina based on particle size S (sifting), SSS (three sifting passages), E (export), F (fine), M (medium), C (coarse), and D groats (durum groats)*

Semolina is classified based on the maximum semolina yield with regard to quality objectives, regulatory constraints (types of semolina) and production costs from cleaned and tempered wheat to the final semolina product.

4.3.2. Bread making

Bread making, whether it be on a household, small-scale or industrial level, has seen a significant technological change due to technical progress (mixing, shaping, baking, refrigerating, freezing, etc.) and greater knowledge of raw materials and the biochemical and physicochemical mechanisms involved in bread production.

4.3.2.1. Baking quality

Baking quality is defined as the utilization value of flour in the production of a bakery product. Each production process has a technological value, for example type 55 flour has a good technological value for making bread, an average value for making puff pastry and a poor value for making biscuits. Technological quality is determined by the implementation of a standardized small-scale production test.

Baking quality (French norm NF V03-716) incorporates various concepts:

– *dough yield*: the amount of water absorbed by flour for a given consistency;

– *dough machinability*: ease of workability until the baking stage. This qualitative characteristic takes into account dough stickiness, elasticity, stability and deformability;

– *dough and bread development*: gas retention capacity and deformability;

– *sensory quality of the crumb*: color, smell, texture.

All these aspects, apart from dough yield, are used to rate the baking quality, which has a maximum score of 300 points. The technological value of flour strongly depends on the wheat variety, the characteristics of which are linked to the quantity and quality of proteins and enzyme activity. However, identifying the variety is not enough to determine its technological value; indirect methods of assessing flour quality include rheological (alveograph), chemical (protein content), enzymatic (Hagberg falling number) and physicochemical analyses (particle size, damaged starch and Zeleny index).

4.3.2.2. Criteria for the baking quality of common wheat

Protein content

The minimum protein content in relation to the amount of gluten formed during kneading to obtain a satisfactory bread is around 10% (w/w) for flour type 55, corresponding to around 11% for the whole grain Approximately 80% of wheat protein can form gluten when hydrated. It is, however, difficult to determine an optimum protein content: excessive levels often cause defects in French bread. Therefore, the protein content can only give an indication of the wheat quality; alone it cannot, for example, be used to differentiate between suitable or unsuitable wheat for bread making.

Protein quality: glutenin/gliadin ratio

The proportion of glutenin and gliadin varies considerably in flour, giving it different rheological properties. In 1896, [FLE 11] associated this ratio with baking quality; he gave a ratio of 0.3 as a reference of good quality and added "when flour contains an excess of gliadin, it produces a soft dough that rises well during fermentation (good extensibility), but which collapses under the effect of heat in the oven. If, however, flour contains an excess of glutenin, the lack of extensibility means that the gas pressure is not able to stretch the dough, which slowly rises during yeast action".

The glutenin/gliadin ratio is characteristic of the wheat variety, but it changes with cultivation techniques (influence of nitrogen fertilizers) and growth conditions, both of which affect protein content; it can be determined by chromatography techniques. The ratio for French wheat is currently around 0.6. When it falls below 0.4, gluten becomes very elastic and conversely, when it increases above 0.8, the dough has low elasticity.

An increase in the gliadin content is proportional to an increase in the protein content, with glutenins remaining relatively constant for a given variety of wheat; these variations change the baking quality.

Zeleny index

This value is based on the absorption of water by gluten and swelling in the presence of lactic acid. A dilute solution of lactic acid is added to flour. After mixing, the mixture is allowed to stand and sedimentation gradually occurs at the bottom of the container. The height of the deposit depends on the quantity of water absorbed and the swelling of proteins, and therefore on the quantity and quality of the latter.

The sedimentation value or Zeleny index can vary between 15 ml for weak flour and 80 ml for strong flour. Wheat used in general baking should have a range of between 30 and 40 ml. Wheat with an index below 20 ml is used for biscuit making.

Baking strength (W)

Baking strength (W) is measured using a dough of flour and salt water formed in a laboratory mixer. The mixing time and the water content are standardized. After mixing, the dough is stamped into circular pieces. After a rest period of 20 min during which swelling occurs, the deformation energy required to inflate the dough until rupture is measured (Figures 4.24 and 4.25).

Deformation energy W related to the area below the curve is relatively well correlated with the amount of gluten. Maximum resistance P, the elasticity index I_e and the swelling index G are quality indicators of the rheological properties of dough (resistance, extensibility)

Figure 4.24. *Dough deformation based on the Chopin alveograph ICC standard N° 121*

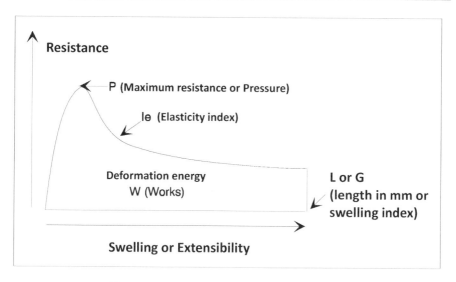

Figure 4.25. Alveograph (resistance curve of a bubble as a function of expansion)

The deformation energy W of a bread-making flour should not be too high or too low (Table 4.12); wheat with a high W value can be disadvantageous when used in pure form, but is commonly used to improve soft wheat varieties. The average W value of bread-making flour (Table 4.13) ranges between 180 and 220, before the optional addition of additives.

Rating	P	G	I_e	W
Too low	<40	<20	<35	<150
Average	40–60	20–22	35–45	150–180
Good	60–80	22–24	45–55	180–220
High	>80	>24	>55	>220

Table 4.12. Analysis of Chopin alveograph parameters based on French bread (without additives)

Type of bread	Average flour strength in alveograph values (W) (flour without ascorbic acid)
Shortcrust pastry	120–140
Puff pastry	180–200
Sweetcrust pastry	150–170
Traditional bread (with long bulk fermentation)	150–180
Traditional bread	200–220
Common French bread, pizza	180–220
Retarded dough, croissant	200–250
Baguette from frozen dough	220–270
Rusk, white sliced/sandwich bread	200–240
Brioche	250–300
American burger buns	>350

Table 4.13. *Indicative values of W for different types of bread*

Amylase activity

Amylase activity, which determines the fermentation capacity of dough, depends on climatic conditions before harvesting and storage conditions after harvesting. Moist wheat quickly passes from a pre-germination stage to germination leading to enzymatic activity. Hydrolase plays an important role in breaking up storage compounds (protein, starch, lipids, etc.) into basic elements used to fuel the growth of the new seedling. If this degradation process continues to an advanced stage, the wheat is no longer suitable for baking or other applications. Too much amylase activity leads to an acceleration of fermentation and excessive reddening of the bread crust. However, a lack of amylase activity slows down the fermentation process, resulting in flatter bread with a pale crust; this defect can be corrected by adding malt flour or fungal amylases.

Amylase activity can be determined using two methods based on the level of starch gelatinization: Hagberg falling number and Brabender amylograph (Figure 4.26 and Table 4.14). These methods are based on measuring viscosity, which decreases once gelatinization begins during the amylase hydrolysis of starch granules.

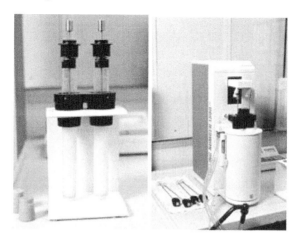

Figure 4.26. *Hagberg device*

Activity	Hagberg falling number(s)	Brabender amylograph (brabender unit)
Low	>300	>400
Normal	250–300	300–400
High	<250	<300

Table 4.14. *Analysis of Hagberg falling numbers*

4.3.2.3. Bread-making processes

Mixing

This first operation has a major impact on the quality of bread. It has three main purposes: to obtain a homogeneous mixture of different ingredients (flour, water, salt, leavening agent, etc.), to form gluten giving bread its texture, and to incorporate air into the dough. There are many different mixing techniques (Figures 4.27 and 4.28).

	Formulation		
Compounds	Conventional kneading	Improved kneading	Intensified kneading
Flour	100	100	100
Salt	1.8 - 2	2	2 – 2.2
yeast	1 – 1.5	1.5 - 2	2 - 3
Water	60	60	60
Ascorbic acid	0	0 - 20 ppm	20 - 80 ppm

	kneading (mixer using a fork mixer)		
Slow (40 tr / min)	10 – 15 min	2 - 4 min	2 - 4 min
Fast (80 tr / min)	0	8 - 12 min	18 - 22 min

	Bulk fermentation (1st fermentation)		
Settling time	2 - 3 h	30 - 150 min	0 - 40 min

Shaping

	Proofing (2nd fermentation)		
Settling time at 27°C	1 h - 1 h 30	1 h 30 - 2 h	2 h - 3 h

Scarification

Baking

Figure 4.27. *Overview of French bread production by direct yeast fermentation based on the type of mixing*

Figure 4.28. *Mixing using a fork mixer*

The intensity of mixing varies according to the desired characteristics of the final product. Its efficiency for the same energy input depends on the properties of the dough, which vary with the type of flour. It is possible to program the mixer to switch off based on the energy dispersed in the dough or a rise in temperature, since there is a relationship between temperature increase and energy input by mixing.

If mixing is carried out in a closed vessel, it is possible to alternate between positive and negative pressure (partial vacuum). In positive pressure conditions, the increased introduction of air ensures good aeration, thereby facilitating oxidation reactions conducive to the creation of air bubbles; however, an excess can lead to a slight irregularity in the alveolar structure. Under vacuum, the dough is "degassed", and the texture becomes very heterogeneous; air bubbles drop in number, become very irregular and develop thick walls. However, slight negative pressure at the end of mixing optimizes structural regularity without decreasing the number of air bubbles. The effects of positive pressure followed by negative pressure are sought in the production of sliced sandwich bread in particular.

Fermentation

Fermentation includes both microbiological activity (yeast and bacteria) and physicochemical changes in the dough. In the presence of yeast, dough gains strength corresponding to an increase in elasticity and a decrease in extensibility and relaxation. This phenomenon is largely due to the oxidation of proteins that form disulfide bonds between gluten molecules: fermentation, which creates constant movement in the dough, produces favorable conditions for reactive molecules to meet. This strengthening of the dough is essential to ensure sufficient stability until it is baked. The dough relaxes unless a product like ascorbic acid is added, which accelerates gluten oxidation; this change is primarily controlled during the first stage of fermentation, also known as primary fermentation or "*pointage*".

Fermentation also causes additional deformation to mixing and thus contributes to gluten development, the enlargement of pre-existing air bubbles and their irregularity (Figure 4.29). In the absence of ascorbic acid and insufficient mixing (low oxygenation, limited gluten development), a long bulk fermentation stage is necessary to complete gluten formation and ensure sufficient strengthening.

Figure 4.29. *Dough expansion measurement device to quantify fermentation activity*

When the stabilization conditions of the dough have been defined (length of bulk fermentation or the presence of oxidative products), the dough is shaped before the final fermentation stage during which the dough rises. This "proofing" depends on the expansion capacity of the dough and its stability throughout baking.

Baker's yeast added at the mixing stage produces carbon dioxide, ethanol and some secondary aromatic compounds from sugars such as glucose in an anaerobic medium. If no yeast is used, the endogenous flora of flour is used to develop a sourdough (Table 4.15). Flour, preferably whole wheat flour containing bran, contains yeast and bacteria, but the concentration of these microorganisms is insufficient to bring about fermentation in the dough within a few hours. They must therefore be developed in dough that is allowed to spontaneously ferment. Old or fermented dough is replenished by the regular addition of flour and water (called "refreshments"), which stimulates microbial activity and growth under aerobic conditions. This mixture is then used as a starter culture in bread once microbial activity stabilizes.

The stability of a starter culture often depends on the skill of the baker. It can vary considerably in quality (e.g. volume, aroma and flavor). On an industrial level, it is difficult for producers to ensure consistency due the lack of flexibility of automated production systems.

Yeast		Lactic acid bacteria		
Species	Size	Species	Size	Type of fermentation
Saccharomyces cerevisiae	10 µm to 50 µm	Lactobacillus brevis	1–10 µm	Heterofermentative
		Lactobacillus san francisco		Heterofermentative
Saccharomyces exiguus		Lactobacillus buchneri		Heterofermentative
Candida holmii, Candida tropicalis		Lactobacillus fermentum		Heterofermentative
		Lactobacillus casei, Lactobacillus plantarum		Homofermentative
Candida krusei				Homofermentative
		Pediococcus cerevisiae		Homofermentative

Table 4.15. *Main yeast and bacteria in natural leavening agents and starter cultures*

The aromatic diversity created by the different fermentation pathways is obtained by the action of the various microorganisms in the flour through long fermentation times and the addition of sufficient amounts of baker's yeast.

Fermentation diagrams

The process of pre-fermentation that produces a "natural" sourdough is very restrictive; it is therefore preferable to initiate pre-fermentation with a small amount of yeast (Figure 4.30) or lactic acid bacteria (starter). Even though these cultures produce less aroma and acidity than natural sourdough, they still improve the sensory quality of bread compared to the direct addition of yeast.

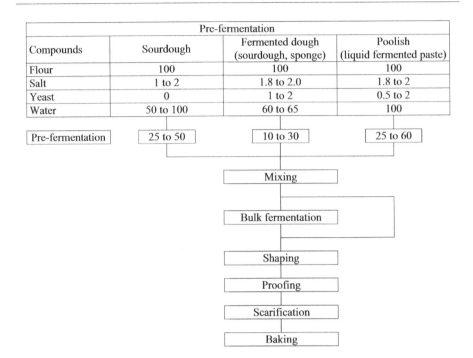

Figure 4.30. *Production of French bread with pre-ferments*

Pre-fermentation, which is more restrictive than direct fermentation, can be controlled by reducing the temperature; this method is used at different stages of the bread-making process, in particular during final fermentation or proofing (Figure 4.31). It is possible to slow down fermentation for up to 48 h at refrigeration temperature and to halt fermentation completely for up to a few months by freezing the dough after shaping. This process offers the possibility of selling the product to purchasers who are not bakers but who own a retail store with a bakery section. This technique has enabled the export of French bread.

Yeasts behave in such a way during deep freezing that it is necessary to take a number of precautions that include decreasing the dough strengthening and using ascorbic acid. If the dough is highly hydrated and if the fermentation begins before freezing, the gluten structure is damaged. The alveolar structure of frozen dough is more irregular with thicker cell walls.

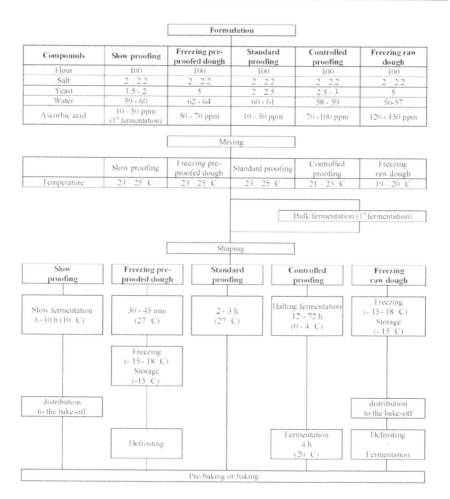

Figure 4.31 *Production flow diagram of French bread obtained by increased mixing and direct yeast fermentation at a low temperature*

The technique of freezing pre-proofed dough works well for bread rolls and pastries but less so for French bread. It has the advantage of halting dough development after the very short bulk fermentation, which allows it to maintain its development potential after defrosting at around 0°C and baking. This technology is almost as flexible as the use of baked products but involves lower production and storage energy costs. For many years, it was thought that this technique could not produce satisfactory results, mainly because of the effects of deep freezing on the significant loss of yeast activity, which impacts development during baking, and the destabilization of gluten structure. Nevertheless, under certain conditions it is possible to obtain high quality by:

– using very high protein flour (increased gas retention);

– adding extra yeast;

– making a soft dough without it being too wet;

– having a short proofing time;

– adding a large quantity of ascorbic acid;

– adding a combination of enzymes and emulsifiers.

The loss of yeast activity and the weakening of gluten can be limited by the stabilization of water using hydrocolloids like guar or carob gum.

Shaping

Shaping dough is necessary to:

– produce products corresponding to consumption or packaging units;

– ensure accuracy in terms of weight for regulatory purposes (Figure 4.32);

– produce aesthetically-appealing products (marketing aspect);

– form small portions of mixed dough, thereby facilitating the process until the packaging of baked products;

– ensure faster and more regular baking conditions;

– facilitate the orientation of the cell structure (differences in sliced pan bread, or country loaves).

Manual or mechanical shaping involves a number of steps (Figure 4.33), from the initial division of dough pieces to final shaping; the mechanical process does not differ much from the manual one, apart from the level of stress applied to the dough and the replacement of manual folding by rolling in a dough-molding machine.

Figure 4.32. *Mechanical division of dough*

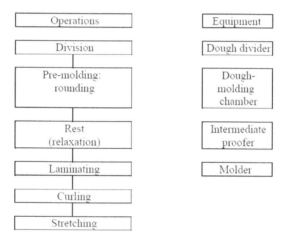

Figure 4.33. *Different stages of dough shaping*

Knowing the deformation modes during shaping as well as controlling the deformation stresses and rates make it possible to better understand their behavior during shaping and optimize their quality.

Dough molders shape and gradually extend the dough by alternating stretching with short rest periods (relaxation), which causes deformation without risk of tearing. This reduction in stress in the dough provides a more regular alveolar structure without excessive degassing. A series of more but shorter periods also increases output and reduces the length of time from division to shaping while ensuring production of thin elongated dough pieces. The introduction of calibration and centering systems to automated production lines ensures excellent regularity in length and cross-section.

Baking

Baking results from an exchange of heat between the oven's atmosphere and the product. Energy can be transferred directly to the product (direct heating) or indirectly. Under the effect of heat, the dough expands and undergoes physicochemical changes (protein coagulation, starch gelatinization in hydrated dough, Maillard reactions and caramelization). These changes ensure improved sensory quality, greater storability and better digestibility. Qualitative variations depend on the conditions and type of baking.

It is important to control dough expansion in order to obtain light, soft or crumbly textures. The factors influencing this include gas production and expansion, dough elasticity and deformation resistance as well as gas retention capacity. There is also a drying phase during which water, as it evaporates, helps to maintain the crumb at a temperature close to 100°C. In the case of pre-baking, the level of dehydration is low; however, controlling residual moisture is important during final cooking since it affects Maillard reactions and the sensory properties of bread. Pre-baking halts the production process at a stage where bread can be stabilized by storing at room temperature, using neutral gas packaging techniques (pre-baked refrigerated) or freezing (pre-baked frozen), while preserving the quality characteristics of bread after final baking.

Baking quality can be assessed based on sensory characteristics (color, crustiness, hollow sound) and weight (degree of water loss). Industrial ovens equipped with sensors that measure moisture and color can automate baking.

The principle of convection baking using baking trays to hold the dough is very common both in artisanal (rotary ovens) and industrial (tunnel ovens) production. Nevertheless, deck ovens produce the best results due to heat transfer by conduction. Transferring the dough onto a loading device (peel), which is necessary although restrictive with this type of baking (Figure 4.34), is increasingly integrated into the design of new bakeries that can be viewed by consumers. On an industrial scale, tunnel ovens could be equipped with a stone deck.

Figure 4.34. *Bread just released from a hearth deck oven*

4.3.3. Pasta making

The generic term "pasta" includes a wide range of products.

Dried pasta

Dried pasta comes in many different shapes and sizes, from long and thin to short and thick (Figure 4.35). It consists mostly of durum wheat semolina and is typically served with sauces or in soups.

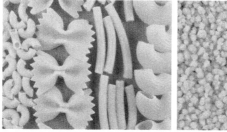

Figure 4.35. *Dried pasta and couscous*

Fresh pasta

Fresh pasta is consumed relatively quickly after production, given its short shelf life due to minimal drying. Boiled in water, it is used as a base or accompaniment for culinary dishes, and is often stuffed with a variety of fillings. Unlike dried pasta, fresh pasta is generally made from common wheat flour. It is usually made on an artisanal scale, often by hand: the instrument used is a dough-brake comprised of a wooden bar with an articulation system attached to the end of a low table. The handmade pasta is placed on the table and pressed with the bar and then cut and shaped manually.

Stuffed pasta

This category includes varieties such as "ravioli" and "tortellini". Marketing regulations in France require the use of durum wheat semolina only. However, this requirement does not apply to culinary specialities prepared by caterers or restaurateurs. The stuffing can be made of pork, veal, beef, chicken, lamb, etc.

Couscous

Couscous is not well defined from a regulatory perspective and "raw couscous" is similar to pasta. Couscous, made from a mixture of water and durum wheat semolina of varying particle size, is rolled, cooked, dried and sifted. Commercial couscous is a pre-cooked "semolina" grain that requires hydration with warm or hot water before being steamed in a couscous pot.

Traditionally, couscous was prepared as follows: "Women prepare the grain by rolling it by hand with great dexterity. Durum wheat semolina and flour are placed in a wooden or clay dish with a little cold salted water; in rolling the semolina, the flour gradually clumps around each grain. The grains obtained are then sifted, making it possible to sort by particle size. Once completed, the grains are then ready for cooking" [COU 84].

Tables 4.16 and 4.17 give the characteristics of pasta made from durum wheat semolina and water; the regulatory aspects of the preparation of pasta as well as its denominations are defined by French decree n°55 1175 of 31/08/1955.

"Superior durum wheat semolina" or "durum wheat semolina of superior quality"	SSS E	Maximum ash content 0.80%, tolerance 10% (percentage on a dry matter basis)
"Durum wheat semolina" or "standard durum wheat semolina"	SSS F	Maximum ash content 1.30%, tolerance 20% (percentage on a dry matter basis)

Table 4.16. *Types of durum wheat semolina*

Quality characteristics	Superior pasta	Standard pasta
Ash content (on a dry matter basis)	Minimum: 0.55% Maximum: 0.80% (tolerance 10%)	Maximum: 1.30% (tolerance 10%)
Acid content (on a dry matter basis)	≤ 0.05%/DM (expressed as sulphuric acid)	≤ 0.07%/DM (expressed as sulfuric acid)
Protein content (on a dry matter basis)	Minimum: 10.5%	Minimum: 11%
Moisture content	Maximum: 12.5% for dried pasta; not applicable to fresh pasta	

Table 4.17. *Main characteristics of pasta made from durum wheat semolina*

4.3.3.1. *"Pasta value"*

"Pasta value" can be defined as:

– the ability of semolina or flour to be processed into pasta (ease of mixing, shaping (drawing through a die) and drying);

– the appearance of the pasta (homogeneous yellow color, glossy, smooth surface);

– the plasticity of raw and cooked products;

– ease of cooking (no stickiness or disintegration).

Ability of durum wheat semolina to be processed into pasta

There are no major differences between the various wheat varieties currently available on the European market, apart from sprouted wheat semolina. New durum wheat varieties are constantly being researched and developed, which could be suitable on a global scale or used for certain quality criteria such as color or strength.

Physical aspects of pasta

Ideally pasta should be amber yellow with minimal brown coloring, glossy, smooth and have no holes, white spots or cracks.

The yellow color is strongly influenced by the amount of carotenoids, but the quantification of these pigments is unable to predict the final intensity of the yellow color of pasta. In fact, lipoxygenase in semolina, in the presence of oxygen, may cause partial discoloration during manufacture. Pasta-making conditions (vacuum pressing, drying conditions) can slow down or accelerate these oxidation reactions. The addition of wheat germ to semolina during milling causes higher lipoxygenase activity.

The brown color is caused by two factors:

– the higher the protein content, the greater the level of browning; an excess of nitrogen fertilization during cultivation can therefore be detrimental. The presence of albumin, a naturally colored protein, could also add to this browning effect;

– polyphenol oxidase is highly involved in enzymatic browning. Advances in genetics have made it possible to obtain varieties with reduced polyphenol oxidase activity.

The browning index increases as the extraction percentage of semolina reaches a maximum. High temperatures at the start of the drying process reduce enzymatic activity, resulting in a decrease in the browning index.

Cracks in pasta result from excessive tension caused by poor drying. This can cause pasta to break during packaging and reduce mechanical strength. The smooth glossy appearance is mainly due to the type of dies used in the screw press (Teflon or bronze dies).

Strength of uncooked pasta

The tensile strength of uncooked pasta is correlated to its resilience to breakage during handling. This is essential to minimize losses during packaging and ensure whole and acceptable products for the consumer. It depends both on the pasta-making conditions and the intrinsic quality of the wheat. At approximately equal gluten quality, the protein content affects the strength of the pasta.

Cooking suitability and product quality

This includes:

– *The cooking time*, which can be considered as the time required to completely gelatinize the starch. It can be measured by pressing cooked pasta between two glass plates and observing the center of the pasta: the presence of a white line or white dots indicates that the pasta is not yet fully cooked (Figure 4.36).

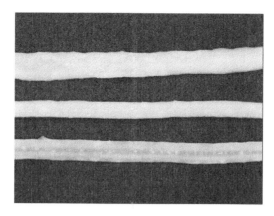

Figure 4.36. *Cooking time for pasta. From top to bottom: overcooked, optimal, undercooked*

The minimum time for cooking spaghetti measuring 1.5 mm in diameter is between 8 – 9 min in boiling water to obtain *"al dente"* pasta. The optimum cooking time corresponds to the optimum characteristics for consumption and varies according to taste. The difference between the minimum and optimum time (2–5 min) increases with the protein content (e.g. egg pasta). The maximum cooking time is the time beyond which the pasta structure falls apart (Figure 4.37); the difference between it and the minimum time reflects the resistance of pasta to overcooking, and should be as high as possible. There is also a decrease in color with cooking time.

Figure 4.37. *Cooking tolerance of pasta: a) pasta with good cooking tolerance; b) sticky pasta that disintegrates (poor cooking tolerance). (Photos: J Abecassis, INRA)*

Swelling or water absorption during cooking results in significant changes in the pasta (starch gelatinization and protein coagulation) and is measured by weighing the pasta before and after cooking. In general, 100 g of dried pasta absorbs 160–180 g of water during cooking. pH plays an important role; it is optimal at a value of 6 (close to the pI of pasta proteins) where the pasta expands less and is less sticky regardless of the wheat variety. At pH 4 and 8, the quality is lower, and at pH 2 the pasta does not swell.

The texture of cooked products is characterized by firmness or resistance to the bite and elasticity; longer spaghetti can be cooked without losing its elasticity, as it is more resistant to overcooking.

The surface of cooked products: Cooked pasta can be sticky and/or "deliquescent". Deliquescence (in this case, the dissolution of pasta and the release of starch into the cooking water) is a serious defect if it occurs shortly after being placed in the cooking water (this is largely influenced by the type of wheat).

The sensory characteristics (aroma and flavor) depend on the drying conditions and the quality and quantity of proteins.

In summary, the differences in culinary quality are due to the varying ability of proteins to form a network capable of holding other constituents. Pasta-making conditions (mixing, compression, shear) significantly impact the quality of the protein network formed. A loose protein network allows starch granules to escape, resulting in the formation of a paste on the surface of the pasta, which becomes sticky and in some cases deliquescent.

4.3.3.2. Pasta-making process

Pasta

Figure 4.38 shows the different stages involved in the industrial production of pasta.

Pressing

The *pressing* stage is preceded by hydration and mixing. Hydration should be sufficient to form a gluten network during pressing without being too high, which would otherwise result in sticky pasta that deforms after shaping. In addition, excessive hydration means longer drying times resulting in higher costs. The hydration range is very narrow (30 – 33%) and should allow the mixture to reach the desired consistency. Water should be optimally distributed to ensure homogeneity of the pasta during pressing. In this case, semolina is of suitable particle size while ensuring an even distribution of water on the surface. The time required for semolina to fully hydrate is between 15 and 20 min in standard mixers. This time can be reduced by using rapid mixers or finer semolina.

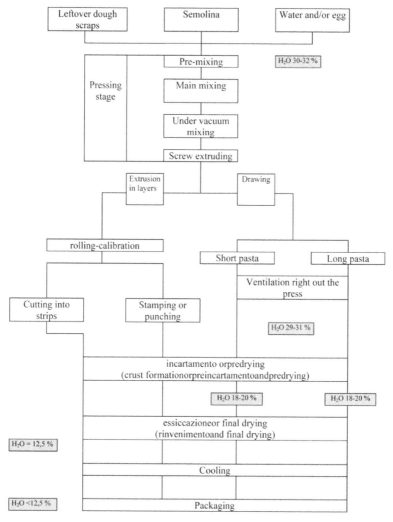

Figure 38 ■ General diagram of the industrial production of pasta

Figure 4.38. *General diagram of the industrial production of pasta*

Hydrated and mixed semolina is gradually compressed in a screw extruder reaching maximum pressure (10 MPa) at the screw head; the pasta becomes more homogeneous and the gluten structure continuous. It is essential to control temperature during this operation as it affects viscosity, residual enzyme activity and the regularity of the starch-protein structure (gelatinization, coagulation). The geometric configurations of the dies allow a variety of different pasta shapes to be created.

Drying

Drying generally consists of two main phases:

– pre-drying or *incartamento* (giving the dough the stiffness of cardboard) removes 30 (short pasta) to 40% (long pasta) of water from the dough within a short space of time; this varies depending on pasta size and weather conditions. Evaporation during this drying phase is irregular, since the outer layer is drier than the center resulting in a cardboard-like texture. This drying of the surface prevents sticking. The resulting firmness helps the pasta to maintain its shape and prevent flattening or elongation;

– final drying or *essiccazione* should be gradual, alternating if necessary between drying and recovery (*rinvenimento*) periods, thus avoiding major contractions causing cracks or breaks, particularly at low temperatures (70–75°C). At high temperatures (90–110°C), there is an increased diffusion of water throughout the pasta, which retains its plastic state up to a moisture level of around 12%; there is therefore no need for *rinvenimento*.

Batch drying is carried out on racks (short pasta) or canes (long pasta). Hot air a few degrees above the outside temperature is needed (35–40°C for *incartamento*, 20–25°C for final drying). The kinetics of evaporation vary depending on size (average duration 24–72 h at 30–40°C, H_R 50–70%).

Continuous drying involves moving the pasta through a dryer (Figure 4.39) consisting of active and non-active ventilation areas. There are three types of drying:

– normal temperature (NT) drying at 40–60°C for 10 – 20 h;

– high temperature (HT) drying at 60–80°C for 4 – 10 h;

– very high temperature (VHT) drying at 90–130°C for 1.5 – 2.5 h.

Figure 4.39. *Long pasta entering a dryer*

The most recent technique is VHT drying, which has the following advantages:

– an increase in the yellow index and a decrease in the browning index by inactivation of lipoxygenase and polyphenol oxidaze above 70°C;

– better pasta stability due to the physicochemical change in proteins (insolubilization), which reduces the qualitative differences among wheat varieties;

– a reduced loss of vitamins A, B_1, B_6 and D compared to HT and NT drying; however, increased Maillard reactions reduce the availability of lysine, which is already lower in cereals.

Couscous

Couscous semolina (Table 4.18) should have a uniform light cream color and have no holes, impurities or acidic or rancid aftertaste. Despite being intended for the preparation of traditional couscous, it is increasingly used in the production of tabbouleh and as an accompaniment to other dishes.

Characteristics	Standard couscous			Superior couscous		
	Coarse	Medium	Fine	Coarse	Medium	Fine
Particle size in mm		1.0 to 1.8	0.50 to 1.25		1.0 to 1.8	0.5 to 1.25
Moisture content	< 12.5%			< 12.5%		
Ash content	1 to 1.3%			0.9 to 1.1%		
Protein content	> or = 12%			> or = 12%		
Acidity	< 0.06%/DM			< 0.06%/DM		
Swelling index (French norm 50.001)	> 220			> 220		
Rehydration rate or capacity	No hard grains after hydration			No hard grains after hydration		

Table 4.18. *Main characteristics of couscous made from durum wheat semolina*

From a processing point of view, the main difference between the shaping of pasta and couscous is the level of cohesion between the hydrated semolina particles. In the case of pasta, the high pressure applied to the semolina binds it into a smooth dough; in the case of couscous, the level of agglomeration is low. Through the process of rolling, semolina is only weakly bound; agglomerated couscous granules must be stabilized by pre-cooking (Figure 4.40).

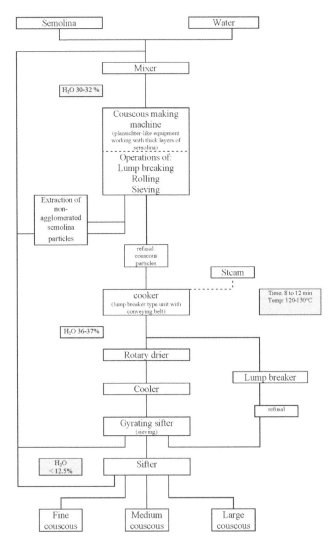

Figure 4.40. *Diagram of the industrial production of couscous*

5

From Barley to Beer

5.1. Biochemistry and structure of barley and malt

Barley (*Hordeum vulgare*), a member of the grass family originating in the Middle East, is a major cereal grain. Cultivated by man as far back as Neolithic times, it is low in gluten and therefore not as suitable as wheat for making bread. It is mainly used as animal fodder or in soups and stews, for example. It is said that "bread drink" or beer originated in Sumeria from the accidental fermentation of wet barley or bread. Over the centuries, barley and beer have remained closely linked, with the quality of barley strongly influencing that of beer.

It is possible to make beer from any cereal grain containing starch, but barley has advantages that make it ideal for beer production. Barley is a covered or hulled cereal: it has a husk that protects it from mold during growth as well as during handling and malting. Husks also act as a filter in the mash tun to separate the wort from the grain during brewing.

Two-row (*Hordeum distichum*) or six-row barley (*Hordeum hexastichum*) is commonly used for making beer. Two-row barley has grains (or spikelets) arranged in triplets, which alternate along the stem (rachis), and only the central spikelet is fertile; thus it has a flat spike. In contrast, six-row or multi-rowed barley has grains arranged all around the stem, six on each level, and all the spikelets are fertile; thus it has a dense spike.

Chapter written by Romain JEANTET and Ludivine PERROCHEAU.

Figure 5.1 shows an ear of six-row barley. Barley can be defined depending on the time of sowing: winter barley is sown in October and requires exposure to cold temperatures to promote subsequent growth, while spring barley is sown in March and does not require low temperatures to trigger growth. Winter barley often includes six-row varieties, whereas spring barley tends to include two-row varieties.

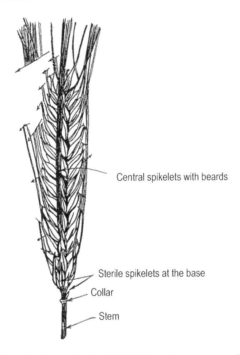

Figure 5.1. *Ear of six-row barley (J. Barloy, Agrocampus Ouest Rennes)*

5.1.1. *Morphology of barley grain*

The barley grain has many distinct parts (Figure 5.2):

– the embryo: where the vital activities of the grain take place;

– the starchy endosperm: storage tissue containing starch granules and storage proteins;

– successive layers surrounding the embryo and the endosperm: the pericarp fused to the testa as well as glume layers.

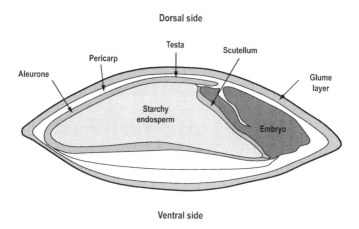

Figure 5.2. *Cross section of a barley grain*

5.1.2. *Biochemical composition of barley*

Each part of the barley grain has a particular biochemical composition:

– the pericarp is composed of 20% cellulose, 6% protein, 2% ash, 0.5% lipids and pentosans;

– the testa is predominantly composed of lipid;

– the aleurone layer contains starch and protein in its outermost portion and lipids (30%), proteins (20%), phytic acid, vitamin B, cellulose and pentosans in its center;

– the endosperm is composed of 65% starch, 7–12% protein, 6–8% cellulosic material in the cell walls (70% β-glucans, 20% pentosans, 5% proteins, 2% glucomannans, 2% cellulose, 0.5% phenolic acid and 0.5% uronic acid) and 2–3% lipids.

Table 5.1 shows the biochemical composition of barley. It is high in starch and low in fat (responsible for the rancid taste of beer). Proteins in malting barley provide yeast with amino acids essential for growth.

At harvest, barley contains 11–16% water. Excess moisture results in a lower yield and limits grain storage, as the endosperm loses its vitality under these conditions.

Carbohydrates	78–83
Starch	63–65
Sucrose	1–2
Reducing sugars	1
Other sugars	1
Pentosans	8–10
β–glucans	3–5
Proteins	9–12
Albumins	8.5–12
Globulins	2.5–5
Hordeins	3–4
Glutelins	3–4
Amino acids and peptides	0.5
Lipids	2–3.5
Nucleic acids	0.2–0.3
Minerals	2
Polyphenols	0.5–1.5
Other substances	4–6

Table 5.1. *Biochemical constituents of barley (% w/w dry matter)*

5.1.3. Composition and structure of starch and protein

5.1.3.1. Starch

Starch is the main constituent of barley. There are two types of starch granule:

– large granules measuring 10–25 μm in diameter, which represent 10% of starch granules and 90% of starch weight; they have a gelatinization temperature of 61–62°C;

– small granules measuring 1–5 μm in diameter, which represent 90% of starch granules and 10% of starch weight; they have a gelatinization temperature of 75–80°C [HEN 88].

The gelatinization temperature is the temperature at which starch granules swell and hydrate, allowing enzymes to act and hydrolyze starch to fermentable sugars (glucose, maltose, maltotriose) and α-dextrins.

Starch consists of 25% amylose and 75% amylopectin. Amylose is a linear polymer of 500 – 2,500 glucose units linked by α(1, 4) bonds. Amylopectin is a branched chain of 10,000 – 100,000 glucose units linked by α(1, 4) and α(1, 6) bonds, with each chain branching every 20 – 25 glucose units.

The other polysaccharides are mainly cell wall components (cell wall polysaccharides) and include β-glucans and pentosans (or arabinoxylans). β-glucans are composed of glucose units linked by β(1, 4) and β(1, 3) bonds. In the cell walls, β-glucans, which are insoluble polymers, are linked to peptides and other compounds. Pentosans or arabinoxylans are polymers containing pentoses (C5), the xylose of which can be substituted by arabinose. In general, the higher the level of substitution in the molecule, the greater the solubility.

5.1.3.2. Proteins

Malting barley contains 9 – 12% nitrogenous substances (proteins, polypeptides, peptides and amino acids). Barley proteins include water-soluble albumins (16% of total protein), salt-soluble globulins (4% of total protein), alcohol-soluble prolamins or hordeins (10% of total protein), glutelins (33% of total protein) soluble in dilute acids or bases, gel proteins (20% of total protein) and insoluble proteins (18% of total protein).

Most enzymes are albumins and globulins, while hordeins are storage proteins. The glutelin fraction includes both storage and structural proteins.

Some proteins are located in the aleurone layer where they do not undergo any change, but most are found in the starchy endosperm cells where they surround the starch granules and are degraded during malting.

5.1.4. Effect of malting

The barley used in malting has well-defined botanical, biological and biochemical characteristics. Currently, about 10% of all harvested barley is converted into malt, particularly for production of alcoholic beverages (beer and whiskey) and some foods (as it provides fermentable sugars and enzymes). Most malt is produced from barley, but it is possible to produce malt from wheat, rye, oats, triticale, corn, sorghum or rice. Processing barley into malt involves three main steps: steeping, germinating and kilning.

5.1.4.1. *The purpose of malting*

Malting, which is the controlled germination of barley, plays an important role because barley in its original state cannot be used in brewing since the nitrogen and carbon substrates are unsuitable for alcoholic fermentation.

First of all, the barley is cleaned to remove any impurities, and graded to select only whole grains of a certain size. Steeping activates the barley from its dormant state and provides the embryo with the moisture and temperature necessary for germination. During this stage, the aleurone layer produces the enzymes needed to modify the grain (α-amylase, β-glucanase, pentosanase and protease). These enzymes break down the starchy endosperm, necessary for subsequent brewing, and release an optimal amount of sugars for the yeast. Kilning, which is the final stage, stabilizes the malt by preventing biological activity. It involves drying the barley so that the moisture content is reduced from 45% to 2–5%. The amount of heat supplied influences the color of the malt, similar to a roasting process. Thus, malting makes the grain friable, which is necessary for its subsequent use in brewing, improves fermentability and contributes to the sensory properties of beer (aroma and color).

5.1.4.2. *Steeping*

Purpose of steeping

Steeping is the most delicate step of the process. The embryo is exposed to moisture and temperature conditions that activate it and trigger germination. An alternating series of wet (immersion in water) and dry ("air rest") periods allows the grain to acquire a moisture content of approximately 45% while avoiding suffocating the embryo. This moisture level is necessary for germination. Chitting (visible white tips of developing rootlets protruding from the base of the kernels) should be triggered to maximize the duration of germination and ensure the absorption of additives. Steeping marks the start of enzymatic synthesis and reactions that continue during germination.

Changes in the grain during steeping

Steeping conditions depend on the type of barley used (variety, size, protein content, etc.). Barley types can be differentiated by the amount of water they need to germinate, which varies between 42 and 46%. The aleurone layer must be sufficiently hydrated both for the production of enzymes and their migration throughout the endosperm. In addition, the amount of water absorbed by the grain should allow homogeneous change in the tissue while minimizing the growth and respiration of the embryo.

The transfer of water in the grain is slowed down by its impermeable external layers, or more precisely the pericarp. After a few hours of steeping, the moisture distribution in the grain is regularized. The saturated pericarp allows water to penetrate to the endosperm and embryonic region by diffusion. Abrasion or scarification of the pericarp facilitates imbibition (the absorption of fluid by the grain which results in swelling).

Barley also requires oxygen to avoid suffocation of the embryo. But oxygen cannot penetrate the grain when submerged in water for prolonged periods. Consequently, the barley is allowed to drip dry (air-rest) after each immersion.

Some materials dissolve in water during steeping. One of the membranes surrounding the endosperm, the testa, acts like a semi-permeable membrane and prevents the penetration of salts from water into the endosperm. The solvent action is therefore limited to the outer layers. During steeping, barley loses 0.6–1.5% of dry matter, mostly in the first 6 h. Dissolved substances include tannins, nitrogenous substances, gum, sugars and minerals.

5.1.4.3. Germination

Purpose of germination

The purpose of germination is to produce as many extractable substances as possible by causing the disintegration of the endosperm through the synthesis and distribution of enzymes throughout the grain. Under the action of gibberellins (plant hormones that regulate growth and influence various developmental processes) produced by the endosperm, the aleurone layer synthesizes the various enzymes necessary for the hydrolysis of starch (amylases), proteins (proteases) and cell walls (β-glucanases).

Changes in the grain during germination

Germination is accompanied by several transformations linked to the development of the embryo, that is the synthesis of new tissue, the respiration and growth of the plumule and radicals. For this, the embryo requires growth (nitrogen, hydrocarbons and minerals) and energy substances (carbohydrates and lipids), which are drawn from the reserves of the endosperm. These reserves are solubilized from the endosperm by the action of enzymes synthesized in the scutellum, which diffuse to the endosperm. Storage proteins in the grain are hydrolyzed into polypeptides, peptides and amino acids. The free flow of nutrients required by the embryo cause it to grow and radicles or rootlets to quickly form. Chitting or sprouting of the rootlets occurs after steeping, on the first day of germination. Radicles "fork" on the third day. At the same time, the plumule develops within the pericarp and if germination takes too long, the plumule breaks through the grain causing overgrown barley grains.

The β-glucanase–pentosanase complex modifies the network of cellulose membranes, which loses its rigidity. Matrix proteins are therefore hydrolyzed, releasing starch granules that are digested on the surface by amylases. Regions of high enzymatic digestion appear in the interstitial material between the starch granules causing disintegration. This starts in the region adjacent to the scutellum and propagates from the proximal end to the distal end of the grain. Disintegration of the endosperm occurs about 18 h after the start of germination. Non-germinated barley kernels are hard and compact, whereas malt kernels are floury and friable.

The degree of germination is therefore based on a compromise between a sufficient level of protein hydrolysis and amylase biosynthesis as well as a limited loss of reserves by the respiration and growth of the embryo. These reserves are substrates for subsequent stages in the production of beer.

5.1.4.4. Kilning

Purpose of kilning

The main aim of kilning is to halt the growth of green malt (the name given to malt after the germination stage) by dehydration in order to facilitate storage and milling: this stage causes the "death" of the grain. Kilning also determines the color and flavor of malt, depending on the intensity of heat treatment.

Changes in the grain during kilning

Kilning results in a gradual increase in the temperature of green or sprouted malt causing changes, the magnitudes of which depend on the heat applied.

The biological activity of the grain must be halted when the production of enzymes and changes in the endosperm have reached their optimal levels. In addition, the water content of the grain must be reduced to 2–5% to allow storage of the malt without any biological change.

In the early stages of drying, the grain temperature is lower than room temperature, given that the air at the (grain/air) interface is fully saturated (H_R = 100%). At this point, the air temperature is around 50–70°C and the grain temperature is around 25–30°C, close to the wet bulb temperature of the air. Under these conditions, the moisture content of the grain is still sufficient for germination and changes in the endosperm to occur. If the grain temperature exceeds 50°C, this activity is rapidly brought to a halt. The substances formed (sugars, dextrins, etc.) then accumulate in the albumen.

After drying, the grain temperature gradually increases, reaching the dry bulb temperature of the air since the relative humidity of the surface decreases. The water in the barley is then drawn to the surface, where it evaporates. As a result, the endosperm dries and its enzymatic activity drops. However, it is important to avoid excessive denaturation of enzymes, which have an essential role to play in the brewing process.

Enzymes gradually become inactive above 60°C. They are denatured above 110–120°C or by a prolonged phase of purely chemical reactions leading to the formation of colored and aromatic substances during non-enzymatic browning reactions (Maillard reactions; see Volume 1). These reactions are the result of condensation between an amino group and a carbonyl group [MAI 12]. The parameters that influence these reactions are temperature, moisture, pH, oxygen, and metal ions. The final stage of the Maillard reaction is the formation of melanoidins, insoluble brown pigments [JAC 78] and volatile aromatic molecules.

5.2. Biological and physicochemical factors of processing

Barley, water, hops and yeast are the basic ingredients of beer. These elements are gradually transformed into beer during four main stages – malting, brewing, fermentation and packaging.

Not all barley varieties are suitable for beer making. Malting barley must meet certain criteria (germination capacity, varietal purity, quality grading, protein content). The most suitable barley is rich is starch (around 65% of its weight), large, uniform, pale yellow, and well protected by its smooth husk.

5.2.1. *Enzymatic degradation of starch and protein*

5.2.1.1. *Enzymatic degradation during malting*

Biochemical changes during malting are complex [BRI 02]. During germination, the embryo needs nutrients to grow, which it draws from reserves in the starchy endosperm. As previously mentioned, these reserves must first be dissolved. The embryo secretes enzymes (protease, amylase) that diffuse from the scutellum to the endosperm, which is where starch and protein hydrolysis takes place. Degradation products can then diffuse through the grain and reach living tissues, where they are metabolized. The synthesis of new complex molecules, such as proteins or polysaccharides in the growing embryo, partially conceals the degradations that occur elsewhere in the grain, in particular in the endosperm. In addition, material loss during malting makes it difficult to analyze these production and degradation phenomena.

Changes in carbohydrates

Barley sugars consist of starch, non-starch polysaccharides (hemicellulose, β-glucans, arabinoxylans), soluble sugars such as mono- or disaccharides, and some oligosaccharides [HEN 88]. Sugars are also found in the form of glycoproteins, glycolipids and nucleic acids.

Malting leads to an increase in soluble sugars due to the degradation of starch and polysaccharides in the cell walls of the endosperm (Figure 5.3), and the synthesis of sucrose from triacylglycerol in particular. Changes in sugars vary depending on the barley used and the malting conditions [CHA 99].

The degradation of cell wall polysaccharides is an important phenomenon in the disintegration of barley grains during malting. Degradation begins with the dissolution of hemicellulose by β-glucanase. Different β-glucanases are synthesized during germination by the action of gibberellic acid (Figure 5.3). They hydrolyze β-glucanes and enable the release of glucose polymers and oligomers. The hydrolysis of arabinoxylans by xylanase and arabinofuranosidase releases xylose polymers, xylose and arabinose.

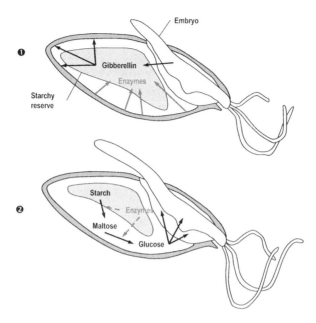

Figure 5.3. *Changes in barley grain caused by the action of hydrolases. For a color version of this figure, see www.iste.co.uk/jeantet/foodscience.zip*

After this stage, cell membranes are permeable to amylolytic enzymes, α- and β-amylases, limit-dextrinase and α-glucosidase. The action of these enzymes results in the partial hydrolysis of starch and the release of glucose and large amounts of maltose and maltotriose [ALL 99]. Hydrolysis also continues during the brewing process.

Changes during kilning largely depend on the conditions. If temperatures are low, the amount of simple sugars increases. If temperatures are high, this amount decreases, probably due to Maillard reactions.

Changes in proteins

Protein degradation is an important part of the malting process. Insufficient degradation causes problems during brewing as well as defects in the final beer [EVA 99]. Starch granules are embedded in a protein matrix, which, if not degraded, hinders the action of amylases during brewing. Moreover, yeast requires assimilable nitrogen for growth during fermentation.

The ratio of different protein fractions changes during malting. Nitrogenous substances are lost by dissolution during steeping. During germination, the amount of hordeins drops significantly and kilning can lead to protein denaturation. The change in hordein concentration during malting has been well documented. Fully disintegrated malt contains half of the hordeins present in non-germinated barley.

During the malting process, storage proteins in the endosperm are degraded due to an increase in the quantity of proteolytic enzymes, an increase in water-soluble non-protein nitrogen and a biosynthesis of nitrogenous substances in the aleurone layer and the growing embryo.

Proteolytic enzymes act from the start of germination and attack the hordein fraction, in particular in the endosperm. They are synthesized in the scutellum of the embryo. Endoproteinases are the main enzymes; they cleave the peptide bonds within the proteins, releasing peptides and polypeptides. Exoproteinases also exist, which hydrolyze proteins at both ends of the protein chain releasing one amino acid at a time. Exoproteinases are called carboxypeptidases when they act on the carbonyl-terminal (C-terminal) end of the protein and aminopeptidases when they act on the amino-terminal (N-terminal) end. Carboxypeptidases are abundant throughout the malting process.

5.2.1.2. Enzymatic degradation during brewing

Brewing is the production of beer through steeping the malted barley in water and then fermenting it with yeast. It aims to:

– dissolve compounds that are formed during malting;

– use enzymes synthesized during malting to convert starch into fermentable sugars;

– activate the residual proteases of kilned malt to continue the degradation of proteins into smaller components (amino acids, peptides, polypeptides);

– add bitterness to the wort (the liquid extracted from the mashing process during brewing) by adding hops.

Carbohydrate compounds, representing around 90% of the solid content of the wort, undergo significant change during brewing; the starch in the malt undergoes the following degradations: hydration of the starch granules, gelatinization with an increase in viscosity, enzymatic degradation (liquefaction) and saccharification.

Mashing

Mashing allows biochemical changes to take place, which produce the elements necessary for the fermentation of the wort by using enzymes that are synthesized or activated during malting. The crushed malt and water are mixed in a large vessel called a mash tun to form the mash: this is known as mashing. The water–flour mixture is subjected to different temperatures in the mash tun: this allows the brewer to control cell wall degradation (by β-glucanases), proteolysis and finally amylolysis by adjusting the optimal activity conditions of the enzymes involved. Temperature and pH are therefore key factors in the mashing process, since these two parameters determine the optimal activity of amylolytic and proteolytic enzymes. If the mash is diluted, amylolytic activity is favored over proteolytic activity, since amylases are less heat sensitive. However, proteolytic activity is favored in a thick mash because peptidases are usually more heat-resistant in concentrated mediums.

The conversion of starch during brewing begins with the breakdown of starch granules under the effect of heat and the release of amylose and amylopectin chains: this is called starch gelatinization. Gelatinized starch is in the form of a highly viscous and compact mass. The action of α-amylase, an endoenzyme that cleaves α(1, 4) bonds, results in the scission of amylose and amylopectin molecules. Dextrins are consequently released and the viscosity of the starch solution decreases, which is known as liquefaction. When all the starch has been gelatinized and liquefied, α- and β-amylases act on the dextrin chains to form fermentable (80%) and non-fermentable (20%) sugars: this is known as saccharification. Unlike α-amylase, β-amylase is an exoenzyme that cleaves α(1, 4) bonds from the reducing end to produce maltose and glucose. The optimal activity of β-amylase occurs at 62–65°C at pH 5.4–5.6. Its denaturation temperature is 70°C. α-amylase shows optimal activity at 70–75°C at pH 5.6–5.8, and is denatured at 80°C. Limit-dextrinase, which only attacks the α(1, 6) bonds of branched amylopectin chains, and maltase, which converts maltose to glucose, do not play a significant role in brewing and therefore do not influence the amount of fermentable sugars. Not all the starch is converted during brewing and starchy materials represent a third of the dry matter content of the spent grains.

Nitrogenous substances are initially converted during malting and then brewing by enzymes such as endopeptidases or carboxypeptidases. Proteases have an optimal temperature of 45–50°C, which explains their essential role in mashing and results in relatively long polypeptide chains. A compromise must be found between the production of amino acids required for yeast growth and the degradation of polypeptide chains favorable to colloidal stability, but detrimental to head retention.

In brewing, pH plays a key role since it affects enzyme activity, protein precipitation and hopping yield. The pH in the mash tun is generally 5.6. Malt has a buffer capacity, and brewing water is often adjusted by acidification to obtain a pH close to the optimum pH of α- and β-amylases. Either sulfuric, hydrochloric, lactic or phosphoric acid is added.

Mashing results in a liquid rich in sugars, amino acids and peptides (wort), which are fermented by the brewing yeast.

Boiling the wort

Once filtered, the wort is boiled in a copper kettle to which hops are added (Figure 5.4). The aim of this process is to sterilize and stabilize the wort by destroying all microbial flora, isomerize α-humulones in the hops to extract bitterness, develop the color and flavor of the wort by evaporating unpleasant sulfur odors, and concentrate the wort.

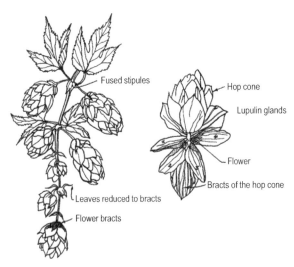

Figure 5.4. *Hop cones (J. Barloy, Agrocampus Ouest Rennes)*

Stabilization is primarily biological, since any bacteria still present in the wort are destroyed. During boiling, enzymes, in particular α-amylase, are inactivated, preventing a change in the sugar composition of the wort during fermentation. This could affect biological stability and head retention in the final product. Finally, the colloidal stability of beer is improved because nitrogenous substances and protein–polyphenol complexes are eliminated.

Flavor and color development is affected by the reduction of undesirable volatile substances, such as dimethyl sulfide and its precursor S-methyl-methionine, and by Maillard reactions that result in a caramelization of reducing sugars (maltose, fructose, glucose) and the formation of melanoidins; these two reactions enhance the color and flavor of the wort.

Hopping is the term given to the use of hops in beer; the hops are boiled to give beer its characteristic flavor and aroma. This process involves the dissolution and conversion of the main active ingredients of hops, including lupulin, which decomposes into slightly bitter and weakly soluble α acids (humulones) and β acids (lupulones). The latter dissolve when heated and α acids are converted into their much more soluble and bitter isomers called isohumulones, which are the bitter substances in beer (Figure 5.5). Isohumulones significantly affect the physical properties of wort and beer because of their surface-active properties.

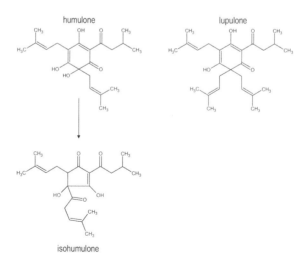

Figure 5.5. *Hop-derived bitter compounds*

The wort composition affects fermentation and the properties of the final beer product. Wort must contain the necessary amounts of fermentable sugars, nutrients for yeast growth and flavor compounds. It contains mainly glucose, maltose and maltotriose. Dextrins make up a large fraction of non-fermentable sugars. Wort proteins (around 400 mg L^{-1}) and lipids (30 mg L^{-1}) play an important role because they are conducive to yeast growth but detrimental to flavor stability and head retention in beer [BAM 85].

Wort clarification and cooling

The wort is clarified and cooled to remove all solid substances. It contains two types of debris: "hot break", the coagulation of substances during boiling, and "cold break", the precipitation or flocculation of substances during cooling. These substances consist of proteins, hop matter and polyphenols.

5.2.2. Fermentability of the wort

The clarified wort is rich in fermentable sugars and nitrogenous substances that can be used by yeast during fermentation (Figure 5.6).

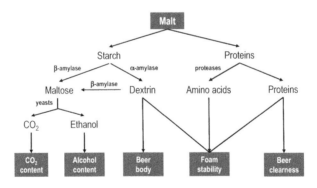

Figure 5.6. *Development of malt starch and proteins during brewing and fermentation*

5.2.2.1. Purpose of fermentation

Beer making is based on alcoholic fermentation, an anaerobic and exothermal process whereby sugar is converted to alcohol by yeast. Yeasts, such as *Saccharomyces*, use the fermentable sugars in the wort as substrates to form alcohols, fatty acids, esters and carbon dioxide.

5.2.2.2. Biochemical changes

During fermentation, sugars and amino acids in the wort are assimilated by yeast. Sugars (glucose, fructose, maltose, maltotriose, etc.) are mainly converted into ethanol and carbon dioxide (Figure 5.6). Once carbon dioxide builds up in the beer, the pH drops from 5.5 to 4.5. A large number of compounds that influence the flavor of beer are also synthesized during fermentation: higher alcohols produced from amino acids, aldehydes (acetaldehyde), ketones (diacetyl), organic acids, and sulfur compounds (hydrogen sulfide, dimethyl sulfide, sulfite). The wort is thus converted into beer. At the end of fermentation, the yeast excretes various compounds (amino acids, peptides, nucleic acids and phosphates) that contribute to the body of the beer.

5.3. Brewing technology

5.3.1. *Stages of malting*

Figure 5.7 summarizes the different stages involved in malting.

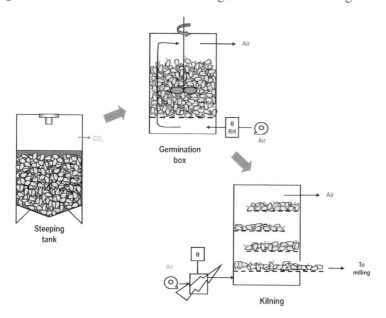

Figure 5.7. *Different stages of malting. For a color version of this figure, see www.iste.co.uk/jeantet/foodscience.zip*

5.3.1.1. Steeping

Initially, water is added to a steeping tank and barley that has been cleaned, sorted and graded is poured on top. The barley is then submerged for 6–8 h; this is the first wet period. The water is then drained off through the bottom of the tank and barley is allowed to rest (air rest); this is the first dry period. An alternating series of wet and dry periods allows the grain to obtain the required moisture level for the subsequent stages. Steeping generally involves three wet periods. The purpose of this method is to allow imbibition, required by the barley for the following stages, while also providing it with the necessary oxygen. To avoid suffocating the grain during dry periods and inhibition during germination, carbon dioxide released by the respiration of the barley is removed; the carbon dioxide/oxygen ratio should not exceed 1. Modern steeping techniques last approximately 40 h.

Steeping is carried out in cylindro-conical or flat-bottomed tanks; it is important that the thickness of the grain bed is uniform, with sufficient aeration and carbon dioxide extraction.

5.3.1.2. Germination

Germination takes place in a device where humidity, temperature, agitation and ventilation are constantly monitored. The basic processes involved in germination are enzymatic hydrolysis in the endosperm and synthesis in the embryo. These processes are strongly influenced by temperature, aeration and humidity conditions throughout the entire procedure. To obtain optimal grain disintegration, germination generally takes 5 h at 16°C for spring barley and 5 days at 18°C for winter barley.

The germination temperature is a difficult parameter to optimize. Low temperatures induce a high level of enzymatic activity, whereas high temperatures can accelerate germination and enzyme production. Thus, a moderate temperature should be maintained. Insufficient moisture can result in germination being slow to start and insufficient disintegration. Finally, aeration, which affects respiration and therefore synthesis, should be limited to avoid material loss.

Two types of malting techniques exist: floor malting, which is hardly used anymore, and pneumatic malting. The different stages take place in germination boxes. The layer of grain is aerated regularly to control its temperature following exothermal reactions associated with disintegration.

The product obtained at the end of this stage is called green malt.

5.3.1.3. Kilning

Kilning involves spreading out the green malt in an even layer on perforated metal plates (single- or two-floor kiln), through which hot air of varying temperatures is circulated. This process lasts between 24 and 48 h. It results in the termination of biochemical changes in the malt, which can then be stored until the milling stage.

Kilning affects color and aroma development in malt. The quality of the final beer thus depends on the control of the last stage of kilning, which is similar to a roasting process. The level of intensity and length of this stage directly influences the aroma and color of the malt (ranging from pale yellow to dark brown), which in turn will be used to produce a yellow, red or brown beer. The kilning of green malt involves five steps:

– initial drying until a moisture content of 23% is reached;

– intermediate drying until the moisture content is reduced to 12% (50–70°C, 12–22 h);

– removal of bound water until the moisture content is reduced to 6% (65–75°C);

– rapidly raising the temperature to reduce the moisture content to 2–5% (85–110°C, 4–5 h). The temperature of the malt intended for pale beer does not exceed 85°C, whereas malt intended for dark beer can have temperatures of 110°C;

– cooling by cold air ventilation to reduce the temperature to 30–35°C.

The survival of heat-sensitive enzymes requires a low initial temperature (50°C for not more than 24 h). Since the heat resistance of enzymes increases as the moisture content decreases, the temperature can be increased to promote the development of the desired aroma and color.

5.3.1.4. Deculming and storage of malt

Deculming removes the malt rootlets (also known as culms) that would give beer a very bitter taste. The rootlets contain non-protein substances, protein compounds, fats and vitamins. The final product after this process is malt. It takes 133 kg of barley to produce 100 kg of malt. Freshly kilned malt contains 1.5–4.5% moisture. Malt should be stored below 25°C and the minimum amount of time between kilning and brewing is 1 month.

5.3.2. *Stages of beer production*

Brewing generally takes about 3 weeks. It is followed by maturation, which gives the beer its character and aroma, and can last several months in some cases. Figure 5.8 shows the various stages involved in brewing.

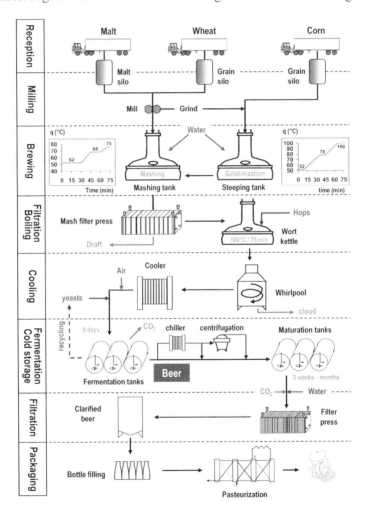

Figure 5.8. *Different stages of beer production. For a color version of this figure, see www.iste.co.uk/jeantet/foodscience.zip*

5.3.2.1. Milling

Before brewing, the malt and sometimes raw grains (corn, rice and wheat) are crushed into grist, usually by roller mills. The husks protecting the grains should not be destroyed, as they subsequently act as a filter bed to strain out the wort. The level of filtration is therefore determined by the degree of grinding, depending on the type of raw grain added during brewing. When a high percentage of raw grain is used, the grind is coarser and filtration is faster.

Two techniques are used: dry milling and wet milling. Wetting the grain prior to milling protects the husks during the crushing process. Hammer mills are also used if the filter technology allows the use of a finer flour.

5.3.2.2. Mashing

Raw grains (corn and rice) are usually gelatinized in the presence of a small amount of malt in the steeping tank. Malt enzymes promote the liquefaction of starch at around 75°C and gelatinization is achieved by boiling.

The crushed malt, water and contents of the steeping tank are mixed together in the mash tun to form a mash: this is called mashing. Various temperature stages activate different enzymes from proteases to β- and α-amylase.

Different brewing methods exist:

– infusion involves mashing at around 65°C. The temperature is then lowered or increased by the addition of water;

– decoction involves successive heating stages at temperatures ranging from 50 to 80°C.

The mash can also be boiled to increase yield and enable the removal of proteins and tannins by coagulation.

5.3.2.3. Lautering

Lautering or filtration involves separating the wort from the spent grain (husks and insoluble kernel fragments, called draff) by extracting the maximum amount of dissolved solids. It is at this point that the husks, kept during milling, act as a filter bed. This operation involves two steps: filtration itself, where the first wort run-off is separated from the spent grains, and sparging, in which extracts, in particular sugars, are rinsed off from the grains with hot water at around 80°C.

Lautering is carried out in a lauter tun or a mash filter press. A lauter tun consists of a boiler with a perforated bottom that allows the first wort run-off to pass through while holding back the spent grains. Hot water is added to sparge the residual grain. Some lauter tuns have rotating rake arms that unclog the grain bed and accelerate lautering. A mash filter press consists of frames (covered in filter cloth) and hollow shafts (through which the wort is filtered). The wort passes through the filter cloth and into the frame, from where it is discharged by a valve or collecting pipe. The residual grain, which cannot pass through the cloth, accumulates in the hollow shafts where it is sparged.

5.3.2.4. Boiling the wort

The filtered wort is boiled in a copper kettle to which hops are added. This operation lasts 90 min at pH 5.2 (pH of the wort) and has the following objectives:

– to sterilize and stabilize the wort by destroying microbial flora;

– to develop the color and flavor of the wort by evaporating unpleasant sulfur smells;

– to isomerize the α-humulones in the hops to isohumulones to extract their bitterness;

– to concentrate the wort; boiling concentrates the wort by evaporation, giving the desired density.

Hops should be adapted to the type of beer, taking into account the sensory quality of the final product. Hops are used in different forms: flower cones (female only, the pollen of which is aromatic), pellets (compressed cones) or extracts.

5.3.2.5. Clarifying and cooling the wort

After boiling, the wort is 100°C and contains suspended solids such as hop particles. Before yeast is added, the wort is clarified to remove these suspended substances (break) and cooled to between 7 and 10°C for bottom fermentation and 15 – 20°C for top fermentation.

Clarification involves the removal of two types of debris: "hot break" or "hot trub" and "cold break" or "cold trub", which gradually appears as the wort cools. Up to 70–80% of the cold break is removed, as it is detrimental to foam head stability and flavor. This is done by centrifugation (whirlpool) or filtration.

The wort is cooled in a counter-flow heat exchanger (using cold water). It is then aerated so that the oxygen has time to dissolve prior to fermentation.

5.3.2.6. Fermentation

After cooling, the wort is inoculated with yeast (15×10^6 to 25×10^6 cells mL^{-1}). The yeast strains are selected based on technological (optimal fermentation temperature, ability to flocculate) and sensory (e.g. low diacetyl production) criteria. Two types of yeast are used in brewing: *Saccharomyces cerevisiae* and *Saccharomyces carlbergensis*, and their different technological properties result in two types of fermentation:

Top fermentation occurs between 15 and 25°C for 2 – 5 days using *Saccharomyces cerevisiae*. At the end of fermentation, the yeast rises to the surface. It is recycled by taking the foam that accumulates on the surface. The cells are washed and stored at a cold temperature before reuse.

Bottom fermentation occurs between 7 and 12°C for 5 – 10 days using *Saccharomyces carlbergensis*. Fermentation is followed by cold storage or maturation at variable temperatures depending on the process. At the end of the cycle, the yeast flocculates at the bottom of the tank. This phenomenon is linked to the properties of the strain and the process conditions (the final fermentation temperature is often lowered to 4°C to promote flocculation). The yeast is recovered for reuse. This type of fermentation is used in the production of lager beer (e.g. Pils).

5.3.2.7. Cold storage

The beer produced following primary fermentation, known as "green beer", does not yet have the required sensory properties for consumption. It must undergo cold storage or maturation. This stage, during which fermentation continues, promotes the dissolution of carbon dioxide, the sedimentation of suspended particles (yeast, precipitates formed by proteins and bitter substances from hops) and flavor development.

Cold storage or maturation is carried out at 5–8°C, especially for non-flocculating yeast. Beer is then cooled from fermentation to maturation temperature within 24–36 h. This temperature is maintained for 5–8 days until the complete reduction of diacetyl, which is assimilated by the yeast. Other compounds such as hydrogen sulfide and acetaldehyde are also removed. At the end of the maturation period, the beer is cooled to 3–4°C, then rapidly lowered to a temperature of –1°C for at least 48 h to precipitate the polyphenol–protein complexes. The beer must not be heated prior to filtration as this would dissolve the cold trub that has formed.

5.3.2.8. Filtration

Filtration is the final step in beer making. Its purpose is threefold:

– clarify the beer to make it translucent;

– decrease the microbial count by removing most of the bacteria and surviving yeast at the end of cold storage;

– stabilize the beer by removing polyphenol–protein complexes and colloids that would rapidly form trub, as well as poorly soluble proteins and polysaccharides.

During filtration, yeast and bacteria are removed based on size exclusion mechanisms. Dissolved bitter substances are removed by chemical affinity (adsorption, precipitation) or ion exclusion (repulsion).

Cooling the beer to 0 to –1°C before and during filtration is the key to good colloidal stability. A centrifuge is used when the yeast content in the beer is high or irregular. The carbon dioxide content is adjusted before the beer is sent to the filtered beer tank.

Beer can be filtered in two ways [FIL 99]:

– by frontal filtration using filter plates (by filter press) or Kieselguhr filters. Kieselguhr is a sedimentary rock composed of the fossilized remains of brown one-celled algae. It is added to beer as a filtration aid (it constitutes the filter cake). The use of Kieselguhr is controversial for environmental (highly polluting effluents) and public health (the powder product can cause respiratory lesions among operators) reasons;

– by cross-flow microfiltration (see Volume 2, Chapter 3), which is the main alternative to Kieselguhr filtration and could help do away with the pasteurization step and all associated sensory changes. In addition, the automation and continuous nature of this method results in significant improvements in productivity. The use of cross-flow filtration in brewing over the coming years will largely depend on the whether the Kieselguhr method remains viable and the development of renewable filtration material.

5.3.2.9. Stabilization

The deterioration in the translucency of beer during storage is mainly due to colloidal instability. Trub can form following a reaction between proteins and polyphenols (tannins or proanthocyanides). These molecules come from malt and hops and have an affinity to certain proteins in malt, especially hordeins, which are storage proteins with a high proline content. The interaction between tannins and hordeins is based on the formation of hydrogen bonds, which is facilitated at low temperatures. Although proanthocyanides are physically separated from these proteins in the malt, they are capable of forming complexes during brewing, which then precipitate. This phenomenon continues after the wort has been boiled and during fermentation, as the temperature is lower.

To control colloidal stability and therefore reduce the amount of compounds that cause trub, brewers remove either polyphenols or proteins. They can use PVPP adsorption; polyvinylpolypyrrolidone is an insoluble synthetic polymer of high molecular weight that forms very stable complexes with polyphenolic substances in beer, which thereby reduces polyphenol–protein complexation. Treatment with bentonite (aluminum silicate), silica gel or proteolytic enzymes helps reduce the amount of proteins and thus prevents the formation of complexes with tannins. The colloidal stability of beer is difficult to achieve. Selecting the raw materials, adhering to the principles of brewing, ensuring cold storage, filtration and airtight filling all help reduce the need for processing aids. Apart from colloidal stability, the brewer also aims to achieve sensory and microbiological stability as well as good head retention [LUS 95, OHA 87, SEG 67].

5.3.2.10. *Packaging*

After filtration, the beer is poured into kegs (counter-pressure filling) or bottles. To ensure the proper preservation of beer, pasteurization can be carried out in bulk (flash pasteurization) or after packaging (tunnel pasteurization).

6

From Fruit to Fruit Juice and Fermented Products

Fruit, whether fleshy, dehiscent or dried, is an important part of the human diet. The four main fleshy fruits are bananas, grapes, apples and citrus fruit, representing 59% of total fleshy fruit crops worldwide [FAO 12]. Fruit can be consumed fresh or as processed products (purees and slices, juices and concentrates, wines and spirits). The proportion of fruit that goes towards processing is large for grapes and citrus fruit, less for apples and only marginal for bananas.

Processing varies depending on the fruit used and the desired final product (clear juice, cloudy juice, nectar, concentrates or fermented beverages, etc.). Methods are adjusted based on the structural and biochemical characteristics of the plant. The challenge facing the food processor is to ensure colloidal and microbiological stability while maintaining the nutritional and sensory quality of the beverage.

6.1. Fruit development

6.1.1. *Stages of development*

On an anatomical level, a true fruit is a mature ovary, thus developed solely or partly from carpel tissue. However, several species of fruit develop

---

Chapter written by Alain BARON, Mohammad TURK and Jean-Michel Le QUÉRÉ.

from other pericarp tissue. For example, in pineapples, blackberries or apples, the largest part of the fruit is respectively derived from bracts, calyces or floral tubes (fused base of floral organs). Strawberries develop from the floral receptacle bearing the achenes or true fruit. Even fruit derived exclusively from carpel tissue can vary significantly in development from the expansion of a single carpel resulting in stone fruits (or drupes) to the differentiation of tissue resulting in the albedo (fleshly middle layer of the pericarp), flavedo (outermost layer of the pericarp) and multicarpel flesh of citrus fruit or bananas.

Upon the completion of pollination and fertilization, the flower ovary begins to enlarge into a developing fruit. This is "fruit set" and it marks the beginning of growth and development, which can be divided into three stages (Figure 6.1).

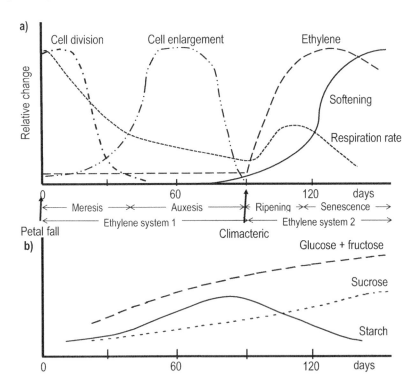

Figure 6.1. *Stages of fruit development; a) main changes during the development of an apple; b) changes in carbohydrate*

The first stage named "meresis" is a period of intense cell division with a differentiation of tissues and an accumulation of metabolites such as polyphenols or organic acids. Depending on the species and cultivar, this can take two to eight weeks. The second stage "auxesis" (3 to 10 weeks) is a phase of cell expansion responsible for the enlargement of the fruit during which there is an accumulation of carbohydrates (starch and/or simple sugars) and a synthesis of wall compounds. The third stage is the ripening and senescence phase. Compositional changes during this final stage are best understood and described in terms of physiology.

6.1.2. *Fruit ripening*

Depending on whether or not fruits have a peak in respiration during ripening, they are classified as climacteric (apples, pears, bananas, melons, etc.) or non-climacteric (citrus fruits, grapes, strawberries, etc.) fruits. The term climacteric means that the fruit is at a critical period of its development when it passes from an unripe to a ripe state, prior to senescence (biological aging).

Climacteric crisis begins by the increase in ethylene production which accompanies the respiratory peak during ripening. It reflects intense metabolic activity and it corresponds to modifications of the nitrogen fraction (proteins, amino acids, RNA), sugars (starch, sucrose, wall polysaccharides), organic acids, aromatic compounds and so on. The role of ethylene in causing this process has been known since the 1930s [KID 33, GAN 34, GAN 35]. Since then, the development of molecular biology has helped to define the mechanisms regulating the ripening of climacteric fruit [WEI 10, ZHE 13]; less data is available on non-climacteric fruit.

6.1.2.1. *Role of ethylene*

Ethylene is a plant hormone that is synthesized from S-adenosyl-L-methionine (SAM), a major metabolite in cells, central to several metabolic pathways. SAM is converted by ACC synthase to 1-aminocyclopropane-1-carboxylic acid (ACC) that is then oxidized by ACC oxidase (ACO) with the release of ethylene (carbon 3 and 4 of methionine), CO_2 and hydrocyanic acid.

Maturing fruit, such as all plant tissues, produce small quantities of ethylene. This production can be inhibited by exogenous ethylene and corresponds to system 1 ethylene biosynthesis. It is the only metabolic pathway present in non-climacteric fruits. During the ripening of climacteric fruit, system 2 ethylene biosynthesis is established (Figure 6.2); it is characterized by an intense production of ethylene, initiated by small quantities of ethylene and self-catalyzed. Thus, the two systems differ in their susceptibility to ethylene: system 1 is inhibited and system 2 is stimulated. System-2 ethylene synthesis is initiated and maintained by ethylene-dependent induction of one of four regulation genes of ACA synthase [TAT 10].

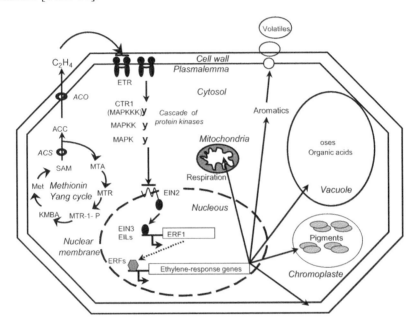

Figure 6.2. *Diagram of metabolic pathways during the ripening of climacteric fruit*

On the left, ethylene biosynthesis. Abbreviations: SAM, S-adenosyl-L-methionine; ACC, 1-aminocyclopropane-1-carboxylic acid; ACS, ACC synthase; MTA, 5'-methylthioadenosine; MTR, 5'-methylthioribose; MTR-1-P, MTR phosphate; KMBA: 2-keto-4-methylthiobutyric acid; Met, Methionine; ACO, ACC oxidase.

In the middle, ethylene signal perception and transduction. Abbreviations: ETR, ethylene receptor; CTR1, CTR1 gene products, that is kinases: Mitogen-Activated Protein Kinase Kinase Kinase (MAPKKK), Mitogen-Activated Protein Kinase Kinase (MAPKK) Mitogen-Activated Protein Kinase (MAPK); EIN2, membrane receptors; EIN3 and EILs, transcription factors; ERF, ethylene response factors (induction of gene expression).

In climacteric fruit, ethylene binds to a set of membrane receptors (ETR), probably located in the endoplasmic reticulum and having kinase activity. In the absence of ethylene, they actively repress the expression of ethylene response genes via a series of protein factors. In the presence of ethylene among ETR receptors, protein factors in turn bind to gene promoters regulating the ethylene response factor (ERF and EDF) and induce their transcription. These proteins, which are themselves transcription factors, activate or repress certain genes that control the metabolic pathways involved in ripening. Among them, the modification in cell walls resulting in fruit softening [IRE 14] and the formation of volatile compounds [SOU 11] are the most visible responses.

In non-climacteric fruit, the role of ethylene seems to be secondary in the ripening process. In citrus fruit, this hormone contributes to the color change of the flavedo by stimulating carotenoid synthesis and chlorophyll degradation. In grapes, the ethylene content, still low, decreases after "veraison" or the onset of ripening. Ethylene does not seem to have any effect in other non-climacteric fruit. In strawberries, ripening could be initiated by the decrease in the amount of auxine produced by the achenes.

6.1.2.2. Cell wall changes

During ripening, fruit softens due to changes in the cell wall structure. Changes can also occur among each family of polysaccharides (pectins, hemicellulose, cellulose) as well as their interactions. A large number of enzymes contribute to these changes (see section 6.2.2.): pectin methyl esterases (PME), polygalacturonases (PG), β-galactosidases, β-D-xylosidase, α-L-arabinofuranosidase, endo-β-mannanase, endo-β-1, 4-glucanases (EGases)and xyloglucan endotransglycosylase (XET). The role of expansins, non-catalytic proteins, is still not clearly understood; they can promote the access of polysaccharidases to their substrate by generation of a mobile conformational defect on the surface of cellulose microfibrils, which in consequence disrupts the bonds between xyloglucans and cellulose microfibrils, thereby affecting the wall structure. Although transgenesis has demonstrated the involvement of some of these enzymes, such as β-galactosidase in tomatoes or pears, many questions still remain unanswered: are all the enzymes involved known, what are the substrates of certain enzymes and what are their functions? EGase is an example of an enzyme in tomatoes, avocados and peppers in which the substrate is neither cellulose nor xyloglucan. In addition, in ripening strawberries, could the accumulation of gene products homologous to those encoding pectate lyase (PAL, an enzyme that degrades low methylated pectin) account for the hydrolysis of the middle lamella, while PG activity is still very low and PAL activity is not detectable *in vitro*?

6.1.2.3. Synthesis of aromatic compounds

Aroma is an important factor in the quality of fruit and fruit juice. It involves a complex mixture of volatile molecules. For example, more than 300 compounds have been identified in the flavor profile of different apples, nearly 200 in oranges, 140 in strawberries, over 80 in blackcurrants and around 50 in bananas. Some aromas contain large amounts of terpenes and terpenoids (e.g. blackcurrants, mangos and citrus fruit). This family of compounds includes linear, cyclic or polycyclic polymers of isoprene (Figure 6.3).

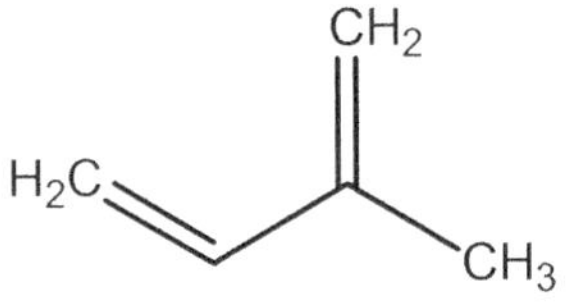

Figure 6.3. *Structure of isoprene, the basic structural element of terpenes*

However in fleshy fruit, the most abundant aromatic compounds are generally esters (78 to 92% in apples) and alcohols (6 to 16%). These compounds are generally formed by the esterification of alcohols and acyl-CoA derived from the metabolism of fatty acids and amino acids; this reaction is catalyzed by alcohol O-acyltransferase (AAT). Several studies on bananas, strawberries and apples have shown that the limiting factor to the biosynthesis of esters is the availability of precursors and not the level of AAT activity. This could indicate that the formation of esters is regulated at an earlier stage. However, recent studies on apples with antisense genes for ACO or ACS, or in other words with reduced ethylene production, have shown a significant reduction in the synthesis of esters (-70%). If the synthesis of precursors of these esters is not affected, ethylene would influence the regulation of AAT. Other key enzymes responsible for specific flavor profiles have been identified such as O-methyltransferase capable of forming dimethyl methoxy furanone, a major compound in the aroma of strawberries, alpha-farnesene synthase in apple skin or carotenoid cleavingdioxygenase1 in tomato that contribute to the formation of different volatile compounds (β-ionone, pseudoionone and geranylacetone).

6.2. Biochemistry of fruit juice

The most abundant constituent of fruit juice is water, which represents 75 to 90% of the total weight. Solutes can be divided into three groups based on their weight.

The first group contains compounds weighing several grams to several hundred grams per liter. They constitute the bulk of the solids content of juice and contribute to the balance of flavor:

– soluble sugars (100 to 200 $g.L^{-1}$), most often glucose, fructose, sucrose in varying proportions depending on the fruit;

– organic acids (2 to 15 $g.L^{-1}$), citric acid in citrus fruit, tartaric acid in grapes, malic acid in apples and grapes, etc.

The second group includes less abundant compounds, but with a high technological impact:

– pectins (0.1 to 2 $g.L^{-1}$), which play a role in colloidal stability and juice clarification;

– amino compounds (0.05 to 0.5 $g.L^{-1}$) involved in non-enzymatic browning reactions (see Chapter 5 of Volume 1 [JEA 16a]) after heat treatment and certain enzymes that may have a positive or negative effect on product development and sensory qualities;

– phenolic compounds (0.1 to 5 $g.L^{-1}$), enzymatic browning substrates (see Chapter 6 of Volume 1) and responsible for the color, bitterness and astringency of juice.

The last group includes low-abundance solutes such as volatile compounds and vitamins that contribute to the aromatic and nutritional qualities of fruit juice.

6.2.1. *Pectins*

Pectic substances are polysaccharides, exclusively of plant origin. They are one of the main constituents of the plant cell wall (middle lamella and primary cell wall). During pressing or maceration with water, they are partially extracted in the juice but in varying quantities depending on the fruit and the state of maturity of the fruit. Thus, the pectin content of raw apple juice steadily increases from 150 to 500 mg.L^{-1} during the season from the beginning of October to the end of December in the Northern hemisphere. The juice from resoaking treatment of pomace contains 75 to 250 mg.L^{-1}.

The pectin backbone is formed by alternating homogalacturonan (so-called "smooth regions") and rhamnogalacturonan regions substituted with side chains rich in neutral sugars (so-called "hairy regions") (Figure 6.4). A critical review of the structure of the pectic substances was done by [YAP 11].

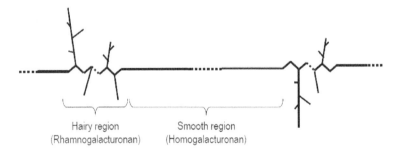

Figure 6.4. *Structure of pectin*

6.2.1.1. *Homogalacturonan regions (HG)*

The monomer unit in this region is α-D-galacturonic acid. Homogalacturonan is a linear chain of (1→4)-linked α–D-galacturonic acid residues with a degree of polymerization (DP) of 70 to 100 (Figure 6.5). The carboxyl groups in position 6 of galacturonic acid residues can be neutralized by cations (Ca^{2+}, K^+, Na^+) or esterified with methanol. Pectic substances are categorized according to their degree of methylation (DM: mole percentage of carboxyl groups esterified with methanol):

– *pectic acid*: DM < 5;

– *low methoxyl (LM) pectins*: 5 < DM < 50;

– *high methoxyl (HM) pectins*: DM > 50.

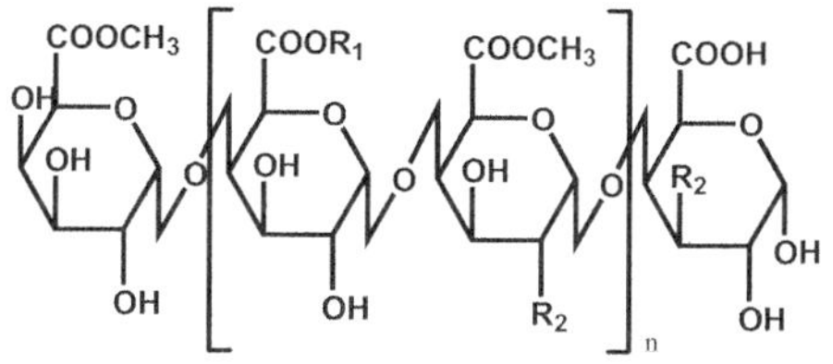

Figure 6.5. *Homogalacturonan sequence of pectic substances; R_1 = -CH_3 or-H, H+, Na+, K+ , etc.; R_2 = -$OCOCH_3$ or –OH*

The secondary alcohol groups at positions 2 and 3 in galacturonic acid can be esterified with acetic acid. The degree of acetyl esterification is generally lower than the DM, but can reach 10 to 15 in the case of pears.

6.2.1.2. Rhamnogalacturonan regions (RG)

Regions rich in neutral sugar side chains with (1→2)-linked α-rhamnose residues arranged in repeating units [→4)-α-D-Gal*p*A-(1→2)-α-L-Rha*p*-(1→] are interspersed between the smooth homogalacturonan regions. These rhamnogalacturonan sequences (Figure 6.6) form the backbone of rhamnogalacturonan I (RG-I), which has additional side chains of L-arabinose and D-galactose linked to rhamnose residues.

Rhamnogalacturonan II (RG-II) consists of a main chain of galacturonic acid with four types of side chains (Figure 6.7). These contain residues typical in pectin compounds such as L-rhamnose, L-fucose, L-arabinose, D-galactose and glucuronic acid. The main characteristic of RG-II is that it also contains less common sugars such as 2-*O*-methyl-fucose, 2-*O*-methyl-xylose, apiose, aceric acid, 2-keto-3-deoxy-D-manno-octulosonic acid (Kdo) and 3-deoxy-D-lyxo-2-heptulosaric acid (Dha). In the cell wall, RG II exists predominantly as a dimer covalently cross-linked by borate, which forms diester bounds with apiose residues of side chains A.RG-II is abundant in juice obtained from the enzymatic liquefaction of the pulp [PER 03].

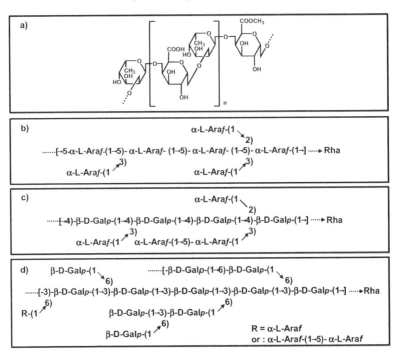

Figure 6.6. *Rhamnogalacturonan I sequences of pectic substances: a) RG I backbone; b) Arabinan; c) Arabinogalactan I; d) Arabinogalactan II*

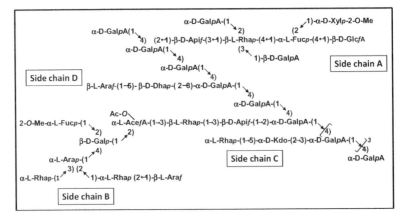

Figure 6.7. *Structure of rhamnogalacturonan II (adapted from [ORT 09])*

6.2.1.3. *Properties of pectins*

Pectins are colloids that form pseudo-solutions of low viscosity. In acidic media, their solubility increases with DM. Pectins can form two types of gels depending on their degree of esterification:

– HM pectins form gels in acidic media (2.8< pH<3.4) at 63 to 80% soluble solids. Gelation conditions vary depending on the DM. The most esterified pectins require less sucrose and gel at higher pH levels than less methoxylated pectins. The pK_a of pectins is close to 3.3; at acidic pH, a drop in the ionization of free carboxyl groups reduces electrostatic repulsion between chains. Sucrose decreases water activity and therefore the hydration of pectin. Under these conditions, low energy bonds can develop between galacturonic residues of the different chains, which can be achieved during the concentration of undepectinized juice;

– LM pectins form gels in the presence of calcium ions according to the "egg box" model [GRA 73, BRA 01]. Hydrated calcium ions form two electrovalent bonds with two carboxylgroups of two galacturonic residues in two pectin chains as well as hydrogen bonds with different oxygen atoms in these units. The junction zone is stable when the structure contains about 15 successive uronic units.

6.2.2. *Pectinolytic enzymes*

Pectins are the substrates of many enzymes, which can either change the degree of esterification (saponifying enzymes) or decrease the degree of polymerization (depolymerizing enzymes).

6.2.2.1. *Pectin methylesterases*

Pectin methylesterases (EC 3.1.1.11) are pectinolytic enzymes that catalyze the demethylation of pectin. They therefore release methanol and produce low methoxy pectins (Figure 6.8). They are classified into carbohydrate esterase family 8 in the CAZy database. Pectin methylesterase (PME) activity occurs in many higher plants, especially in fruit where several PME isoforms can exist (tomatoes, oranges). Their molecular weight ranges from 23.7 (tomatoes) to 57 kDa (kiwis) and their optimum pH is close to or greater than 7. They are activated by monovalent or divalent cations. They display processive behavior along the pectin chain, i.e. multiple attacks on a single chain [DEN 00]. PMEs bind to the cell wall by low-energy bonds and therefore a portion can pass into the juice during extraction. Some bind to debris particles and the rest dissolve.

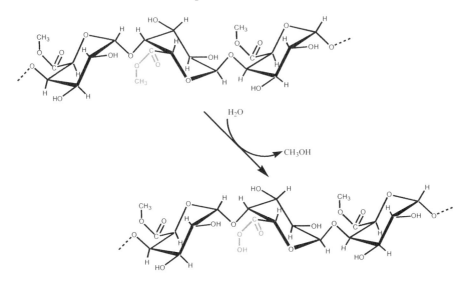

Figure 6.8. *Reaction catalyzed by pectin methylesterases (PME). For a color version of this figure, see www.iste.co.uk/jeantet/foodscience.zip*

PMEs are also produced by microorganisms, in particular fungi. They are exogenous enzymes and thus active in acidic media with an optimum pH similar to that of fruit juices. However, in France, only PMEs extracted from culture media of *Aspergillus niger* and *A. wentii* are authorized in fruit juice. These enzymes demethylesterify pectins in a random manner [DEN 00].

6.2.2.2. Polygalacturonase

Polygalacturonases (PG) hydrolyse the α-(1→4) bond between two non-esterified galacturonic acid units. They are classified into carbohydrate hydrolaze family 28 in the CAZy database. A distinction is made between exopolygalacturonases (exo-PG) and endopolygalacturonases (endo-PG). Most exo-PG act at the non-reducing end of the pectin chain by releasing galacturonic acid (EC 3.2.1.67) or digalacturonic acid (EC 3.2.1.82). Endo-PGs (EC 3.2.1.15) randomly attack homogalacturonic acid (Figure 6.9) and the final degradation products are monomers, dimers and sometimes trimers of galacturonic acid.

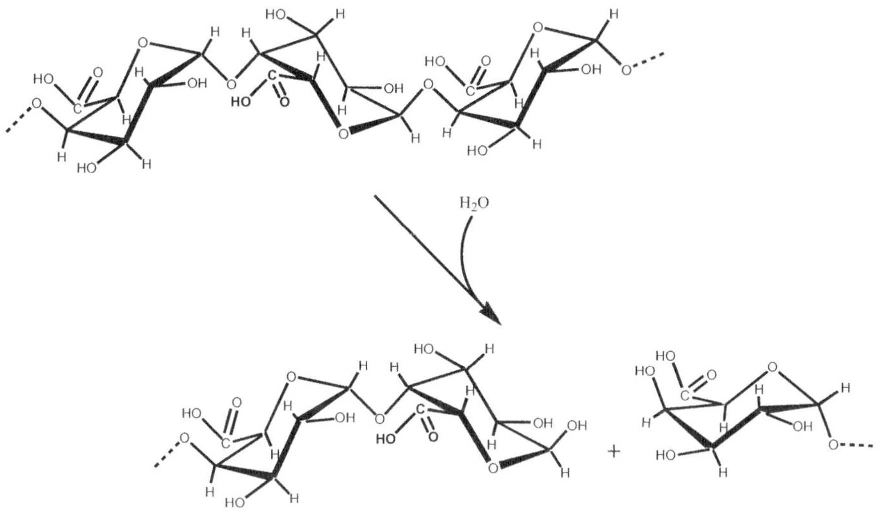

Figure 6.9. *Reaction catalyzed by endopolygalacturonases (endo-PG)*

Polygalacturonases are found in plants, fungi, bacteria and animals (insects). They all have an optimum pH of between 3.5 and 5.5. Their preferred substrate is pectic acid. Endo-PG activity, which is non-existent in pectin with a DM of 75, increases as DM decreases. Thus, PME and PG act synergistically on natural pectins: PME creates demethylated sequences allowing hydrolysis by PG: in return PG degrades homogalacturonan sequences. Only PG extracted from culture media of *A. niger* or *A. wendii* are permitted in the production of fruit juices.

6.2.2.3. Lyases

Lyases act on pectic substances by catalyzing the rupture of two galacturonic acid units by β-elimination. They are classified into polysaccharide lyase family 1. A distinction is made between pectin lyase (PL, EC 4.2.2.10), endo-pectate lyase (endo-PAL, EC 4.2.2.2) and exo-pectate lyase (exo-PAL, EC 4.2.2.9). PALs, produced by microorganisms, are inactive at the acidic pH of fruit juice; they have no processing application.

PL cleaves the bonds between methylated galacturonic acid residues (Figure 6.10). All identified pectin lyases are endo-enzymes. They are produced by fungi (*Aspergillus* species) and bacteria (*Erwinia* species). They have never been detected in higher plants. Their activity decreases with the degree of pectin esterification. In addition, the optimum pH drops from 6.5 to 5 when the DM of the initial substrate decreases from 95 to 65%.

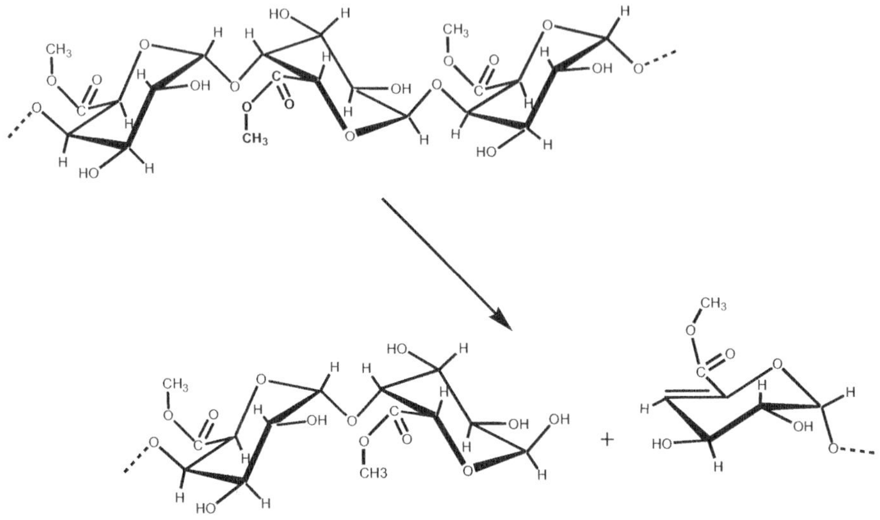

Figure 6.10. *Reaction catalyzed by pectin lyase (PL)*

6.2.2.4. Arabinanases

A number of enzymes have been identified and purified from culture media of microorganisms, which hydrolyze different glycosidic bonds of rhamnogalacturonans and xylogalacturonans. Only arabinanases with activity in pectinolytic preparations from *Aspergillus sp.* are used in industrial apple juice production to prevent arabinan hazes. These enzymes are mainly classified in family GH43 and GH51.

Branched arabinan chains are linearized after the hydrolysis of (1→2)- and (1→3)-α–L-Arafbonds by endogenous or exogenous exo-α-L-arabinofuranosidase (activity II; EC 3.2.1.55) (Figure 6.11). The linear chain of (1→5)-α-L-Araf gradually insolubilizes to form a characteristic haze, in particular in concentrates. Endo-arabinanase (EC 3.2.1.99) prevents the formation of such haze by reducing the degree of arabinan polymerization. The end products are arabinose dimers and trimers. (1→5)-α–L-Ara*f* oligomers are then hydrolyzed by the non-reducing endexo-α-L-arabinofuranosidase (activity I; EC 3.2.1.55), with arabinose being the only reaction product. The optimum pH values of enzymes extracted from *A. niger* are 4.1, 3.7 and 5.0 respectively for arabinofuranosidases activities I and II and endo-arabinanase. They are also active on side chains of arabinogalactan.

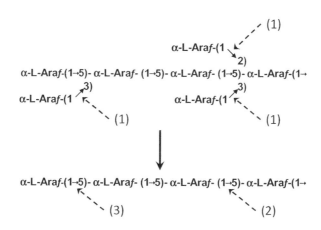

Figure 6.11. *Reactions catalyzed by arabinanases: (1) α-L-arabinofuranosidase II (exo-enzyme); (2) Endo-arabinanase; (3) α–L-arabinofuranosidase I (non-reducing end exo-enzyme)*

6.2.3. *Bitter and astringent compounds*

Depending on the fruit juice, different compounds can add bitterness and astringency. The main compounds are terpene derivatives and polyphenols.

6.2.3.1. *Limonoids*

In citrus juice, the main limonoid, monocarboxylic limonoic acid, is an oxidation derivative of tetracyclic triterpene (apotirucallol). By itself, it is not bitter, but in an acid medium, it cyclizes to form a second lactone group (Figure 6.12).

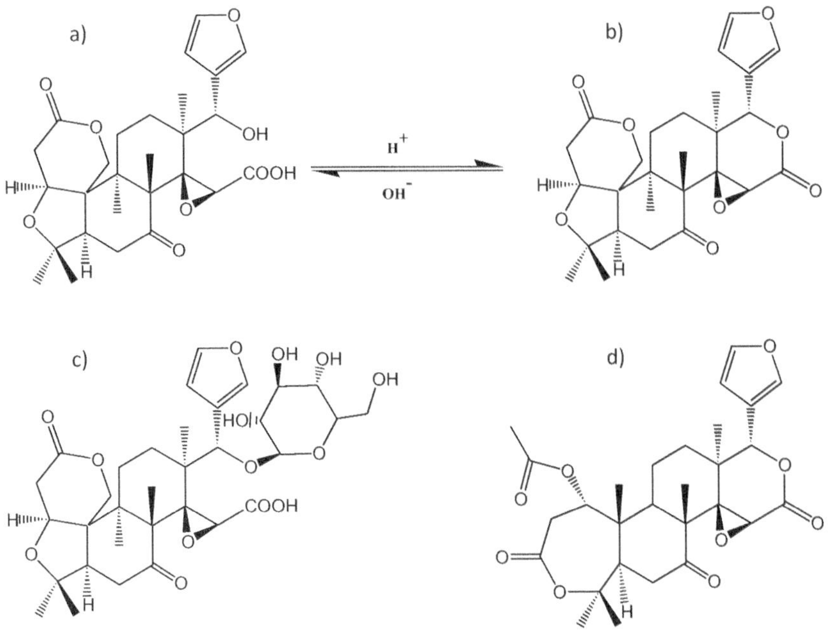

Figure 6.12. *Structure of some limonoids in citrus fruits:*
a) limonoic acid; b) limonin; c) 17-β–D-Glup-limonoic acid; d) nomilin

Limonin is a very bitter compound. Heating accelerates its formation. Its detection limit ranges from 0.075 to 5 mg.L^{-1} in water and from 0.5 to 32 mg.L^{-1} in orange juice, depending on the taster. Nomilin is another limonoid with almost the same detection threshold. It can account for up to 40% of bitter limonoids in grapefruit juice (*Citrus grandis* L.). The average content of these compounds in juice varies significantly depending on the extraction method, species and maturity. Fruit harvested early in the season has higher levels. The average content of grapefruit juice is around 8 mg.L^{-1}. Values of 4.2 to 14.2 mg.L^{-1} have been reported for lemon juice (*C. limon* L.). In sweet orange juice (*C. sinensis* L.) the limonin content decreases rapidly as the fruit ripens. The presence of this compound poses real sensory problems in Navel orange juice due to its content. Limonin is poorly soluble in aqueous solutions (16 mg.L^{-1} of boiling water). However, glucosyl-limonoate found in citrus fruit juice is water-soluble and tasteless. [FON 89] recorded an average content of 120 mg.L^{-1} in grapefruit juice. The enzymatic or chemical hydrolysis of glycosylated limonoids has never been reported.

6.2.3.2. Flavanones

Of the flavanone glycosides (Figure 6.13), only those with a neohesperidosyle residue [α–L-Rha-(1→2)-β-D-Glc-(1→] are bitter.

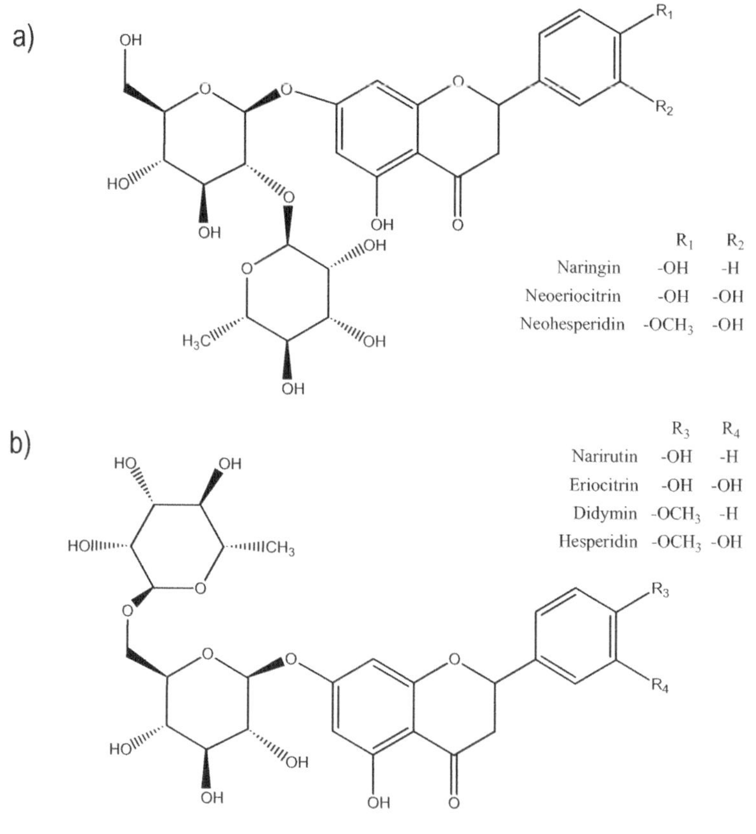

	R_1	R_2
Naringin	-OH	-H
Neoeriocitrin	-OH	-OH
Neohesperidin	-OCH$_3$	-OH

	R_3	R_4
Narirutin	-OH	-H
Eriocitrin	-OH	-OH
Didymin	-OCH$_3$	-H
Hesperidin	-OCH$_3$	-OH

Figure 6.13. *Structure of some flavanones in citrus fruit; a) bitter compounds; b) easily crystallizable compounds*

Naringin, the main flavonoid glycoside compound of citrus fruits occurs principally in the peel. It is only present in the juices of grapefruit, pomelos (*C. paradisi* Macf.) and bitter oranges (*C. aurentium* L. subsp *amara* L.). Depending on the variety, the content in juice varies from 0.05 to 2.3 mg.g^{-1}; neohesperidin and neoeriocitrin are sometimes present in these citrus fruit juicesup to a level of 0.4 mg.g^{-1}. Other citrus fruit juices such as lemons, oranges and mandarins (C. reticulata Blanco) do not contain any bitter compounds. The detection threshold of naringin (less than 10-4 M) is 5 to 10 times greater than that of limonin.

In all of these citrus fruit juices, there are flavanone glycosides, whose glycoside group is rutenose [α–L-Rha-(1→6)-β–D-Glc-], an isomer of neohesperidose. They have no bitter flavor. Hesperidin, the most abundant compound, can reach a concentration of 3 mg.g-1 in some lemons (e.g. Santa Teresa variety). In orange juice, its content does not exceed 0.6 mg.g^{-1}. The other derivatives of rutenose (didymin, eriocitrin, narirutin) are present in significant amounts in lemon juice (0 to 0.1 mg.g^{-1}). In grapefruits and pomelos, these compounds are generally present in trace amounts. Less soluble than bitter compounds, they can precipitate as needle-like crystals during storage.

6.2.3.3. Proanthocyanidins

Proanthocyanidins refer to a class of flavan-3-ols, which can be oligomeric or polymeric (see Chapter 6 of Volume I). The term "condensed tannins" is sometimes used to refer to highly polymerized forms. Proanthocyanidins can have very different structures depending on the plant. These can vary based on their monomeric units (catechins, gallocatechins esterified or not with gallic acid), the type and number of interflavan bonds between monomers (one or two; C_4-C_8 or C_4-C_6 in series B; C_4-C_8 or C_4-C_6 and C2-O-C5 or C2-O-C7 in series A) and the stereochemistry of each unit with the interflavan bond (type B) generating a new asymmetric carbon in C4 in addition to those in C2 and C3. Procyanidins consist of (+)-catechin and (-)-epicatechin units; prodelphinidins are composed of gallocatechin and/or epigallocatechin units.

In apples, only type B procyanidins consisting of epicatechin units have been observed, with catechin located at the terminal position. Their degree of polymerization ranges from 2 to over 250. They are the most abundant phenolic compounds in this fruit in terms of weight. Prodelphinidins, absent in apples, are present in large quantities in grapes. Proanthocyanidins have never been identified in citrus fruit or juice.

During must extraction, these compounds that accumulate in the vacuoles are partially adsorbed on the cell wall compounds (in particular pectins), especially the most polymerized [REN 11]. They add bitterness and astringency to juice. As regards apple procyanidins in water, DP 4 has more bitterness while DP 7 and 8 are responsible for astringency [SYM 14]. However, the matrix (acid, sugar, alcohol, other phenolic compounds, polysaccharides) in which the polymer is dissolved may modify the sensory perception.

In addition, proanthocyanidins are not polyphenoloxidase substrates (see Chapter 6 of Volume 1) but can be involved in redox reactions combined with o-quinones of other phenolic compounds or in addition reactions. The sensory properties of these newly formed compounds are not known or whether or not they affect the bitterness or astringency of juice.

6.3. Fruit juice processing

6.3.1. *Preparation of fruit*

After delivery to the factory, fruit is generally stored for a number of days under optimal conditions to inhibit spoilage. The fruit is then washed and any damaged items are manually or mechanically removed. Apples are stored for up to one week on concrete floors in piles of no more than one meter in height. They are then floated along water-filled supply channels to the preparation hall; this constitutes the first sorting since only healthy apples float. They are then sent in a tilted washer, which includes a cylindrical rotatable conveyor equipped with nozzles for spraying water at high speed. The power of the spray jets removes any damaged parts, leaves and vegetation from the orchard. Citrus fruit is stored for a maximum of 5 to 6 days, usually less than 12 h, in inclined silos to limit crushing due to height. They are transported on conveyor belts to sorting tables where sorting is carried out manually. The fruit is then washed and in some cases disinfected using chlorine agents. Red berries (strawberries, raspberries, blackberries, blueberries, blackcurrants, grapes, cherries, etc.) are much more fragile. They must be treated as soon as possible after harvesting to avoid fermentation and spoilage. They can sometimes be refrigerated for a few days. They must be transported and delivered in small containers (6 to 20 kg) due to their sensitivity to impact. Impurities such as leaves or small branches are removed using blowers. Generally berries and stone fruit are not washed, or if so, this is done using low pressure boom sprayers.

6.3.2. *Pre-treatment*

The pre-treatment of fruit varies with each species. Apples and pears are shredded using graters; a rotary blade propels them onto knives or interchangeable perforated grids. The fineness of the grating depends on the knife or grid chosen, which in turn is selected based on the ripeness of the fruit; the grated fruit is pumped to the press or falls directly into the feed hopper. Citrus fruits are graded into two or three classes and dispatched to extractors adapted to each class of fruit. Cherries are stemmed in machines with counter-rotating rubber rollers, which pluck the stems off. Stone fruits are pitted either by pitting machines or by pulping machines after an initial heat treatment with steam.

6.3.3. *Pressing*

Two main types of presses are used to treat fruit pulp: batch presses (pack press or cylinder-piston hydraulic press) and continuous presses (belt press). Today, the crew press is no longer in use for the extraction of fruit juice. Specific machines have been developed for citrus fruit. Pulping machines, centrifugal extractors and refining machines are used to treat certain products: exotic fruits, tomatoes, apricots and so forth.

6.3.3.1. *Batch presses*

"Rack and cloth" presses are still used on a small scale for the production of apple juice, but have disappeared from factories in the western world due to their low level of productivity. Layers of pulp are enclosed in woven synthetic cloths and alternated with wooden or stainless steel racks to form a press cake or 'cheese', which is mechanically pressed.

Cylinder-piston and bladder presses are also discontinuous systems hydraulically or pneumatically driven. In basket press the end of the cylinder and the piston are connected with flexible drainage elements. The drains consist of a flexible fluted core fitted with a woven polypropylene filter sleeve. Batches of grated fruit are placed in the cylinder where pressure is applied using a piston or membrane. In case of horizontal press, the complete cylinder and piston assembly is slowly rotated. The automation of pressurization, compaction and homogenization processes ensures even extraction. Almost any type of fruit can be used in batch presses. The pressure can be adapted to the physical characteristics of the fruit to achieve maximum yield. The system, which is fully enclosed from the moment the fruit is added until the juice is drained, ensures good product hygiene. Some models include inert gas to minimize oxidation of the juice.

6.3.3.2. Conveyor belt press

The design of conveyor belt presses is based on the principle of pack presses or filter presses. The pulp, deposited onto a screen, is pressed into a thin layer to allow the juice to flow out. The first belt presses were only semi-continuous. Many double-belt presses can now operate continuously. The pulp bed is passed between two mesh belts that rotate in opposite directions and gradually move together to exert increasing pressure on the pulp. The juice flows through the lower belt into a collection tray. The residue falls into a discharge hopper at the end of the belt. Beating devices and spray jets are used to clean the belts.

6.3.3.3. Pressing aids

Different techniques are used to increase the extraction yield. In the treatment of apples, counter-current diffusion of pomace by water or the direct addition of water (10 to 20% in volume) to the pomace after the first pressing followed by a second pressing all help to remove as much sugar as possible and to obtain a further yield of juice with lower soluble solid. This weaker juice is intended to be concentrated.

Pre-treating the pulp with specific enzyme preparations rich in PG (see section 6.2.2) or the addition of PME and/or calcium salts (usually as carbonate) promotes the extraction of juice by modifying the texture of the pulp and the viscosity of the juice.

In the last 20 years innovative processes for extracting juice from plant matrix have emerged. Among them, the pre-treatment of the whole fruit or fruit mash by pulsed electric fields (PEF) appears the most promising [VOR 08].

The efficiency of PEF is related to the phenomenon of electroporation that can be described as a dramatic increase in membrane permeability caused by externally applied short and intense electric pulses (Figure 1.14).

The external electric field E induces a transmembrane potential on the membrane. At a critical electric field (E_c), the transmembrane potential exceeds some threshold value (typically about 0.2-1.0 V), semipermeability of cell membranes could temporarily (reversible electroporation) or definitely occurs (irreversible electroporation). In the second case, the turgor pressure of the cell is lost causing plasmolysis and a drain of the cell into the extracellular space causing a change in tissue.

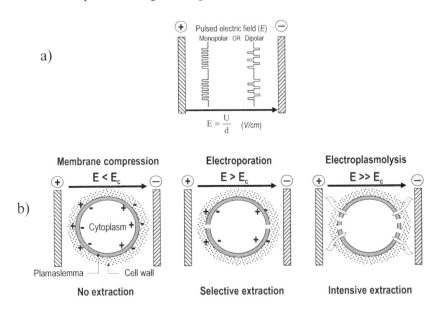

Figure 6.14. Principle of pulsed electric fields applied to the extraction of juices; a) modulation of the electric field (E) between the spaced electrodes (d) by a pulsed electric voltage (U); b) effect of the electric field on the migration of cellular ionic charges depending on its value relative to the critical value (Ec) specific to a given tissue and consequences on plasmalemma permeability

Pulsed electric fields were extensively studied in laboratory within fruit juice production as a pretreatment technique of different raw material such as carrots, grapes and apples. Depending on the plant material and applied electric fields (from 0.5 to 10 kV.cm^{-1}) processing time ranges from 100 ms to 10 µs. In addition to the higher juice yield obtained after PEF-treatment, evaluation of selected quality criteria for both juices showed that application of high-intensity electric field pulses on mash resulted in higher antioxidant content and pigment release in treated juices than from samples processed in a conventional way. Currently, only a few tests were conducted at pilot or industrial scale for apple and grape juice processing. Flow rates did not exceed 4.5 t.h^{-1} in these tests and the efficiency of the treatment is closely dependant on the treated raw material.

6.3.3.4. *Citrus extractors*

The presence of essential oils in the flavedo prohibits the complete pressing of the fruit. The maximum amount of oil in the juice should not exceed 0.015–0.02% (v/v).

The oldest machines are rotary extractors. The fruit is placed in semispherical cavities or cups in the surface of two counter-rotating drums (Colin extractor). The fruit is held while a blade slices it in two. Each half, remaining in its cup, moves in front of plastic extraction reamer. The movement is such (translation and rotation) that the carpels are pressed without bursting the seeds. Otherwise, they would release bitter compounds detrimental to the quality of the juice. Brown extractors (mounted either horizontally or vertically) work on the same principle but the cups are supported along two opposite chains which, after cutting the fruit in half, moves to the extraction reamer. The speed and pressure of the extraction reamers are controlled to adapt to the thickness of the peel.

To avoid the extraction of essential oils from the skin when the pressure increases, the fruit is first passed through an oil extractor: it is carried along on rollers equipped with small needles that puncture the secretory cells of the epidermis. The oils are collected in water flowing in the opposite direction.

The in-line juice extractor developed by FMC is based on a different principle to that of the Colin or Brown extractor: the contents of the citrus fruit are extracted through a system of tubes so that the juice never comes into contact with the oils. For maximum yield, the size of the head should be adapted to the class or size of the fruit.

6.3.4. *Treatment of fruit juice*

Raw juice, or must, is treated to ensure colloidal stability or clarification. Processing methods vary depending on the final product, but they all aim to control biochemical reactions associated with the macromolecules present (polysaccharides, proteins, polyphenols). The introduction of enzymes to this sector has radically changed production techniques. A good knowledge of each processing stage is essential. Controlling the production of these enzymes by microorganisms and their purification offers new opportunities for product development and may help to overcome technological challenges using new mix of enzymatic activities.

6.3.4.1. Stabilization of cloudy juice

Type and composition of cloudiness

Cloudiness in apple or pear juice is due both to the opalescence of pectic colloids in pseudo-solution (diameter <0.1 µm) and to particles in suspension, stabilized by insoluble pectins that act as protective colloids. The size of these particles ranges between 0.5 and 10 µm. The largest elements, consisting of cell debris or cell clusters are considered to form the pulp but do not contribute to the cloudiness in the juice. The stability of particles has been studied extensively during the 1960s [YAM 64]. The outer layer of the particles, consisting of pectins, is negatively charged; the nucleus, composed of proteins, is positively charged at the pH of the juice. Such hydrated particles give stable suspensions due to their small size and electrostatic repulsions. Their sedimentation is also reduced by the viscosity of the medium due to the soluble pectins. Blanching the mash before extraction or immediately heat treating the juice results in smaller, more stable particles.

In citrus fruit juice, the size of suspended particles varies from cells ("sacks of juice") and cell fragments of more than 100 µm to macromolecular complexes less than 0.1 µm. The largest particles hardly affect turbidity, but contribute to the pulpy appearance of the juice. Only particles measuring between 0.4 and 5 µm constitute cloudiness. They consist of around 34% protein, 32% pectin, 25% lipids and small amounts of cellulose and hemicellulose. Hesperidin crystals also contribute to cloudiness; pectin acts as the nucleation site for hesperidin crystal formation. Such interaction is specific to pectins de-esterified by enzymes up to a DM of around 10%.

Destabilization mechanism

The most frequent defect in naturally cloudy fruit juice is spontaneous clarification during storage. Under the effect of endogenous PME, soluble pectins form a calcium pectinate gel that traps suspended particles. The gel then contracts by syneresis resulting in a clear serum.

In a model system, buffered at pH 3.8, gelation is initiated by the PME of orange peel when the DM of pectin drops from 75 to 64%. As saponification continues, calcium pectinate precipitates, thereby entrapping particles. Clarification is achieved when pectins reach a DM of about 40% for apple juice and 27–38% for orange juice, depending on the initial DM, the pectin concentration, PME activity and the calcium content. Hesperidin – pectin complexes with a DM of 25% – flocculate in the presence of 100 ppm calcium ions. The calcium content of juice (30 to 120 ppm) is sufficient to allow the precipitation of partially demethylated pectins.

Prevention of destabilization

Since the addition of calcium chelating agents to juice is not permitted, the only possibilities to prevent the destabilization of cloudy juice are the degradation of soluble pectin or the inhibition of PME activity. The use of polygalacturonase prevents gel formation by degrading pectin in the demethylated sequences. However, there is a partial flocculation of particles and rapid sedimentation.

Heat treatment

Many studies have been published on the heat treatment of citrus fruit juice. The optimal treatment is a compromise between denaturing PME as much as possible while preserving the sensory qualities of the juice. Treatment can last from 1 min at 70°C for lemon juice at pH 2.3 to 5 min at 92°C for orange juice at pH 3.8. However, low residual PME activity, due to the presence of a heat-resistant isoform, usually remains detectable. To limit the impact of this activity on the stability of cloudiness, juice concentrates are frozen and stored at low temperatures (-18°C). To reduce the effect of concentration and storage on the loss of quality, some fresh juice is added either at the time of packaging (add-back technique) or production (cut-back technique). Another technique also exists whereby the suspended components in the juice (the pulp) are separated from the clear phase by centrifugation or cross-flow microfiltration (Fresh note process). The clear juice is pre-concentrated by reverse osmosis and then concentrated to 65° Brix. After washing with water, the pulp is refrigerated (4°C) or frozen (-18°C) until reincorporation into the juice during packaging (Figure 6.15)

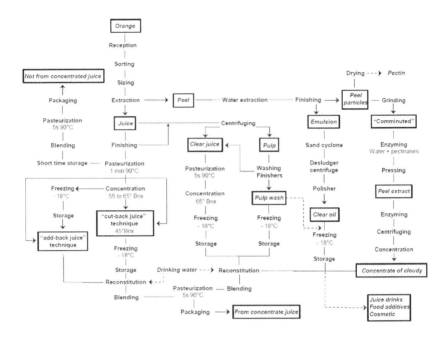

Figure 6.15. Production of citrus fruit juice using different techniques (cut-back, add-back and Fresh note)

Stable cloudy apple juice can be achieved by adding cloudy particles (obtained by crushing and refining heat-treated apple pulp) to clear juice obtained by diluting the concentrate (Figure 6.16).

High isostatic pressure treatment

Research has been conducted since the 1980s to determine the stabilization conditions for fruit juice by high pressure treatment at room temperature. The advantage of this method is that the sensory qualities of the juice are maintained.

PME appears to be relatively pressure-resistant. An 80–90% drop in PME activity occurs in orange juice subjected to 600 MPa for 0.5 to 10 min at initial temperatures below 45°C (sensory changes become more noticeable at higher temperatures). PME isoforms can vary in their pressure sensitivity. Heat-resistant PME is also pressure-resistant. Even though it has been possible to stabilize cloudiness in orange juice on a laboratory scale, juice treated in the pilot stage or industrial-scale production underwent phase separation after a few days since the equipment available does not exceed a pressure of 500 MPa.

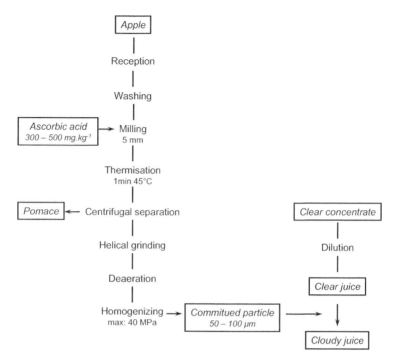

Figure 6.16. *Production of cloudy apple juice using clear concentrate*

The colloidal stabilization of apple juice is achieved under moderate treatment conditions (e.g. 400–500 MPa, 8-12 min at room temperature) although purified apple PME is more pressure-resistant than purified orange PME. In apple juice, the activation of polyphenol oxidase (PPO) between 200 and 300 MPa promotes the production of phenolic inhibitors of PME, thereby improving juice stability.

6.3.4.2. Juice clarification

Many fruit juices (red berries, apple, etc.) are sold in clear form. In addition, their concentration above 60°Bx requires pectin degradation to prevent gelation. Thus, raw juice, fresh from the press, must be depectinized and clarified (Figure 6.17). It cannot be directly filtered on plate filters because soluble pectins quickly clog the filter.

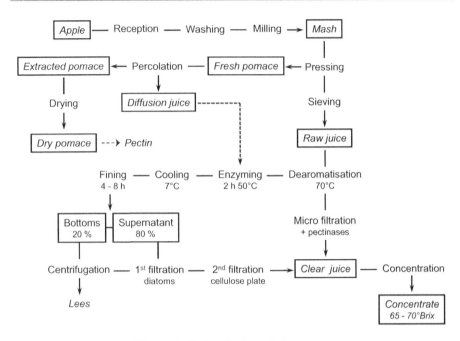

Figure 6.17. *Production of clear apple juice and concentrates*

Clarification by fining and enzyme treatment

Exogenous enzymes of fungal origin are used in the fruit juice industry for clarification. Suspended particles are partially degraded by the enzyme mixture of PME and PG. Pectin, which acts as a protective colloid for particles is solubilized. Its gradual hydrolysis causes a decrease in the viscosity of the juice and the accumulation of final products: monomers, dimers and trimers of galacturonic acid [END 65]. The partial solubilization of pectin partially exposes positively-charged protein compounds at the pH of the juice [YAM 67]. The particles can then flocculate by electrostatic interaction and settle more easily the lower the viscosity of the medium. In juice that is low in protein, flocculation is promoted by the addition of exogenous proteins, usually gelatin. It also reacts with flavan-3-ols and proanthocyanidins; the amount of gelatin used for fining is adapted to each type of juice. Conversely, excess protein in the juice, which can cause cloudiness during pasteurization (thermo-coagulation) and storage (insolubilization, reaction with polyphenols, etc.), can be corrected by bentonite fining (hydrated aluminum silicate).

Enzyme preparations rich in pectin lyase can also be used to treat juice. They restrict the amount of methanol released from pectin. However, their effectiveness is limited in the treatment of juice rich in PME or naturally containing low-esterified pectins (DM between 60 and 70%), such as grape juice.

Originally, clarification was a slow process lasting almost 24 hours. Nowadays, the selection of thermostable pectinase and the use of decanter centrifuges means juice can be enzyme-treated at 50°C for one to two hours in the presence of gelatin; the cooled juice can be decanted three or four hours later and filtered on cellulose plate filters resulting in perfectly clear juice.

Clarification by cross-flow microfiltration

Membrane techniques, introduced for the clarification of apple juice at the beginning of the 1980s, were soon applied to the treatment of fruit and vegetable juices. Ultrafiltration, originally used, was gradually replaced by cross-flow microfiltration (see section 9.1.2.2) due to better flow and less clogging.

In practice, industrial facilities use organic and inorganic membranes with a pore size of 0.1 to 0.2 µm resulting in a permeate flux of 100 to 600 $L.h^{-1}.m^{-2}$ for a transmembrane pressure of 0.1 to 0.2 MPa and an average retentate flow rate of 1 to 7 $m.s^{-1}$. However, the accumulation of pectin in the retentate increases the thickness of the polarization layer and consequently the transfer resistance. As a result, there is a gradual decrease in the flow rate of the permeate. Since the power of the circulation pump is limited, it is not possible to ensure constant hydrodynamic conditions of the system over time. To overcome this limiting factor, pectinolytic enzymes are added to the must. Only half the amounts of enzymes are required compared to the batch technique. The microfiltration module behaves similarly to that of an enzymatic reactor with a high enzyme and substrate concentration close to the membrane, resulting in maximum enzyme activity. Under steady state conditions, only galacturonic oligomers with a DP of less than 10 or 12 permeate.

6.3.4.3. *Debittering treatment*

Plant treatment

The genetic improvement and selection of citrus fruit varieties low in bitter compounds is a first step in tackling the problem of bitter juice. This takes time and requires a major restructuring of orchards.

Treatment trials have been conducted on trees and fruit at different stages of development. The use of gibberellic acid on unripe grapefruit significantly reduces the naringin content of ripe fruit. In the case of lemons, the use of naphtylacetic acid on seedlings or fruit after harvesting reduces the accumulation of limonoids by more than 80%. These methods are however expensive and rarely used.

Modulation of bitterness during extraction

During extraction, some methods can reduce the transfer of bitter compounds to the juice. Placing apple pulp in a vat for about 30 minutes or even one to two hours promotes the oxidation of polyphenols. This operation, combined with slow pressing, promotes the retention of procyanidins on the pomace.

As already mentioned, the use of oil extractors (see section 6.3.3.4) before the extraction of citrus fruit juice limits the transfer of bitter compounds from the peel to the juice.

Enzyme treatment

The naringin content (Figure 6.13) can be lowered using naringinase, a mixture of two exoenzymes, α-rhamnosidase and β-glucosidase. The first, specific to the rhamnosyl-($1\rightarrow2$)-β-glucose bond, releases rhamnose and prunin, a less bitter substance; the second hydrolyses the glucose-naringin bond resulting in aglycone (naringenin) that has no bitter flavor. At pH 3.5–4 and between 20 and 50°C, this enzyme system rapidly reduces the bitterness of grapefruit juice. However, this treatment is not permitted everywhere for "100% fruit juice", for example in the United States.

An exorhamnosidase called *hesperidinase* also exists, which is specific to the rhamnosyl-($1\rightarrow6$)-β–glucose bond. It releases glucosyl hesperidin which, in an acid medium, slowly hydrolyses to glucose and hesperetin, thus preventing the crystallization of hesperidin.

These enzymes do not affect the limonoate content of juice. To reduce this, pilot scale studies were conducted whereby the juice was passed through an immobilized cell column of *Arthrobactus globiformis*, producing limonoate dehydrogenase. This enzyme prohibits the lactonization that results in limonin (Figure 6.18) by oxidizing the alcohol group on carbon 17. However, this technique has not been further developed due to control problems.

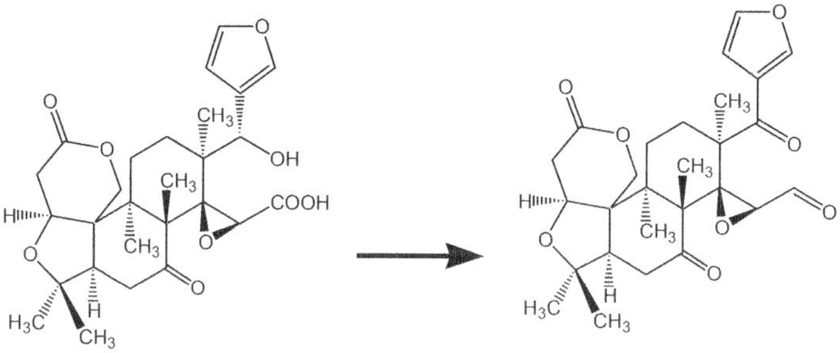

Figure 6.18. *Reaction catalysed by limonoate dehydrogenase*

Resin treatment

Debittering units based on adsorption resins have been developed by several manufacturers since the late 1980s. The juice, de-oiled if necessary, is clarified by centrifugation or cross-flow ultrafiltration. The pulp, which represents about 15% of the volume, is stored, refrigerated or frozen. The clarified juice passes through food-grade divinylbenzene columns. Almost all the limonoids are adsorbed and 40-90% of flavanones are retained depending on the treatment conditions. The pulp is reincorporated into the debittered juice adding some bitterness for a well-balanced flavor. This can be done immediately with fresh clear juice or after its storage as clear frozen juice, or as concentrated and frozen clear juice.

Resins need to be regenerated every 5 to 20 h with dilute alkaline solutions. Some systems are designed to simultaneously or separately deacidify juice, which is often too acidic early in the season, using ion exchange resin.

The use of these resins, known as processing aids, does not affect the other components in the juice and does not have to be declared on the product packaging.

6.3.5. *Pasteurization, high-pressure treatment, pulsed electric fields and concentration*

6.3.5.1. *Pasteurization*

The microbiological stabilization of fruit juice is usually achieved by pasteurization (see section 10.6, Volume 1)

Post-packaging pasteurization

Cold juice is filled into glass bottles or tin cans. Once sealed, it is then heated to 75–85°C in a bath or a hot water or steam shower (tunnel pasteurizer) and then cooled by spraying with cold water. Water used to cool outgoing bottles is used to heat incoming bottles, thereby saving energy.

Flash-pasteurization

Flash pasteurization involves rapidly heating the juice to 95–97°C, maintaining it at this temperature for 10 to 50 seconds and then rapidly cooling it. It is carried out in plate or tube heat exchangers (small diameter tubes). Tube heat exchangers are preferred for the treatment of pulpy products as there is less fouling. Heat exchangers that can operate under pressure have also been developed. They can heat the juice to 130°C with a corresponding reduction in treatment time. It is thus possible to pasteurize orange juice at 107°C in 3 seconds.

– *Hot filling and auto-pasteurization*: this method involves subjecting the juice to flash-pasteurization and then cooling it to 82–85°C. After filling at this temperature, the bottles are closed immediately and shaken so that the hot liquid comes into contact with the entire inner surface of the bottle to sterilize it. After 3 to 4 min, they are cooled by immersion or spraying.

– *Cold filling*: after pasteurization, the juice is cooled to 5–10°C in the last section of the heat exchanger. Subsequent packaging must be performed under aseptic conditions. Aseptic cold filling was initially developed for multilayer packaging that cannot withstand hot products. This concept was recently extended to the packaging of liquids in glass and plastic bottles. Strict compliance with hygiene rules is crucial to avoid contamination after pasteurization. Companies using this system are equipped with clean rooms.

6.3.5.2. High pressure treatment

Not all orange juice can be stabilized by isostatic high pressure. Only a small number of microbes are pressure-resistant. The values D_P and z_P (similar coefficients to those defined for heat treatment; see section 10.6, Volume I) are rarely known. However, experience shows that in orange juice treated at 400 or 500 MPa, a residual flora is detectable in all cases. This proliferates during storage at 4°C for products treated at 400 MPa. It decreases only in juices treated for 5 min at 500 MPa. Pathogenic strains of bacteria such as *E. coli* or *Salmonella* sp. are sensitive to pressure close to 400 MPa.

6.3.5.3. Pulsed electric field pasteurization

Pulsed electric field is a non-thermal juice preservation technique that can be used for microorganism inactivation (see section 6.3.3.3). The treatment is very effective for the inactivation of all kinds of vegetative spoilage and pathogenic microorganisms while limited effect is observed on spores and viruses. To cause irreversible electroporation, treatment requires energy input all the more important that the cells are small. Electric fields from 15 to 60 $kV.cm^{-1}$ are applied for very short times (<1 ms) to avoid overheating.

Pulsed electric field is an alternative for heat pasteurization especially for heat sensitive pumpable juices. The electric treatment does not affect vitamins, antioxidants and other nutritional components of juices. The use of PEF for food preservation is not limited to mild pasteurization purposes. Food sterilization is possible with PEF in combination with higher temperatures to eliminate spores. The first commercial application of PEF for fruit juice preservation has been reported in the United States processing 200 $L.h^{-1}$ of apple, strawberry, and other juices and is showing potential for industrial exploitation.

6.3.5.4. Concentration

Concentration reduces the volume and therefore storage and transportation costs. It provides microbiological stabilization by a reduction in a_w.

Fruit juice is concentrated using a three-effect evaporator (apple juice) or five- to seven-effect evaporator (orange juice) (see section 10.2, Volume 1). The *Taste*® (thermally accelerated short time evaporator) system was developed to concentrate citrus fruit juice: a convergent-divergent nozzle optimizes the wetting of the tubes (expansion of the product in liquid/vapour form) and accelerates evaporation (when exiting the diverging section, the expanded product is slightly above boiling point), which promotes the sensory qualities of the product.

Volatile compounds can be collected in the condensate at the bottom of the evaporator if not previously extracted by stripping with an inert gas.

The concentration unit is degrees Brix (°Bx). Based on the measurement of the refractive index, it is the percentage by weight of the soluble solids in solution expressed in terms of sucrose equivalent. Clear concentrated apple juice is close to 70°Bx, and can be stored at room temperature. Concentrated orange juice (between 60 and 65°Bx), grapefruit juice (55°Bx) or lemon juice (45°Bx) are stored under aseptic conditions and frozen.

6.4. Cider

6.4.1. *French cider*

Cider is the traditional beverage in the west of France and is obtained by a series of different methods ranging from the simplest (household production) to the more elaborate (industry). Despite the wide variety of techniques, some inherent characteristics, either related to the raw material or the choice of processor and/or customer, can be considered as general constraints that influence development and give French ciders their particular features.

Specific varieties of apples are used for the production of cider. They differ from eating apples by having higher levels of phenolic compounds that are responsible for the sensory properties of cider; color formation in cider is mainly due to their enzymatic oxidation by polyphenol oxidase (PPO) while bitterness and astringency are directly linked to the presence of procyanidins. The acidity of juice and cider is also determined by the raw material; due to malic acid, it is usually between 40 and 100 meq.L^{-1} at a pH of 3.5 to 4.3. The pH is higher compared to wine and therefore less selective in terms of spoilage flora, since the direct effect of pH is less efficient and the effectiveness of sulfur dioxide (SO_2) as an antimicrobial agent is reduced.

The choice of fermentation is also specific: most French ciders ferment under the action of "spontaneous" flora consisting of several species of yeast giving ciders their particular aroma: the search for fruity aromas explains why systematic inoculations by *Saccharomyces sp.* are very rare compared to other fermented beverages. In addition, cider fermentation is slow compared to wine and beer. The relatively slow rate is desired for sensory purposes: cider producers feel that a sugar consumption above 4–5 g. L^{-1}.day-1 has a negative impact on cider aroma and that it is better not to exceed a rate of 2 $g.L^{-1}$.day-1. It is important to consider the maximum yeast population reached, so it may be more accurate to talk about "low biomass fermentation" than slow fermentation. In any case, cider producers use a number of methods to slow down fermentation.

Cider should contain a considerable amount of residual sugar to counteract the bitterness and astringency. This sensory constraint means that the final product contains an energy source for microorganisms, thereby making microbiological stabilization difficult. To reduce the risk of yeast growth in the finished cider and to ensure commercial stability, cider producers wishing to avoid pasteurization can greatly reduce the level of a common nutrient. Nitrogen deficiency can achieve sufficient stability for local distribution.

There is a significant risk of microbial growth associated with cider production: high pH, spontaneous inoculation, low alcohol content, and the presence of residual sugar. As the microbiological instability of cider remains one of the main concerns for producers, most industrial ciders are heat-treated to overcome this drawback.

Given these characteristics and the diversity of possible techniques, the process of making cider will be briefly described with particular emphasis on the main stages and constraints.

6.4.2. Fermentation process

6.4.2.1. Extraction of the must

This stage involves different operations: transport, washing, sorting, pressing and extraction of pomace by water percolation to which pomace drying can be added.

Pressing has evolved considerably over the past 40 years going from labor-intensive rack and cloth presses to automated semi-continuous basket presses or continuous belt presses (see section 6.3.3). Depending on the type of press, the extraction yield varies from 600 to 800 kg of "pure juice" per ton of apples.

In industrial plants, *fresh pomace* (residue obtained after pressing, ranging from 200–400 kg per ton of apples) is extracted on a diffusion apparatus that functions continuously by counter-current percolation. This operation yields about 200-300 liters of "diffusion juice" per ton of apples, with a sugar concentration of around 60% of that of the pure juice. The extracted pomace is then either pressed, dried and used to produce pectin or immediately discarded for animal feed.

It is mainly during this stage that phenolic compounds undergo enzymatic oxidation, resulting in color formation in must. Phenolic compounds in apples are located in the cell vacuoles while the enzyme that catalyses the reaction (PPO) is located in the plastids. Macerating and pressing the pulp breaks down the tissue and brings these two components into contact, which in the presence of oxygen triggers a series of browning reactions. Some apple polyphenols, cafeoylquinic acid in particular, are oxidized to quinones by PPO. These highly reactive molecules are also involved in other reactions (oxidation-reduction coupled with other polyphenols, condensation, etc.), which form compounds responsible for the brown color (see Chapter 6 of Volume 1).

6.4.2.2. *Pre-fermentation clarification*

Apple must is generally clarified before fermentation. In the past the method used, called "keeving", involves the "spontaneous" gelation of pectin in the must in the presence of calcium (see section 6.3.4.1) within 2 to 5 days. The gel, containing must deposits and growing yeast, contracts and is lifted to the top of the vat by microbubbles of carbon dioxide produced at the start of fermentation. This forms a brown foam, the so-called *"chapeau brun"*. The foam is separated from the clear juice, which is put into another vessel. Today, the current practice of adding a fungal enzyme (pure PME) to ensure sufficient demethylation and soluble calcium salts (4 mM $CaCl_2$) makes it easier to control this stage.

To separate the gel and the must deposits, some industrial plants use continuous flotation with nitrogen microbubbles. Resulting from the treatment of waste water, this method involves the saturation with nitrogen (gas) at 5 bar pressure of a part of clarified must and added calcium salt. Then the expansion of this pressurized juice at the bottom of the flotation device, which is continuously fed by the previously enzymed (PME) raw juice, causes the rise of the gel to the surface. A scraper eliminates the solid phase on top of the clarified must, which is pumped into a vat. The clarification effect is similar to the static process while avoiding the risks associated with an uncontrolled rapid start to fermentation.

These clarification processes are considered favorable to the sensory quality of cider since fermentation is always slower than that of crude must or otherwise clarified must: depectination alone, centrifugation, cross-flow filtration. The effect of the process on fermentation is attributed to the depletion of nutrients in the must due to the growth of the first generation of yeast. The exact type of nutrient is still uncertain; the drop in nitrogen has long been given as the reason, but given the low yeast population reached during clarification, assimilable soluble nitrogen is hardly affected and the drop observed is mostly linked to nitrogen of the must deposits, which is not assimilated by yeast. This operation not only allows the must to be depleted of nitrogen, but also of some vitamins like thiamine, known to be rapidly consumed by the growth of secondary yeasts (non-*Saccharomyces*). Moreover, while clarification removes the must deposits, it also removes the adsorbed compounds including plant sterols necessary for yeast membrane synthesis in the absence of oxygen. Clarification also removes PPO, which is mostly insoluble and plays a major role in the anaerobiosis of must due to its involvement in enzymatic oxidation reactions.

In reality, the process is more complex than the description given and a number of effects, sometimes contradictory, influence fermentation during this operation.

6.4.3. Action of microorganisms

6.4.3.1. Succession of yeast species and fermentation

Cider fermentation occurs under the effect of a succession of flora introduced by the fruit and/or equipment. It can be artificially divided into three phases based on the dominant flora: oxidative phase, alcohol-producing phase and maturation.

Oxidative phase

The first phase takes place after extraction and lasts from a few days to two weeks. It is called the *oxidative phase* because of the presence and growth of yeasts with low fermentative activity, also known as oxidative species; *Hanseniaspora valbyensis* being the most active. Although facilitated by the presence of air, this species has a fermentative metabolism and under cider fermentation conditions, its activity is considered favorable to the production of fruity aromas. Products developed experimentally by a strain of *Saccharomyces* sp. alone are very neutral while the presence of *H. valbyensis* brings out fruitiness and ester notes linked to the production of acetate esters and/or an interaction with *Saccharomyces* sp. Another species, *Metschnikowia pulcherrima*, commonly present and even sometimes dominant after pressing is quickly eliminated by the absence of oxygen, and does not seem to have an impact. This phase also corresponds to a latency step of *Saccharomyces* sp. growth, but at this stage, their population is still insufficient to have any biochemical impact.

Alcohol-producing phase

The second phase, the alcohol-producing phase, is characterized by the activity of fermentative yeast species of the genus *Saccharomyces*, which ferment most of the sugars with a yield of around 1% vol. for 17 g of sugar consumed. In cider, the most common species is *S. uvarum* although the species *S. cerevisiae*, responsible for wine and beer fermentation, may also be present. Low fermentation temperatures (between 5 and 15°C) and the prior growth of *Hanseniaspora* species explain this selectivity of fermentation conditions in favor of *S. uvarum*, provided mostly by the equipment during pressing. At equivalent populations, this species quickly dominates over non-*Saccharomyces* yeasts; as a result, the oxidative phase can only exist if the population of *S. uvarum* or *S. cerevisiae* is initially low. It is also these species that consume most of the soluble nitrogen in the must during their growth. As a result, the growth or successive growth of these fermentative floras depletes the medium of nitrogen and helps achieve microbiological stability without heat treatment.

Maturation

Maturation follows the alcohol-producing phase and involves very slow fermentation resulting from the elimination of the fermentation biomass by sedimentation/racking, filtration or centrifugation. It lasts until bottling or consumption if the cider has not been microbiologically stabilized. Maintaining residual fermentation has the advantage of preventing the growth of aerobic flora (acetic acid bacteria, film forming yeasts as *Candida* sp., *Pichia* sp. etc.). However, yeast species with low growth levels such as the *Brettanomyces* species can develop after a few months of storage. This occurs as a result of warmer temperatures in spring if the fermenting room is not refrigerated. It is also during this period that bacteria (mainly lactic acid) invade the medium. They include *Œnococcus œni*, which causes malolactic fermentation (MLF), that is the conversion of malic acid (diacid) to lactic acid (monoacid). This reaction is particularly important in cider since malic acid is the main acid and MLF drastically reduce the acidity of the cider.

Secondary fermentation in the bottle or closed tank can be considered a particular feature of maturation. It involves carrying out the last stage of alcoholic fermentation in a sealed container (pressure-resistant vessel or corked bottle) to preserve the endogenous carbon dioxide and make the cider fizzy. The carbon dioxide content of cider ranges from 4 to 6 $g.L^{-1}$, which corresponds to the fermentation of 9 to 13 $g.L^{-1}$ of sugar. When secondary fermentation occurs in the bottle, the cider should be bottled with a higher level of sugar than desired in the final product. This type of fermentation takes about two months. It is the most common method among small cider producers given its operational simplicity. In addition, the absence of additives and the possibility of adding the label "naturally effervescent" are beneficial to the image of the product. This technique also avoids the formation of off-flavors associated with aeration during bottling because in secondary fermentation, yeast consumes any oxygen introduced. However, some uncertainty surrounds this method because it is difficult to control the length and intensity of fermentation while the yeast remains active in the bottle.

6.4.3.2. Yeast and bacterial spoilage

Apart from the positive flora, cider is also prone to the growth of spoilage flora, which can damage its sensory quality. Most undesirable microorganisms are bacteria capable of growing at a pH below 4 but are more active the higher the pH. Some yeasts, in particular *Brettanomyces sp.*, are also classified as spoilage flora.

The production of acetic acid – acesence – by acetic acid bacteria is no longer a serious problem: cider undergoes light fermentation in the tank so that the constant release of CO_2 is sufficient to prevent the dissolution of air. However, lactic acid bacteria, especially *Œnococcus œni* responsible for MLF, grow well under anaerobic conditions, with some even facilitated by the growth of *Saccharomyces sp*. MLF is not spoilage as such, but it negatively affects cider by raising its pH; since this species is heterofermentative, it metabolizes sugars to D-lactic acid and acetic acid – known as lactic acid spoilage – which is detectable when the bacterial population becomes too large. Since MLF in cider always occurs in the presence of sugars, this spoilage exists but is likely to be limited by the absence of other nutrients such as nitrogen.

Another type of spoilage that remains difficult to control is "ropiness" or oiliness. It is caused by a production of exopolysaccharides that surround the bacteria making the cider highly viscous with an oil-like consistency. Other sensory perceptions are not necessarily affected unless another type of spoilage such as lactic acid spoilage develops at the same time. Ropiness is mainly attributed to lactic acid bacteria, but recent studies have shown that bacteria of the *Bacillus* (*B. licheniformis*) and *Pediococcus* genera could be responsible. It poses a real problem as the high viscosity makes it difficult to filter.

Another type of spoilage described as "cider sickness" or "*framboisé*" in French can also be highly problematic: it is characterized by an excessive production of acetaldehyde (ethanal), which gives a raspberry aroma and which reacts with flavan-3-ols to produce a white turbidity. Two mechanisms have been proposed to explain it. According to the first, ethanal results from the action of acetic acid bacteria, which under anaerobic conditions, metabolize lactic acid generated mainly during MLF. However, as these bacteria are strictly aerobic and therefore can only grow in the presence of oxygen, this mechanism is rarely referred to given that cider is seldom in contact with air. According to the second, ethanal is directly produced from sugars by *Zymomonas mobilis* ssp. *Francensis* strains according to the Entner-Doudoroff pathway. This mechanism is currently seen to be the main cause of spoilage. Once started, it is very difficult to control, especially by SO_2, which loses its effectiveness by combining with excess ethanal. When the pH is below 3.7, the preventive use of SO_2 on a small population before the production of large amounts of ethanal is however effective to a certain extent. Whatever the case may be, the depletion of available nitrogen in the medium is the main way of preventing bacterial growth.

Cider can have "animal", "leather" or "earthy/woody" odors especially when stored at high temperatures (>15°C) for a long time. These odors are linked to the occurrence of volatile phenols and are mainly attributed to yeast strains of the *Brettanomyces* genus and certain bacteria. Some consumers value these odors provided they are not overpowering; otherwise the cider can become unpalatable. However, most consumers find these odors unpleasant, which leads us to consider that they constitute a spoilage. In general, recent technological developments have had a positive impact as the number of ciders with defects caused by this spoilage flora seems to be declining.

The widespread control of keeving, filtration, refrigeration and inoculation with dry yeast for secondary fermentation helps to limit the impact of bacteria. In the case of artisan producers and farmers, the input of technical advisors has led to a progressive and significant reduction in the main defects, which could be totally eliminated in the future. Tests have been developed for particularly problematic types of spoilage to predict risks and avoid serious situations by adapting production methods.

6.4.4. *Fermentation and post-fermentation*

6.4.4.1. *Fermentation*

Traditionally, cider was racked (or siphoned) during fermentation to remove the settled biomass and gradually slow down fermentation, with two aims: to avoid rapid fermentation, considered detrimental to the sensory quality, and to gradually deplete the medium of nitrogen, sterols and vitamins to stop the fermentation at the right time while maintaining the residual sugar. This method is inefficient; if the initial rate of fermentation is too high, and the only method is then centrifugation.

However, controlling fermentation by reducing the biomass can only be achieved at a certain time since its effectiveness depends on the nutrient content of the fermentation medium. If intervention takes place when the yeast has consumed only a small amount of nutrients, it will be ineffective because yeast growth will resume and the population (as well as the rate of fermentation) will quickly reach a level similar to that without treatment. On the other hand, if centrifugation is carried out when all the available nutrients in the medium have been consumed, the yeast cannot resume growth and, as a result, the rate of fermentation will be reduced to almost zero. To achieve this objective, it has recently been proposed to systematically carry out an early decimal reduction of the biomass before yeast growth is complete, that is at the very beginning of the alcohol production stage. To do this, the fermenting cider is filtered or centrifuged, then blended with 10% untreated cider to prevent a delay in fermentation resumption. In practice, the operator does not know the nutritional potential of the must at any given time; it is therefore based on the corresponding change in density, which is an indirect indicator. It is estimated that for most ciders, the operation should be performed after a drop of about 5 to 10 points in density, that is after the consumption of around 10 to 20 g L^{-1} of sugar. While the role of nitrogen may be important in pre-fermentation clarification, it is crucial for the effectiveness of this operation: yeast quantities of between 5×10^6 and $1 \times 10^7 CFU.mL^{-1}$ must be removed during this operation, the growth of which has consumed large amounts of nitrogen. The effectiveness of reducing biomass therefore depends on the amount of residual nitrogen in particular, which would be the best indicator to control this operation if the determination of this amount were easier.

While many artisans and farmers commonly use a reduction of the biomass more or less early in the fermentation, industrial cider producers prefer late centrifugation. Unlike the previous method, the procedure takes place at the end of yeast growth, which limits residual fermentation. This low level of fermentation can either be a result of new but low yeast growth if it was previously limited by a factor other than nitrogen, or caused by unremoved residual flora. This option makes it more convenient to manage fermenting rooms, and since cider is often heat-stabilized after blending, the entire method offers greater protection against spoilage (sensory defects and risks of re-fermentation in the bottle).

6.4.4.2. Packaging

The final cider is usually a blend of different basic ciders with the aim of achieving the desired balance of flavors (sourness, bitterness, sweetness, astringency) and aromas.

Depending on the desired level of clarity, cider is then subjected to various clarification treatments: basic sedimentation, centrifugation or varying degrees of filtration with or without the use of fining agents (gelatin, bentonite, etc.). Earth filtration was common but cross-flow microfiltration is now becoming more widespread.

Effervescence (the escape of gas bubbles) can be achieved in two ways: either the cider is saturated by the injection of exogenous carbon dioxide or it undergoes secondary fermentation in a closed container. In physicochemical terms, the first method is based on the diffusion of gas through a gas-liquid interface (gas and cider are mixed together as much as possible to increase the exchange surface) and Henry's law on gas solubility, where the solubility increases with high CO_2 pressure and low temperature. This method is mainly used by artisan and industrial producers. Secondary fermentation is mainly implemented by small-scale producers (farmers) with two variations: in the first, secondary fermentation is carried out by the yeast present in the cider provided that fermentation has been sufficiently slowed down; in the second, the cider is finely filtered to remove as many microorganisms as possible and then inoculated with active dry yeast, both to prevent spoilage and ensure better reproducibility of secondary fermentation. However, since some yeast still remains at the end of fermentation, the final pressure is difficult to control.

The type of bottling is directly related to how effervescence is achieved: sparkling cider (carbon dioxide saturation or secondary fermentation in a closed vat) should be bottled using a counter pressure bottle filler to avoid foaming by maintaining the pressure in the bottle higher than the equilibrium pressure of the cider. Different machines vary in their ability to prevent the introduction of oxygen into the cider, which has an impact on product quality. If cider is intended for secondary fermentation in the bottle, it is bottled at atmospheric pressure by more basic machines.

Some of the cider is stabilized by heat treatment. Due to effervescence, treatment is carried out after packaging in tunnel pasteurizers at low temperatures (around 60–65°C) for long periods (20–60 min). Given the acidic pH of cider, the low growth capacity of microorganisms in the fermented medium and the absence of major health risks with this type of product, the treatment conditions are relatively mild. Calculated with a z value of 7°C and a reference temperature of 60°C, treatment ranges from 70 to 270 PU (pasteurization units) for sweet cider and 40 to 150 PU for raw cider.

The operations carried out during packaging can sometimes have a detrimental effect on the sensory quality of the final product. They include fining, which is required to obtain a clear, bright and physicochemically stable cider, and prolonged heat treatment used for microbial stabilization. In addition, oxygen incorporated during bottling can result in the oxidation of various compounds, which affects the sensory quality of the cider.

7

From Grape to Wine

Like many fermented products, wine is an ancient drink. It has held an important place in many civilizations because of its mythical and sacred nature. Wine is produced from the fermentation of grape juice by yeast such as *Saccharomyces*, which is naturally present on the outside of grapes. Wine has a particular feature compared to many other food products – it can vary in character not only depending on the wine region but also on the plot, the year and even the bottle since many factors affect its development:

– grape variety;

– soil and climatic conditions (soil type and structure affect root development and growth direction; temperature and sunshine vary annually) regulate water supply to the vine, which is crucial. Too much water results in excessive foliage preventing sunlight from reaching the grapes whereas too little water causes drought stress in the vine. All these factors contribute to the "terroir effect";

– the contribution of the winemaker in maintaining the vineyard and making wine;

– the skill of the cellar master in charge of the wine until it is sold.

Chapter written by Thomas CROGUENNEC.

7.1. Raw materials

7.1.1. *Grape variety*

Grape variety is one the main criteria that determines the character and quality of wine. Grapes (around 250 different varieties in France alone) differ in their ability to adapt to climatic conditions, as well as their physical (stem proportion, size and juiciness of the grapes, skin thickness, etc.) and chemical characteristics (amount of acid, sugar, phenolic compounds, nitrogen compounds, aromatic compounds, etc.). These differences have resulted in a large variety of technological skills and consequently a wide range of wines. For example, Chardonnay and Cabernet Sauvignon grape varieties adapt well to different climates and are therefore successfully grown across the world. Chardonnay and Pinot varieties have high acidity levels making them suitable for the production of sparkling wine. The Semillon variety, which is used in the production of sweet wines, favors the development of *Botrytis Cinerea* or "noble rot". Grenache and Muscat varieties are particularly suited to dry terrain and produce high-sugar yields, which is ideal for the production of naturally sweet wines. The Syrah grape variety is rich in pigments (anthocyanins) resulting in very dark red wines. Low-sugar grape varieties such as Trebbiano (or Ugni blanc) and Colombard are used to produce low-alcohol wines with limited aging potential. They are used for the distillation of brandy such as Cognac or Armagnac.

7.1.2. *Composition of grapes*

A bunch of grapes is composed of a woody part (the stem) and a fleshy part (the berries) consisting of skin, pulp and seeds (Figure 7.1). About 20% is solids in the form of the stem, skin and seeds and 80% is must, mostly from the pulp. The woody and herbaceous stem contains minerals, acids and astringent tannins. The seeds are rich in lipids and astringent tannins. The skin is a concentrated source of "specific" components from the individual grape variety, such as pigments, flavors and aromas, making it the most important part of the grape. Sugars, organic acids, amino acids and other non-specific components are contained in the pulp.

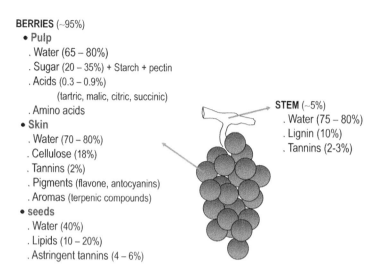

Figure 7.1. *Chemical composition of grapes*

7.1.2.1. Carbohydrates

Pulp cells contain a high concentration of fermentable sugars (glucose, fructose). Sugar is synthesized from the time the vine bursts its buds until the leaves fall off, with a peak in concentration when the grapes ripen. The fermentable sugar content generally varies between 200 and 350 g/L depending on the grape variety and time of harvest. The must also contains non-fermentable sugars (arabinose, xylose) mainly from the solid parts of the harvest (stems and skins). A percentage of carbohydrates is also in the form of starch since the plant synthesizes more sugar than it consumes. There are also substantial amounts of other carbohydrate substances such as pectic compounds in the cell walls of the grape.

7.1.2.2. Organic acids

Large amounts of organic acids are found in the pulp cells, usually in the form of potassium salts. Tartaric acid, which is synthesized in unripe grapes, and malic acid, produced from the metabolism of sugars, are the main organic acids in the must (up to 90–95% of organic acids). Their content in the finished wine is less than in the must due to their insolubility (precipitation of potassium bitartrate) and/or their degradation (malolactic fermentation). The must also contains other organic acids such as citric acid, succinic acid and so on. Maximum acidity is reached at the onset of ripening after which point it decreases under the combined effect of dilution, caused by an increase in grape volume, and the degradation of some acids. In the production of red wines, organic acids promote color development. They also contribute to the freshness of wines, an important characteristic in white wines. If a wine is too low in acid, it can taste flat and dull whereas if a wine is too high in acid, it tastes harsh and sour.

7.1.2.3. Nitrogenous substances

The nitrogenous substances in grapes include proteins, free amino acids, ammonia and so on. Levels are generally low and result from an interaction between the genetic characteristics of the vine, soil, climatic conditions and cultural practices. Free amino acids (mainly arginine, glutamic acid/glutamine, proline and threonine) largely result from the fixation of ammonium to α-keto-acids in the grape cells. Proteins are rapidly produced during ripening with a simultaneous accumulation of sugar in the grape.

Enzymes present in grapes can be highly active once in contact with their substrate. The most common enzymes are oxidoreductases (polyphenol oxidase, lipoxygenase) and hydrolases (proteases, pectinases, glucosidase and invertase):

– proteases degrade proteins and release amino acids necessary for yeast nutrition during alcoholic fermentation. Amino acids and vitamins facilitate the onset of alcoholic fermentation. They also contribute to the formation of higher alcohols (more than two carbons) enriching the flavor of the wine. In addition, they release polyphenols and aromatic compounds adsorbed on the proteins;

– in the presence of oxygen, polyphenol oxidases convert phenolic compounds into quinones capable of polymerizing. Under the action of polyphenol oxidase, grape juice oxidizes (enzymatic browning, see Chapter 6, Volume 1 [JEA 16a]) and its color changes. Polyphenol oxidases therefore have a negative effect on the clarity, aroma and flavor of wine. Their action is blocked by adding sulfur dioxide to the must and avoiding contact with oxygen;

– pectinolytic enzymes (pectinases) hydrolyse pectin, which acts as physical barrier preventing the diffusion of compounds in the skin or pulp cells to the fermentation medium. They therefore have a positive effect on the extraction of juice and contribute to the spontaneous clarification of grape juice from the press;

– lipoxygenases oxidize unsaturated fatty acids, mostly found in grape seeds, to give aldehydes or alcohols. The latter alter the aroma of grape juice giving it a grassy or herbaceous character. This defect is controlled by reducing the intensity of crushing and pressing, which limits the amount of fatty acids released into the grape juice;

– glycosidases facilitate the release of aromatic compounds (mainly terpenes) by hydrolyzing the covalent bond between the aromatic molecule and the carbohydrate derivative. Grapes contain these compounds in the form of flavor precursors; they are formed by covalent association between an aromatic molecule and a carbohydrate residue making them non-volatile.

7.1.2.4. Aromatic compounds

The characteristic aromas of grape varieties develop in the skin during ripening. Aromas exist as free compounds (volatile terpenols, esters, etc.) or as precursors (terpene glycosides). They are released upon crushing and pressing of the grapes. If present in excessive amounts in wine, some aromatic compounds can cause defects. This is the case for example with the phenolic aroma of certain wines containing grapes affected by rot or mercaptan compounds that can mask good aromas and, in more extreme cases, develop a putrid odor.

7.1.2.5. Phenolic compounds

Phenolic compounds or phenols comprise phenolic acids and flavonoids. Phenolic acids include derivatives of benzoic acid and cinnamic acid while the large family of flavonoids can be divided into several sub-classes containing flavones, anthocyanins and tannins. Phenolic compounds significantly affect the color and sensory characteristics of wine. They also contribute to the stability of wine during storage due to their antioxidant properties. Their solubility in water (must) decreases with the degree of polymerization of phenolic compounds. However, they are soluble in acidic hydroalcoholic solutions. Thus the dissolution kinetics of phenolic compounds during the maceration stage of red winemaking accelerates as alcoholic fermentation increases.

Colored phenolic compounds are found in the skin cells of grapes where their concentration increases during ripening. Flavones are yellow pigments present in the skin of all grapes. They do not play any role in the color of white wine, which has a low concentration of flavones, around 20 times less than that of red wine. Anthocyanins are red, purple and blue pigments present in the skin of red grapes and are responsible for the color of red wine. The concentration of different colored compounds depends on the grape variety and the degree of ripeness. The extractability of anthocyanins increases as the grape ripens and softens.

Tannins are a heterogeneous family of colorless compounds formed by the polymerization of phenolic compounds. In grapes, they are primarily in the form of condensed tannins or oligomers. They are found in the skin, seeds and stems. Their importance in wine depends on the grape variety, the length of maceration and fermentation. Tannins have an important antioxidant property; they regulate oxygen activity and stabilize the color of wine during the aging process. Tannins are highly reactive toward proteins and can form co-precipitates resulting in a deposit at the bottom of barrels and bottles.

7.2. Winemaking techniques

Winemaking involves a series of processes starting with the selection of grapes to bottling the finished wine. It can be divided into two main sets of operations:

– the first set includes Physicochemical operations. They involve releasing the pulp from the grapes, transferring the desired components (aromatic compounds, pigments, etc.) from the solid part of the grapes, mainly from the skin, to the liquid part, separating the solid part from the fermented or unfermented must, and controlling enzymatic or chemical oxidation;

– the second set includes biological operations. They involve alcoholic and malolatic fermentation.

Red and white winemaking differs in when the alcoholic fermentation stage takes place with regard to the separation of the solid and liquid parts of the harvest (Figure 7.2). In white wine production, alcoholic fermentation begins as soon as the solid part of the grape harvest has been removed, that is immediately after crushing and pressing. In red wine production, alcoholic fermentation takes place together with the grape skins. This stage of maceration, or fermentation on skins, generally varies from a few days to a few weeks, with the most common period being one to two weeks.

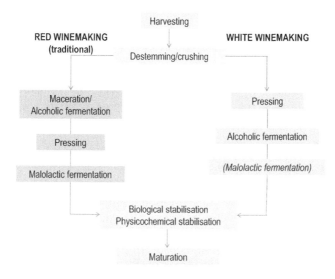

Figure 7.2. *Red winemaking (traditional) and white winemaking*

7.2.1. *State of the harvest and adjustments*

The degree of ripeness is estimated by the sugar/acid ratio (indicator of maturity) in the grape. This ratio, which varies with the type of wine to be produced, the grape variety, climatic conditions and so on often determines the harvest date. Grapes intended for the production of fresh, fruity white wine are harvested with a sugar and acid content of approximately 190 g/L and 5 g/L, respectively. Grapes intended for the production of red wine for aging are picked at full maturity and generally contain 210 to 220 g/L sugar. For naturally sweet wines, grapes must have a sugar content of at least 252 g/L. However, other criteria for grape maturity can come into effect such as the concentration of aromatic and phenolic compounds, which significantly influence the final characteristics of the wine.

Sometimes, bad weather negatively impacts the harvest in terms of the sugar and acid content, aromatic and phenolic compounds and so forth. The perfect balance between the physical and chemical characteristics (concentration of various desired components) of grapes is only achieved during exceptional years or under ideal maturity conditions. Under-ripe grapes are deficient in pectolytic enzymes, which results in a firmer texture and a lack of juiciness, making pressing more difficult. The resulting must contains a large amount of coarse elements, is highly acidic, low in sugar, has an excess of herbaceous and astringent tannins mainly from the seeds, and lacks anthocyanins and aromatic compounds.

To overcome any discrepancies between the optimum maturity of the different components of the grape, which affects the final quality of the wine, the winemaker may employ a number of regulated practices. For example:

– to overcome a lack of sugar, it is possible to add sucrose to the unfermented grape must, which is known as chaptalization or sugaring; around 17–18 g of sucrose is required to increase the alcohol content of wine by 1%. It is also possible to concentrate the must by removing some of the water by evaporation or reverse osmosis; this operation can only be applied to a limited amount of must;

– to correct acidity levels, tartaric acid or "verjuice" (highly acidic juice from unripe grapes) is added to increase acidity while the addition of potassium tartrate or further malolatic fermentation lowers acidity.

7.2.2. *Physicochemical processes involved in winemaking*

Traditionally, grapes are destemmed, crushed, macerated to extract the desired components, pressed, and sulfur dioxide is added. Many operations are common to all types of winemaking, but others are more specific to a certain type of winemaking, such as maceration in the production of red wine.

7.2.2.1. *Destemming and crushing*

Destemming and crushing are often combined operations. Destemming is the process of removing all or part of the stems from the grapes. Crushing is the process of mechanically bursting the grapes to release the juice and allow yeasts naturally present on the skin to come into contact with the sugar-rich must. Destemming and crushing intensity affect the flavor and subtlety of the finished wine. In the case of red wine being, fully destemmed and a long period of maceration of lightly crushed grapes improve red wine quality. In other cases, the grapes are not crushed at all and are instead transferred directly to the press to extract a clear must (e.g. Champagne). Any rotting grapes are removed during the sorting process associated with destemming so as to avoid any sensory defects.

7.2.2.2. Extraction of desired compounds

The maceration and fermentation of crushed grapes is an important process in traditional red winemaking. It allows the components in the grape skins to be extracted and solubilized in the grape juice. An increase in the alcohol content resulting from the fermentation of the grape must by yeast as well as an increase in temperature accelerate component extraction and solubilization. The solubilization of aromatic compounds is faster than that of phenolic compounds, especially with a high degree of polymerization of phenolic compounds; anthocyanins are extracted earlier than tannins. The solubilization of phenolic compounds determines the maceration and fermentation time. From the same harvest of red grapes, it is possible to produce white wine (no fermentation on skins), rosé wine (short fermentation on skins), fruity red wine for early consumption (medium fermentation on skins) or red wine for aging (long fermentation on skins) (Figure 7.3). Thus, minimal solubilization of components contained in the grape skins is required for white wine. Its characteristics are achieved by balancing the aromas of grape varieties and fermentation with alcohol, acidity and residual sugar levels.

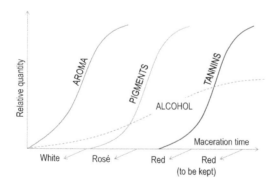

Figure 7.3. *Maceration time to produce white wine, rosé, and red wine to be kept from red skin grapes*

Thermovinification is an alternative to traditional maceration to extract desired compounds. It is either carried out on its own or immediately followed by traditional maceration (combined method). Thermovinification applies heat to facilitate the extraction of various compounds from grape skins, in particular anthocyanins. Heating to 75°C for 20 to 30 min in a maceration vat results in a dark-colored must and increases the ratio of anthocyanins to tannin compared to traditional maceration. Heating promotes the extraction of organic acids in the must, in particular tartaric acid/potassium tartrate, which is responsible for mineral precipitation. Apart from stabilizing the must from degradation by microorganisms present in the grapes, heating also denatures some enzymes that could negatively affect the final quality of the wine: removing oxygenases tends to have a positive effect whereas removing hydrolases, isomerases, transferases and so on decreases the aromatic potential of the wine.

Another winemaking technique involves macerating intact grape clusters in a CO_2-saturated environment, known as carbonic maceration. In this process, the uncrushed grapes are placed in a sealed container and gradually saturate with CO_2 from plant cell respiration. When oxygen levels are depleted, aerobic metabolism is taken over by anaerobic metabolism in plant cells and a limited amount of intracellular ethanol is produced. This production of ethanol occurs without the intervention of yeast as there is no contact between yeast and intracellular sugar. However, changes within the cells promote the diffusion of desired compounds (aromatic and phenolic compounds) from the skin to the vacuolar sap. In addition, intracellular fermentation consumes a large quantity of malic acid whereas the amount of succinic acid increases. After 8 to 15 days, the mostly intact grapes are transferred to the press to separate the solid from the liquid part.

7.2.2.3. Pressing

Pressing involves extracting the liquid fraction (fermented or unfermented) from the grapes. In white wine production, this is carried out on fresh, crushed or whole grapes. In red wine production, usually totally or partially fermented grapes are pressed after the "free run wine" has been drained. The solid part of the grapes is removed from the maceration tank (devatting) and pressed to give "pressed wine". This is either mixed with free run wine or used in subsequent operations such as distillation. In this case, the recovered alcohol can be used in "mutage" (a process to artificially stop or "mute" fermentation of the must by adding alcohol) for the production of sweet wines (see section 7.4.2.2). The extraction yield and final quality of the wine are affected by pressure and pressing time, which must be carefully controlled.

7.2.2.4. Sulfiting

Sulfiting is the process of adding moderate amounts of sulfur dioxide (SO_2) or its derivatives to ensure the microbial and chemical stability of grapes and grape juice during the various winemaking operations. The SO_2 content in white and red wines at the time of sale must not exceed 210 mg/L and 160 mg/L, respectively. This content may be increased by 50 mg/L in wines containing high amounts of sugar (sweet wines). Sulfur dioxide has an antioxidant function by protecting phenolic compounds from oxidation and an antibacterial function making it possible to choose the fermentation medium; yeast is less sensitive than bacteria to SO_2. Added in large amounts, SO_2 can slow down or stop fermentation. This property is exploited in the preparation of must for white wine from red skin grape varieties by extending the extraction time of desired compounds (aromas) before fermentation begins. In addition, must intended for white wine is more susceptible to oxidation due to a lower content of phenolic compounds, which explains the addition of greater amounts of SO_2. Sulfur dioxide is highly reactive to many grape compounds and a high concentration (e.g. a high local concentration) causes reactions that may adversely affect the aromatic balance of the wine.

7.2.3. Biological processes involved in winemaking: fermentation

Two types of fermentation can occur in winemaking:

– alcoholic fermentation consisting of the conversion of sugar to ethanol by yeast;

– malolactic fermentation during which lactic acid bacteria degrades malic acid to lactic acid.

The main challenge is to avoid an overlap of these fermentation processes. Lactic acid bacteria can metabolize sugar into lactic acid, which is known as lactic souring. However, grape must is more favorable to the development of yeast than lactic acid bacteria, the growth of which is limited by acidity and the alcohol content of the medium. Wines are "biologically" stable when they no longer contain any fermentable sugar or malic acid. In red wine production, both types of fermentation usually continue to completion. In white or rosé wines, acidity contributes to freshness and highlights the fruity aromas of the grape varieties. As a result, malolactic fermentation is optional and depends on the acidity of the must and the desired characteristics of the wine.

7.2.3.1. Alcoholic fermentation

During alcoholic fermentation, fermentable sugars in the must are mostly converted to alcohol (and some secondary fermentation products) by yeast. It is generally preceded by a limited phase of aerobiosis (consumption of surrounding oxygen) during which the degradation of one mole of glucose or fructose ($C_6H_{12}O_6$) provides 38 moles of ATP, which is necessary for yeast growth. The degraded sugar molecules do not create alcohol during this phase. Glycolysis (presence of oxygen) proceeds as follows:

$$C_6H_{12}O_6 + O_2 \rightarrow 6\ CO_2 + 6\ H_2O + 38\ ATP$$

Under normal winemaking conditions, oxygen quickly becomes a limiting factor and yeast metabolism changes to ethanol production (alcoholic fermentation). Under these conditions, the degradation of one mole of glucose results in the formation of two moles of ethanol, two moles of CO_2 and two moles of ATP as shown in the alcoholic fermentation reaction:

$$C_6H_{12}O_6 \rightarrow 2\ CO_2 + 2\ CH_3 - CH_2 - OH + 2\ ATP$$

Thus, alcohol is only produced anaerobically, but these conditions are less favorable to yeast growth since fewer ATP molecules are produced. Based on the equation for alcoholic fermentation, the degradation of 188 g/L of fermentable sugars by alcoholic fermentation alone can produce a wine with an alcohol content of 12% or 96 g/L (mass density of ethanol 800 g/L).

During alcoholic fermentation, various phenomena can be observed:

– the mass density of the must, initially around 1,080 g/L, decreases to around 990 g/L in wine. This is an indicator of the changes due to alcoholic fermentation;

– the temperature in the fermentation tank increases due to the exothermic nature of the biological reactions involved. This temperature increase facilitates the extraction of phenolic compounds in the production of red wine but also increases the loss of aromatic compounds by volatilization. It is therefore important to control the temperature in the fermentation tanks depending on the type of wine produced (around 30°C for red wine and 20°C for white wine);

– in red wine production, the solid part of the must (skins and pulp) rises to the top of the mixture under the pressure of escaping CO_2 generated by alcoholic fermentation. This forms what is known as a "cap" on top of the fermentation vessel. If left on the surface, only a small amount of must would come into contact with the skins. Also, if the cap dries out, this would be a perfect breeding ground for spoilage microorganisms. To avoid these problems, the solid matter is kept submerged using a screen or by a process called "pumping over", which involves pumping must out from the bottom of the tank over the top of the cap. Around 50 liters of CO_2 (gas) are generated per liter of wine produced (12% alcohol content). Due to its toxicity, cellars are ventilated to remove the large volumes of CO_2 produced during fermentation.

Alcoholic fermentation is controlled by different conditions or operations:

– inoculation with a selected culture of exogenous yeast can rapidly trigger fermentation. This process is also used to accelerate or revive latent fermentation;

– the oxygen content of the fermentation medium, modified by the aeration of the must during the "pumping over" process for example, is an important control point in alcoholic fermentation. Aerating the must lowers the amount of alcohol produced and promotes the oxidation of certain compounds in the grape;

– temperature influences the fermentation process. Alcoholic fermentation significantly slows down above 35°C or below 10°C;

– the pH of the must also affects alcoholic fermentation. It generally varies between 2.8 and 3.8;

– sulfur dioxide (SO_2) has an inhibitory effect on bacteria above 50 mg/L, but at higher concentrations, it also slows down yeast activity. In addition, it increases the amount of secondary fermentation products such as glycerol;

– when the alcohol concentration increases, yeast activity slows down. This is the principle behind the aforementioned "mutage" process (see section 7.4.2.2) in the production of sweet wines;

– tannins in the stems, seeds and skins of grapes interact with yeast walls and limit exchanges with the must. Tannins thus slow down yeast activity;

– deficiencies in nitrogenous substances (amino acids, vitamins) slow down alcoholic fermentation. The stability of certain sweet wines is based on a deficiency of nitrogenous matter during storage.

7.2.3.2. *Malolactic fermentation*

Malolactic fermentation by lactic acid bacteria occurs under anaerobic conditions and reduces the acidity of wine. During this fermentation, lactic acid bacteria convert malic acid (diacid) to lactic acid (monoacid) and CO_2 (Figure 7.4). This is a desired outcome in red wine production but sometimes avoided in the production of white or rosé wines. It is conducted in white winemaking when grape varieties grow in cool climates due to the high acidity levels of the grapes. It adds a "buttery" or "honey" aroma to the wine.

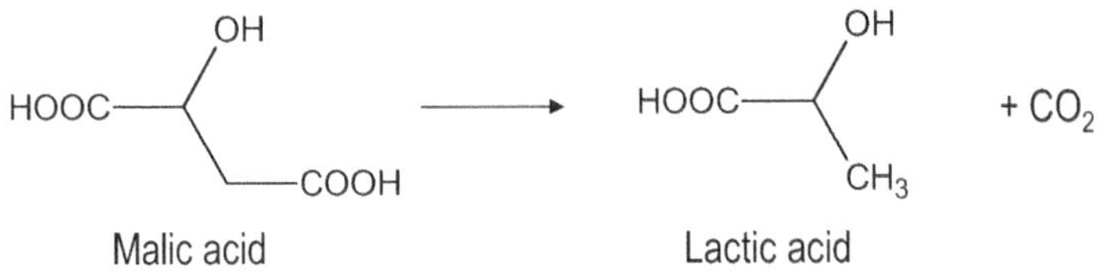

Figure 7.4. *Malolactic fermentation*

Before starting malolactic fermentation, it is important to ensure that all the sugar has been fermented to avoid further lactic acid production. When malolactic fermentation is desired, conditions more favorable to the growth of lactic acid bacteria must be met in terms of pH, temperature and the SO_2 content of the must. At the end of alcoholic fermentation, the alcohol content, the acidic pH and excessive levels of SO_2 have generated adverse conditions for the growth of lactic acid bacteria. Therefore, to promote malolactic fermentation, it is recommended to:

– reduce the acidity of the wine by a precipitation of potassium bitartrate at low temperatures;

– store the barrels in cellars where the temperature is strictly controlled between 18 and 20°C;

– begin malolactic fermentation in sulfur dioxide-free tanks prior to blending when bacteria are in the exponential growth stage.

After malolactic fermentation, sulfur dioxide is added to the wine to avoid any risk of fermentation defects.

7.3. Stabilization and maturation of wine

After fermentation, the new wine is still not ready for consumption. It is cloudy, often unappealing in appearance and harsh in taste (acidic, astringent and bitter). It contains carbon dioxide, suspended solids and residual microorganisms (yeast, lactic acid bacteria), which can spontaneously activate further fermentation and biological degradation. The new wine also contains compounds that are close to solubility limit. After a few days, crystalline or amorphous precipitates can appear, which is a normal development during wine maturation. The aromatic profile of the wine also changes and becomes more refined during aging (maturation) due to multiple biological and chemical reactions. Chemical reactions tend to take precedence over biological reactions. Thus the fruity characteristics of young wines gradually fade while the bouquet becomes richer and more complex.

7.3.1. *Biological stabilization*

Three main types of microorganisms can act on the new wine: yeasts, lactic acid bacteria and acetic acid bacteria:

– yeasts may be responsible for further fermentation, particularly in wines containing residual sugar;

– lactic acid bacteria have many potential substrates in wine: fermentable sugars in sweet wines and residual malolactic acid in some white and rosé wines. By degrading citric acid, lactic acid bacteria give the wine a lactic flavor. At higher pH, they can also degrade tartaric acid, causing changes in wine color and taste defects;

– acetic acid bacteria, in the presence of oxygen, convert ethanol to acetic acid, which is known as acescency (vinegar). They are inactive when the alcohol content is high.

Several Physicochemical methods can be used to biologically stabilize wine:

– cold temperatures slow down or stop fermentation, but do not remove microorganisms;

– successive racking (moving wine from one barrel to another) depletes wine of microorganisms, but only filtration can remove most of the yeast and bacteria;

– sulfur dioxide is added to inhibit microbial growth;

– maintaining sufficiently low oxygen levels prevents the growth of acetic acid bacteria. The oxidative aging of wines is an exception. This method is practiced in regions where wines have a high alcohol content (sherry, port, etc.). The wines are barrel aged, and exposed to air and sometimes heat (in the sun), which enriches the aromatic palette of the wines.

7.3.2. Physicochemical stabilization

The precipitation of a compound is defined as a "casse". This can be caused by metal elements such as iron or copper (iron casse, copper casse) or organic compounds (tartaric acid, proteins).

Tartaric precipitation occurs on a large scale and very rapidly once the tanks are cooled after fermentation. At the pH of wine, tartaric acid is mostly in the form of potassium bitartrate, which is poorly soluble in alcohol and at cold temperatures. Its precipitation is increased by malolactic fermentation, which tends to increase the pH of the wine to its pH of minimum solubility (pH 3.6). The precipitation of potassium bitartrate forms a tartar deposit. To prevent this from occurring in the bottle, any excess tartaric acid is precipitated under cold conditions and filtered before bottling.

Protein casse results from an excess of proteins in wine, which originate either from the grape or yeast. They do not exactly correspond to the protein fraction of the grape since the majority is degraded or precipitated during winemaking. Grape proteins that are resistant to proteolytic activities, acidic pH or the presence of ethanol are often stress-resistant proteins (against fungus attack, water stress, etc.). The more proteins in a wine, the more unstable they are. Thus the quality of the wine depends on the health and maintenance of the vineyard as well as the winemaking method. Thermovinification tends to reduce the protein content of grape must through heat denaturation. At the pH of wine, proteins are positively charged and are therefore able to establish electrostatic interactions with negatively charged polysaccharides and hydrophobic interactions with phenolic compounds in the wine. The precipitates resulting from such interactions affect the clarity and stability of the wine. They are removed by clarification which is achieved by filtration or fining. The most effective fining method involves using cation exchange resins, or bentonite, which establish electrostatic interactions with proteins in the wine, thus removing cloudiness. The major drawback of this method is the non-specific adsorption of molecules of interest: bentonites can adsorb molecules other than proteins and thus remove some of the aromatic compounds and flavor from the wine. Another common alternative is fining by the addition of exogenous proteins (casein or potassium caseinate, egg white, food grade gelatin, etc.). By removing or reducing the concentration of certain compounds such as tannins (hydrophobic interactions), this method also reduces astringency in wine.

7.3.3. *Maturation of wine*

The maturation or aging of wine is a long process that can improve the quality of the wine through a series of chemical reactions. It is carried out in oak barrels (new, recycled or heated on the inside before filling) or in contact with wood chips and continued in the bottle. In some cases, wine is aged on yeast lees (Champagne, Muscadet, etc.). It is one of the most expensive operations in winemaking due to production downtime, but it helps to refine the sensory characteristics of the wine. The change in aroma, color, mouthfeel and taste of wine aged in barrels is due to a change in the composition and concentration of certain compounds resulting from:

– the extraction of wood compounds (phenolic compounds such as tannins, eugenol, etc.), which dissolve in wine;

– the decomposition of macromolecular components of wood (lignin, cellulose) resulting in aromatic aldehydes, furan compounds, etc.;

– esterification, polymerization and other reactions between wood and wine compounds. In addition, many redox reactions occur in response to the redox potential of wine controlled by gas exchanges through the barrel and the cork after bottling.

7.4. Specific technology

7.4.1. *Sparkling wines (traditional method)*

The main grape varieties used in the production of sparkling wines based on the traditional method (*méthode champenoise* or Champagne method) are Chardonnay (white grape), Pinot Noir and Pinot Meunier (red grapes), which are commonly used in the production of champagnes. Harvesting, which is carried out manually, should be early to ensure a high acidity level in the grapes. The whole, uncrushed clusters are gently and gradually pressed to give three pressings (the first pressing is the *cuvée* and the second and third pressings are referred to as *tailles*). In the traditional method, the amount of juice extracted is limited to less than 2/3 of the weight of grapes used. Pressing is a particularly delicate operation since the juice quickly changes color once the grapes burst, and clarity is usually a desired criterion. Coloration increases from the *cuvée* to the *tailles*. After pressing and sulfiting, the different pressings are left to rest for about 12 hours in the tanks to be clarified by the natural sedimentation of suspended solids; this is known as settling or *débourbage*.

Alcoholic fermentation (primary fermentation) is carried out by selected yeasts (yeasting or seeding) at a controlled temperature (18–20°C). They convert all the sugar in the clarified must to alcohol. Malolactic fermentation is not systematic, but when it occurs, it is immediately after alcoholic fermentation. The result is a dry white wine with an alcohol content of around 10–11%. Once fermentation is complete, the wine is racked and lightly clarified (naturally, by centrifugation or filtration) to reduce the quantity of lees. The base wine is stored in tanks at around 10°C until blending.

Blending involves the harmonious mixing of different batches, vintages and reserve wines (stored for several years) to achieve a consistent taste. Blended wine is cooled to -3°C to precipitate excess potassium bitartrate (tartaric stabilization), which is removed by filtration. Before bottling, selected yeasts and sugar are added to the wine (around 24g/L) and the bottles are stored horizontally in a cellar at 10–12°C. During this stage, which can last from several weeks to several months, a secondary fermentation occurs in the bottle, known as bottle fermentation. A yeast deposit (lees) forms at the bottom of the bottles, which remains in contact with the wine for up to seven years.

Before sale, the yeast deposit is removed by riddling and disgorging. In riddling, the bottles are placed on special racks called *pupitres* that hold them at a 45° angle, with the crown cap pointed down. After several quarter rotations of the bottles, the yeast deposit gradually settles in the neck of the bottle. Disgorging involves removing this deposit. The neck of the bottle is immersed in brine at -28°C. The ice formed traps the yeast deposit, which is expelled by natural pressure from the opening of the bottle. To compensate for the resulting loss of product, each bottle is topped up with *liqueur d'expédition*, which is a mixture of wine and possibly sugar. A cork is then inserted with a wire cage known as a *muselet* securing it in place.

7.4.2. Sweet wines

7.4.2.1. Naturally sweet wines

Sugar-rich musts produce naturally sweet wines. In practice, there are several ways to concentrate the sugar content of grapes:

Some grape varieties (Semillon, Furmint, etc.) are conducive to the development of *botrytis cinerea*, a type of fungus known as noble rot. During growth, *botrytis cinerea* perforates the skin of the grape causing water to evaporate and thereby increasing the sugar content. At the same time, it feeds on nitrogenous matter and as a result creates a hostile environment for yeast growth during winemaking. Only grapes affected by noble rot are harvested and the must obtained after pressing is extremely high in sugar. In the production of naturally sweet wines, fermentation is halted when the optimal balance between alcohol and residual sugar has been achieved, which is usually an alcohol content of 13 to 15%. Fermentation is stopped by a combination of three factors:

– a gradual slowdown in yeast activity for an alcohol content of more than 13% (progressive auto-intoxication of yeast);

– a low level of nitrogen in the wine, which kills yeast. To prevent nitrogenous matter in the dead yeast cells from being reused, it is removed by racking or filtration, which considerably limits fermentation activity;

– the addition of a significant amount of SO_2 to block yeast activity.

In other parts of the world, a high sugar content is achieved by leaving grapes on the vine past normal harvest time so that their sugars become concentrated; this process is known by the French term passerillage. Leaves are removed from the vine to expose the grapes to the sun while the bunch stem is twisted to break the sap-conducting vessels. Due to the evaporation of water, the sugar content increases.

Another method involves leaving bunches of grapes to dry on straw racks in a dry and well-ventilated place (straw wine). Grapes with a sugar content of around 300 g/L are pressed and the must is used to make naturally sweet wines. Fermentation stops without human intervention through the auto-intoxication of yeast for an alcohol content of 15-18% even though 40 to 50 g/L of residual sugar remains. The wine is aged for several years in oak barrels.

7.4.2.2. Fortified wines

There is a difference between naturally sweet wines and fortified wines. Wines called "*vins doux naturels*" (Port wine, Banyuls, Rivesaltes, etc.) are produced from grapes that are left to over-ripen on the vine, usually from Grenache (black, white) and Muscat varieties. The grapes must contain at least 252 g/L of sugar at harvest time, which corresponds to a potential alcohol content of more than 14.5%. They are used in the production of red and white wines. Fermentation is abruptly stopped when the alcohol content reaches around 12% by adding 96% neutral alcohol, which is obtained by distilling wine from a previous year; this process is known as *mutage*. The addition of alcohol, amounting to 5 to 10% of the product, preserves the sugar in the must. At equilibrium, the actual alcohol content is between 15 and 18% and the residual sugar content is greater than 50 g/L. In all cases, the sum of the actual alcohol and potential alcohol content, which is calculated based on the residual sugar, must be at least 21.5%. In the production of red wine, *mutage* is sometimes carried out during maceration (*mutage sur grains*) prior to devatting and pressing.

Élevage, the French term for the progression of wine between fermentation and bottling, plays an important role in shaping the quality of *vins doux naturels*. Muscat wines are generally stored under reducing conditions prior to bottling to preserve their young aromas, while other wines only achieve aromatic maturity after *élevage* or maturation under oxidative conditions. White wines develop an amber color and dried fruit aromas while red wines take on a tawny color and a bouquet of cooked fruit, prunes and cocoa. Before bottling, it is not uncommon to blend wines that have undergone different maturation processes.

Grapes used in the production of liqueur wines (e.g. Pineau des Charentes) contain at least 170 g/L of sugar. After harvesting and crushing but before any sign of fermentation, the must is mixed with locally distilled spirits. In the case of Pineau des Charentes, 60% oak-aged Cognac from the same region is added. Then, the mixture is placed to the cellar for fermentation. After some of the sugar in the must has fermented, the fermentation is halted and the mixture is aged for at least one year in oak barrels. The alcohol content of the liqueur wines generally varies between 16 and 20%.

8

From Fruit and Vegetables to Fresh-Cut Products

Fresh-cut fruit and vegetables have undergone minimal processing such as trimming, cutting and washing and are ready for consumption. They are packaged under a modified atmosphere and refrigerated at +4°C or lower. They are referred to as "fourth range" products in commercial terminology. By comparison, fruit and vegetables are fresh in the "first range", canned in the "second range" and frozen in the "third range".

Fresh-cut products were developed in the 1980s in an attempt to boost the consumption of fruit and vegetables, which dropped considerably in the 1970s. Between 1971 and 1982, the consumption of salad greens in France dropped by more than 25%, while during the same period, the consumption of easy-to-prepare vegetables such as tomatoes and chicory slightly increased. These changes in food habits in developed countries can be attributed to urbanization and an increase in the number of women entering the workforce, which led to significant growth in the restaurant and catering industry and less time devoted to the preparation of food at home.

Given these facts, nutritionists and those responsible for the fruit and vegetable sections of supermarkets looked for ways to boost the consumption of fresh fruit and vegetables. The first "ready-to-eat" iceberg lettuce appeared in the United States (1950) and Northern Europe (1970).

Chapter written by Florence CHARLES and Patrick VAROQUAUX.

However, the sensory and hygienic quality of these products, which were not packaged or film-wrapped, was not satisfactory and their shelf-life was limited. To address market requirements and distribution constraints, lettuce was sealed in airtight packages of about four servings per unit to avoid microbial recontamination; product shelf-life was thus extended by around a week. Studies carried out in France between 1981 and 1984 provided a better understanding of the impact of unit production operations on microbial growth and the enzymatic browning of salad leaves, and thus contributed to the development of fresh-cut products. This research led to the development of processing methods that comply with the cold chain (0–4°C) throughout production and distribution until the product reaches the consumer. In France, the industrial production of ready-to-eat salad greens began in 1984–1985 and the first varieties were broad-leaved endive (escarole), curly endive (frisée), Italian red chicory (radicchio), sugarloaf, lamb's lettuce and, later, different types of lettuce such as butterhead, Lollo Rossa, Romaine and Batavia lettuce. Success was immediate: 500 tons in 1985 and 30,000 tons in 1988. Production then underwent a sharp decline due to the non-compliance of production and hygiene conditions. In 1996, the DGCCRF (Directorate General for Competition, Consumer Affairs and Repression of Fraud) proposed a code of good practice for producers of fresh-cut products, which increased production to 40,000 tons in 1999 and more than 100,000 tons in 2005. At the same time, food producers developed production lines specifically for this new industry. Industrial gas companies and film manufacturers offered products designed to optimize active or passive modified atmosphere packaging. National research centers offer the necessary technical assistance.

Since fresh-cut products mainly consist of salad greens, this chapter is limited to the study of these vegetables. The quality control of these products requires a thorough knowledge of the respiratory activity of plants used as well as enzymatic browning caused mainly by mechanical stress. A description of these physiological phenomena is given, followed by the technological stages involved in the production process of fresh-cut products.

8.1. Respiratory activity of plants

Ready-to-eat fresh products are living tissues and must remain so throughout the production chain until consumption. Respiration is the main metabolic activity of post-harvest plants. It involves a series of enzymatic reactions including glycolysis and the citric acid cycle (Krebs cycle) with the consumption of oxygen and the production of carbon dioxide, water and energy (ATP and ADP).

8.1.1. *Measurement and modeling of respiratory activity*

Knowing the respiratory parameters of plants is essential to optimize modified atmosphere packaging. The respiration rate is expressed in millimoles (mmol) of oxygen consumed or carbon dioxide produced per unit of plant weight (kg) and per unit of time (h). It is measured using a static or dynamic method.

In the static method [CHA 89], plants are placed in sealed jars and gas levels are analyzed for three to eight hours. The respiration rate is proportional to the slope of the decrease in oxygen and the production of carbon dioxide. This method is quick and easy, but it cannot be used in a stable atmosphere.

To overcome this problem, [VAR 99] developed a respirometer, based on the dynamic method. In this system, the atmospheric composition in the measuring chambers is maintained by the absorption of carbon dioxide by a sodium hydroxide solution. The oxygen uptake by the plant and the release of carbon dioxide create a vacuum in the measuring chamber. The initial pressure is restored by injecting pure oxygen via a control valve. Thus, the partial pressures of oxygen and carbon dioxide remain constant throughout the measurement. The amount of oxygen delivered by the control valve is measured (opening time x nominal flow rate of the valve) as a function of time. The O_2 respiration rate (RR_{O_2}) can be calculated from the injected O_2 versus time function. The CO_2 respiration rate (RR_{CO_2}) is obtained by measuring the change in conductivity of the sodium hydroxide solution (proportional to the concentration of carbonate) as a function of time.

Several authors have modeled the change in respiration based on the surrounding gas levels. Respiratory activity is currently considered as a series of enzymatic reactions with Michaelian behavior, characterized by the apparent Michaelis constant Km for O_2 (substrate concentration allowing half the maximum reaction rate to be achieved). [PEP 96] studied the effect of carbon dioxide and proposed a non-competitive inhibition equation to model the respiratory behavior of all plants:

$$RR_{O_2} = \frac{RR_{maxO_2} \cdot O_2}{(K_{mappO_2} + O_2)\left(1 + \frac{CO_2}{K_{iCO_2}}\right)} \quad [8.1]$$

with RR_{O2} the O_2 respiration rate (mmol kg^{-1} h^{-1}), RR_{maxO2} the maximum O_2 respiration rate (mmol kg^{-1} h^{-1}), O_2 and CO_2 the partial pressures of oxygen and carbon dioxide (kPa), K_{mappO2} the apparent Michaelis-Menten constant for O_2 (kPa) and K_{iCO_2} the inhibition constant of CO_2 (kPa). From this equation, the double reciprocal plot or Lineweaver-Burk plot (see Volume 2 [JEA 16b]; $1/RR_{O2} = f(1/O_2)$) gives K_{mappO_2} and RR_{maxO2}. K_{iCO_2} is obtained from the graph: $1/RR_{O2} = f(CO_2)$.

The respirometer can also measure the apparent K_m after a slight change in the gas circuit, that is the replacement of oxygen by nitrogen to restore the pressure equilibrium between the measurement and reference cells. In this configuration, the drop in the partial pressure of oxygen is automatically measured in the measuring chamber, under constant carbon dioxide (Figure 8.1). A sufficient polynomial fit, with a coefficient of determination (R^2) of at least 0.99 is used to evaluate RR_{O2} as a function of time.

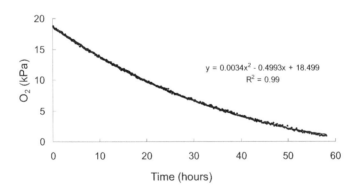

Figure 8.1. Measurement of the oxygen consumption of chicory as a function of time and polynomial fit of the kinetics (20°C) [CHA 05]

Since the partial pressure of oxygen is known at each time, the $RR_{O_2} = f(O_2)$ curve can be drawn. The apparent K_m can be calculated from this data by the double reciprocal plot (Figure 8.2).

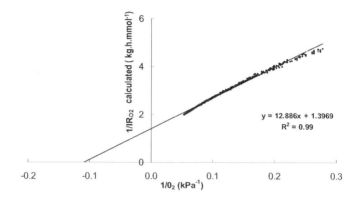

Figure 8.2. *Lineweaver-Burk plot: determination of the apparent K_m (O_2) and RR_{maxO2} of chicory (20°C) [CHA 05]*

8.1.2. *Control of respiratory activity*

The shelf life of plants after harvest is inversely correlated with their respiration rate: if high, the plant quickly consumes its reserves and moves into senescence. To extend product shelf life, it is therefore necessary to reduce respiration. The most common method is to reduce the storage temperature above the point of freezer burn to avoid chilling injury. The effect of temperature on the respiratory chain is characterized by its activation energy (E) based on the Arrhenius equation (see Volume 2) where the natural logarithm of the respiration rate is inversely proportional to the absolute temperature (K). These quantities are difficult to understand, which is why technologists use an approximation, known as Gore's law, based on proportionality between the logarithm of the respiration rate and temperature (°C). According to this law, it is possible to characterize the effect of temperature on respiration by calculating Q_{10} (see Volume 2). The latter corresponds to the ratio of respiration rates obtained at θ and $\theta + 10°C$. Q_{10} is between 2 and 3 for most fruit and vegetables.

The respiratory activity of fresh-cut products can also be reduced by modifying the internal atmosphere of packets and in particular by decreasing the oxygen content. This packaging method is described at the end of the chapter.

8.2. Enzymatic browning

8.2.1. *Mechanism and evaluation*

Wounding of plant tissue results in enzymatic browning, which is chiefly responsible for the reduced quality of ready-to-eat salad greens. The physical stress involved causes cellular relocation and contact between substrates (mainly vacuolar phenolic compounds) and oxidation enzymes (cytosolic or membrane). This change in cellular compartmentalization is essential for browning to occur. The enzymes involved in this process are polyphenol oxidase (PPO) and peroxidase (POD). They enable the formation of highly reactive end products called o-benzoquinones that polymerize to form visible brown pigments called melanins. All the browning mechanisms are covered in Chapter 6 of Volume 1 [JEA 16a]. Phenolic compounds play an important role in these reactions, as well as the key enzyme phenylalanine ammonia lyase (PAL), which forms the basis of the synthesis of phenolic substrates. Studies have shown a strong correlation between PAL activity and the intensity of browning in cut mid-rib sections of iceberg lettuce [TOM 01]. It is now recognized that PAL is induced by the stress of injury and that it catalyzes the conversion of L-phenylalanine to trans-cinnamic acid, ultimately causing browning in these products ([SAL 97], Figure 8.3).

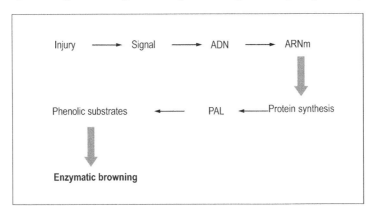

Figure 8.3. *Transcriptional mechanism of the injury signal*

Enzymatic browning can be measured using different methods. Sensory analysis by a trained panel gives an indication of the overall condition of the product in terms of visual quality. However, this method cannot be routinely used and lacks reproducibility. Instrumental measures have therefore been proposed. They can be used on liquid extracts after crushing the tissue, centrifugation and removing the pellet. Pigments, the phenolic content and enzymatic activity can be measured by absorbance, generally between 400 and 490 nm. The color of solid samples can also be analyzed by reflectometry and colorimetry. Thus, image analysis methods were developed for different food products. Samples are scanned or analyzed by digital image or video. The L*a*b* system is the most common method for food since it provides information directly related to the sensory perception of the color of the product (see Chapter 9 of Volume 1): L* (from 0 to 100) represents lightness, a* and b* (from –128 to +127) are the chromatic parameters and describe color respectively from green to red and blue to yellow. Other systems such as RGB (Red Green Black) and CMYK (Cyan Magenta Yellow Black) can also be used.

8.2.2. Prevention of enzymatic browning

In general, browning is controlled by reducing injury, which also promotes the entry and growth of microorganisms. Thus, knife blades and machine discs must be as sharp as possible.

Oxidation is limited by removing and washing the injured area of fresh-cut products. Water jet cutting can significantly reduce browning by automatically cleaning damaged cells.

Other prevention techniques are based on the use of chemical inhibitors. Thus, reducing compounds such as ascorbic acid, citric acid or sulfur compounds (sulfite, cysteine) are used in fresh-cut products for their potential inhibitory effect on PPO. Ascorbic acid is commonly used in the production of fresh-cut fruit while sulfites are used up to a maximum of 50 ppm on certain cooked products such as potatoes and mushrooms. In some cases, the browning of cut iceberg lettuce has been prevented by PAL inhibitors.

Modified atmosphere packaging is an effective method of reducing enzymatic oxidation reactions in fresh-cut products. In the case of lettuce, low oxygen levels (1 to 5%) and high carbon dioxide levels (10 to 15%) inside the bags can significantly reduce browning.

Heat treatment can also be used to prevent browning. Initially used to reduce the microbial count of fruit and vegetables, this technique has proven effective in combating certain physiological defects in plants. In iceberg lettuce, for example, short heat shocks (> 45°C) can reduce browning in the mid-rib section after cutting [SAL 00]. This phenomenon was correlated with an inhibition of PAL synthesis and appears to result from a redirecting of protein synthesis away from stress proteins like PAL to the synthesis of heat shock proteins (HSP). This diversion of PAL synthesis lowers the accumulation of phenolic substrates and therefore tissue browning (Figure 8.4). However, heat treatment is difficult to apply on an industrial scale since it can cause tissue burning and its effect varies depending on the variety and physiological state of the plant.

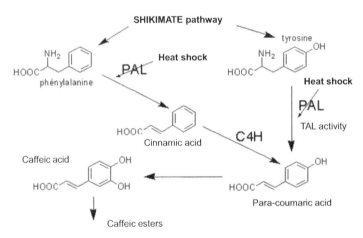

Figure 8.4. *Diagram of the operating principle of heat treatment on enzymatic browning in plants [SAL 00]*

8.3. Unit operations in the production of fresh-cut products: main scientific and technical challenges

The production of fresh-cut products includes a series of defined steps that are shown in Figure 8.5 for ready-to-eat salad greens.

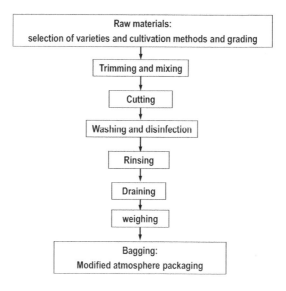

Figure 8.5. *Unit operations in the production of ready-to-eat salad greens*

The first rule in production is the principle of "forward flow". According to this rule, clean products should not come into contact with contaminated products or waste (Figure 8.6). "Forward flow" does not imply linearity: direction changes in the production chain are possible, even necessary. This principle simply removes potential meeting points.

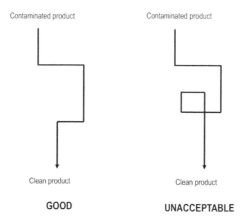

Figure 8.6. *The principle of "forward flow"*

In practice, to avoid cross-contamination, different unit operations are carried out in separate rooms. Since cold temperatures are essential in maintaining the quality of ready-to-eat fruit and vegetables, the temperature in the different rooms must be controlled in accordance with the regulations (Figure 8.7).

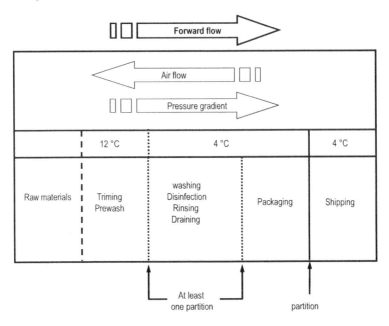

Figure 8.7. *Segmentation and temperature gradient in the processing units*

8.3.1. *Raw materials: selection of varieties and cultivation methods*

The choice of raw material is a key factor that determines the final quality of the product. Varietal susceptibility and cultivation conditions influence physiological behavior and therefore the final quality of the product. For example, "brown stain", which is characterized by pinkish-brown spots on the mid-rib of iceberg lettuce or the leaves of butterhead lettuce, is linked to varietal susceptibility. Enzymatic browning is closely correlated with cultivation conditions.

However, no physiological selection criteria have yet been validated for obtaining well-suited varieties for fresh-cut production. The only selection criteria for butterhead lettuce intended for processing is resistance to downy mildew (Bremia lactucae), morphology (the ability of leaves to separate from each other), weight (so as to obtain the best possible technical yield) and seasonal cultivation. Research is currently being conducted in France to develop tests for detecting raw materials susceptible to enzymatic browning or physiological disorders after processing, and therefore making it possible to reject unsuitable batches. These evaluation tests should be carried out on raw materials during grading, on all batches intended for ready-to-eat processing.

8.3.2. *Raw material quality control: grading*

When raw materials arrive at the factory, they are immediately inspected, i.e. analyzed from a quality perspective according to specifications supplied to producers. The main criteria are general appearance including turgidity of the leaves, absence of foreign material (insects, stones, roots, etc.) and physiological or microbial disease, technical yield (percentage of product that can be used in processing) as well as compliance with regulations on pesticide residues and nitrate levels. Some batches of raw materials intended for longer storage (Italian red chicory, cabbage) are examined for the presence of pathogenic bacteria such as *Listeria monocytogenes*. Depending on the result, the batch may be accepted, partially rejected or refused. All the criteria measured are recorded for the purpose of traceability. The selected lettuce is then placed in cold storage at $2 \pm 1°C$ for up to two days before being processed.

The correct use of all data on varieties, cultivation methods, and quality measures for raw materials as well as end products could be a valuable tool for optimizing the quality of ready-to-eat products.

8.3.3. *Trimming and mixing*

The first processing operation is trimming, which involves removing the inedible parts of the fruit and vegetables. In the case of leaf vegetables, the stem and outer dark green leaves are removed. This is generally carried out manually using a sharp stainless steel knife. However, with broad-leaved and curly endives, trimming may be partially mechanized. In an automatic trimming machine, the product is held at the base by a mandrel. As the device rotates, centrifugal force causes the leaves to spread out; the outer layers are then removed by a rotating disc. The advantage is a reduction by half in the cost of labor; however, this device can only handle salad plants with similar shapes and mechanical behavior.

The remaining leaves are then separated on conveyor or hydraulic belts [VAR 02]. Figure 8.8 shows a trimming table consisting of a hydraulic belt for transporting trimmed fruit and vegetables (top) and a conveyor belt for removing waste (bottom).

Figure 8.8. *Trimming table with hydraulic belt*

Salad leaves are mixed at this stage. The final proportion of each type of salad plant is added to the table, taking into account its yield. To reduce costs, each type of salad plant could be prepared separately before mixing on demand using automatic weighing scales. In this case, salad leaves should be properly washed before cold storage, which should not exceed 24 hours.

8.3.4. *Cutting*

Cutting gives the product its final shape. For example, chicory leaves are cut in lengths of 3 to 6 cm by rotary blades (perpendicular to flow), discs (parallel to flow) or water jet. Water jet cutters (Figure 8.9) position the midrib parallel to flow and reduce the thickness of the product to only two or three layers [BEG 96]. To be effective, the water jet pressure must be between 50 and 100 MPa depending on the type of product. In most production lines, the product immediately drops into a wash tank after cutting.

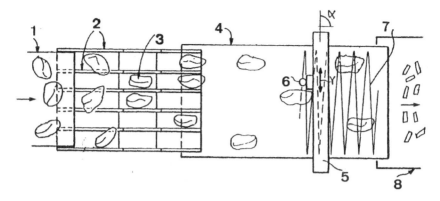

Figure 8.9. *Diagram of a water jet cutting system [BEG 96]*

8.3.5. Washing and disinfection

Washing must be carried out immediately after cutting to reduce browning. This has been highlighted in a study on the browning of apples either cut in air followed by immediate immersion in a wash tank, or directly in water ([KUC 93], Figure 8.10). Browning was measured by reflectometry as a function of storage time in air at 25°C. Unlike slices cut in air that quickly turn brown, apples cut in water remain colorless longer. This phenomenon can be explained by the flushing of cell sap from the injured tissue when cut in water. Thus, enzymes and browning substrates are immediately removed whereas in air, exudates are partially absorbed by capillary action in the internal tissues, thereby inducing browning. For fresh-cut salad greens, hydraulic trimming tables (Figure 8.8) that allow the immediate washing of cut products can be used to limit discoloration and facilitate the separation of butterhead lettuce, the leaves of which are not subsequently cut.

8.3.5.1. Traditional cleaning and disinfection methods

In France, cut products are disinfected using chlorinated water. The concentration of active chlorine, set at 120 ppm in 1988, was reduced to 80 ppm in 1992. The contact time is generally limited to one minute. Chlorine is used in the form of sodium hypochlorite or chlorine gas. The latter is more difficult to use but is more effective. The antimicrobial efficiency of chlorine in aqueous solution is pH-dependent due to hypochlorous acid (HOCl) and not its ionized form (ClO$^-$). The lower the pH, the greater the proportion of hypochlorous acid and therefore the more effective the treatment. In processing units, the concentration of active chlorine and the pH of the disinfectant bath are regularly measured because cell sap released by plants after injury reduces the oxidizing power of the bath. Figure 8.11 gives an example of automatic washing: the disinfectant bath is mixed by a tangential injection of air or by water jets, and the transfer time is controlled by shafts with blades. At the outlet of the machine, a roller cage removes insects and small plant debris. Disinfected leaves are then rinsed with potable water containing less than 0.5 ppm of active chlorine in accordance with regulations.

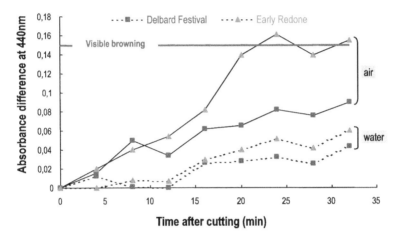

Figure 8.10. *Browning of apple slices cut in air or water, measured by the difference in absorbance at 440 nm as a function of storage time, in air at 25°C [KUC 93]. For a color version of this figure, see www.iste.co.uk/jeantet/foodscience.zip*

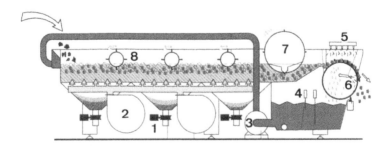

Figure 8.11. *Diagram of a traditional washer. For a color version of this figure, see www.iste.co.uk/jeantet/foodscience.zip*

8.3.5.2. Alternatives to the use of chlorine

In some countries in the European Union (Belgium, Germany and the Netherlands), the use of chlorine for washing fresh produce is prohibited. In France, chlorine is not authorized as a processing aid, but is presently allowed for this particular purpose. The current trend is to eliminate chlorine. As a result, new disinfection methods and materials have been developed.

In collaboration with industry [BEG 98], INRA has participated in the design of a new washer that includes a succession of turbulent and laminar flows (Figure 8.12). In this method, salad plants are fed into a feed hopper filled with a solution containing 5 to 8 ppm active chlorine. The chlorine concentration is adjusted using a specific electrode and the solution is filtered prior to partial recycling. The submerged product makes it way by gravity into the water stream. The bottom of the first section of the channel has protrusions that create a succession of turbulent and laminar flows. At the end of this first section, the product is separated from the chlorinated water by passing over a perforated conveyor belt. The water is then returned to a buffer tank after filtration and adjustment of the chlorine concentration. The salad plants fall into a tank filled with potable water. The base of the conveyor belt in the rinsing section also has protrusions. After draining, the product is then transported to the drying system.

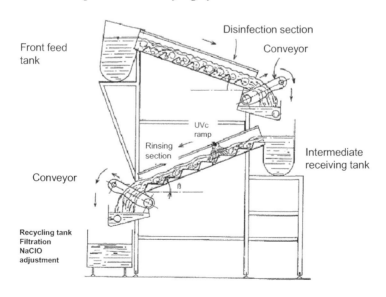

Figure 8.12. *Diagram of a cascade washer [BEG 98]*

Alternatives to chlorine have been tested: ozone is an excellent disinfectant but its use is limited due to its high toxicity and volatility; peracetic acid was also used, especially on grapes, but it has a lower disinfectant level than chlorine. The advantage of this compound is that it degrades into harmless by-products with no long-term toxicity. Hypothiocyanite ion (OSCN⁻), present in saliva and tears, is another potential disinfectant. This natural antimicrobial is produced by an enzyme called lactoperoxidase in the presence of hydrogen peroxide. The effectiveness of OSCN⁻ has been demonstrated in the destruction of several pathogenic organisms like *Yersinia enterocolitica* and *Listeria monocytogenes*. This system does not produce by-products and is very easy to use.

8.3.5.3. Inhibiting microbial growth

The outer surfaces of fruit and vegetables are naturally rich in microorganisms (yeasts, moulds and bacteria), the type and proportion of which depend on the physicochemical characteristics of the product [NGU 94]. During processing stages like trimming and cutting, contamination can derive from the product surface or the environment. It is therefore essential to keep machines and premises meticulously clean. In addition, it is necessary to reduce the contamination of raw materials and processed products to ensure the desired shelf life. Washing products in 80 ppm of chlorine reduces the original flora of salad plants by 90 to 99% and is more effective on yeasts and moulds than on the total aerobic mesophilic flora. Compliance with the cold chain slows down microbial growth. The use of modified atmosphere packaging also affects the development of microbial flora, e.g. moulds are inhibited by low levels of oxygen and carbon dioxide levels above 10%. Current EU regulations (EC 2073/2005) require testing for *Listeria monocytogenes* and *Salmonella*, and impose hygiene criteria for *E. Coli*. However, fresh-cut products do not pose a serious risk to consumers since they quickly degrade if contaminated by phytopathogens. Nevertheless, a risk still exists and hygiene precautions are necessary like in any food sector.

8.3.6. *Draining and drying*

The purpose of this step is to reduce the amount of water on the surface of the plant product so that it can be packaged at optimal humidity. Excess water in bags is highly detrimental to the final quality of the product since it promotes microbial growth, especially on the packaging material. Moreover, the presence of water in salad bags reduces consumer acceptability. The draining method must be optimized so that the salad leaves do not contain more than 1 to 3% residual water compared to the raw material.

Two drying methods are currently used: centrifugation and hot air tunnel drying. The most common centrifuges are semi-automatic and automatic. The rotation speed is low during feeding, increases when loading is complete and slows down again once centrifugation is finished. Meanwhile, the double bottom of the centrifuge drum is gradually driven upwards to discharge the dried salad leaves into a chute leading to a weighing scales. The centrifugal force should be carefully adapted to the type of salad since excessive centrifugation causes bruising to the leaves. Very delicate salad plants like butterhead lettuce or young shoots require very gentle draining.

Air drying tunnels have the advantage of not causing mechanical stress to the product. This is a recent technique developed in Italy and now commonly used in many processing plants in Europe and the United States. The drying tunnel has three sections: in the first pre-drying section, water droplets are mechanically removed by powerful fans; the second section consists of a series of vibrating tables leading the product through a stream of warm air (around 30°C); in the third section, the salad leaves are cooled to between 1 and 4°C in fresh air. However, the equipment used is large and therefore not suitable for all premises.

8.3.7. Weighing

The weighing of products is generally automated. The multihead weighing machine used is not specific to the fresh-cut sector. It consists of a vibrating feeder that regularly feeds a weigh hopper attached to the outside of the machine. A level sensor identifies which combination contains the weight closest to the target weight and controls the opening weigh hoppers and filling of the bags. The salad leaves are then released by gravity.

8.3.8. Bagging

The last stage of the production process is bagging. This stage is usually synchronized with weighing to obtain bags of precise weight. Bagging is carried out in a clean, cold (1–2°C) packaging room, separate from the washing room and storage room.

Packaging ensures several physiological functions, which are:

– to limit drying of the products;

– to reduce respiration without suffocating plant tissue;

– to slow down maturation and senescence;

– to halt physiological and microbial spoilage

In the production of fresh-cut salad greens, vertical or horizontal packaging machines are used. In vertical packaging machines, the first part of the machine consists of a "gooseneck", a cylindrical tube around which the packaging film is attached. Side and end sealing units form the bags, which are filled in the cylindrical tube by the opening of a flap attached to the multihead weigher. Finally, gas may or may not be injected into the bags before being sealed closed.

Horizontal packaging machines (flow pack) are also used for products in trays. In this case, the products arrive on a flat conveyor belt and are film-wrapped or capped as they move along the machine.

8.4. Modified atmosphere packaging

Generally, low oxygen and high carbon dioxide levels improve the preservation of fresh fruit and vegetables [KAD 89]. However, the excessive reduction of oxygen can lead to anoxia and induce fermentation reactions detrimental to product quality. It is therefore necessary to predetermine the optimal atmospheric composition for product preservation.

These levels are estimated by the method of controlled atmospheres. Fruit and vegetables are placed in sealed chambers connected to a control device that maintains a constant atmospheric composition. The quality of products is analyzed over time. Guidelines exist with optimal atmospheres for a large number of products. However, this assessment is difficult because the ideal atmosphere depends on the type of fruit or vegetable, the level of ripeness, the conditions of the external environment and the type of spoilage to be inhibited.

The atmospheric composition in bags of fresh-cut products results from gas flow from the respiratory activity of fruit and vegetables and gas diffusion through the film (Figure 8.13).

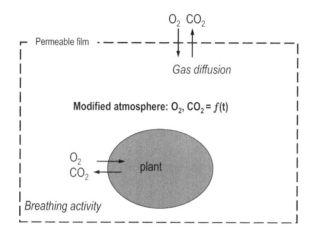

Figure 8.13. *Principle of modified atmosphere packaging*

8.4.1. *Diffusion of gases through packaging*

8.4.1.1. *Measuring and modeling gas diffusion through packaging film*

Knowing the permeability (P_e, usually expressed in mol m^{-1} s^{-1} Pa^{-1}) or permeance (ratio of permeability to film thickness e; mol m^{-2} s^{-1} Pa^{-1} or mL m^{-2} day^{-1} atm^{-1}) of packaging film is essential in determining the choice of film (non microporous or micro-perforated) in order to obtain the desired gas mixture at equilibrium. It is possible using permeability cells to assess the diffusion of gases through a disk of film with a given surface area and thickness, knowing that gas diffusion through a dense non-porous film follows Fick's first law (see Volume 2):

$$\frac{dn_x}{dt} = A \frac{P_e}{e} \Delta P \qquad [8.2]$$

where $\frac{dn_x}{dt}$ is the flow of gas X through the film (mol s^{-1}), P$_e$ is the gas permeability coefficient of the film (mol m^{-1} s^{-1} Pa^{-1}), A is the surface area of the film (m^2) and e its thickness (m), and ΔP is the difference in gas partial pressure between each side of the film (Pa).

The permeability values depend on the polymer structure, but also on temperature, humidity and environmental conditions. The films are also characterized by their selectivity S$_e$ (= P$_{eCO2}$/P$_{eO2}$).

8.4.1.2. Different types of packaging films

To create a specific equilibrium atmosphere that meets the requirements of a certain type of fruit or vegetable, it is necessary to have a wide range of permeability and selectivity of packaging films. Plastic film is commonly used for food such as low density polyethylene (LDPE), polypropylene (PP) and biaxially oriented polypropylene (BOPP). These polymers are generally single layer, with a thickness of 15 to 35 µm. Characteristics include optic quality, mechanical quality, sealability and low cost, and their selectivity S$_e$ generally varies between 3 and 10. However, they are not always suitable for fruit and vegetables. For example, BOPP (35 µm) with a permeance of 900 mL O$_2$ m^{-2} day^{-1} atm^{-1} and 4,500 mL CO$_2$ m^{-2} day^{-1} atm^{-1} can correctly preserve salad greens at low temperatures (0 to 4°C), but not products like spinach, cauliflower and grated carrot, which have higher respiration rates. To overcome this problem, several techniques have been developed to modify permeability.

Physically altering the film, such as perforation, can provide solutions. Four micro-perforation techniques currently exist in Europe:

– electrostatic discharge, which was the first method used but does not provide reproducible results;

– laser, which is the best technique, but is more expensive;

– cold needle, which tears the film, making it difficult to estimate permeability;

– hot needle, which makes overly large holes, contrary to recommendations of a diameter between 20 and 100 µm to prevent microbial contamination (Figure 8.14).

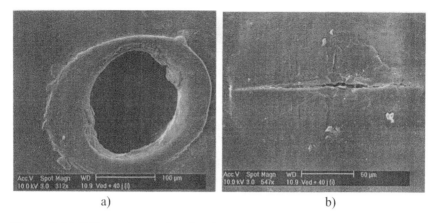

Figure 8.14. *Different types of perforation. a) laser; b) cold needle (photo INRA)*

The permeability of micro-perforated films is modified and selectivity is reduced depending on the density and size of the perforations. Thus, the selectivity of micro-perforated films is close to 0.85 following Graham's law whereby the diffusion rate of a gas is inversely proportional to the square root of its density. When selectivity is close to 1, it is not possible to obtain atmospheric equilibrium, that is simultaneously low in oxygen and carbon dioxide. In addition, micro-perforation also reduces Q_{10}, which can, in the case of rupture of the cold chain, cause anoxia and fermentation reactions in fruit and vegetables since they have higher Q_{10} levels [KAD 86]. Thus, other films have been proposed to diversify the range of permeability. Hydrophilic film, for example, is sometimes used for packing fruit and vegetables. This type of film is made of polyolefins and polymers such as polyester-polyether, polyether-polyamides (Pebax®) or others. Hydrophilic film has high selectivity (above 10) and allows a high level of water vapor diffusion. Moreover, when it is hydrated, selectivity varies. Research is currently underway to investigate the ability of new protein materials to form biopolymers [GON 94].

8.4.2. *Change in gas content in modified atmosphere packaging*

According to the principle of modified atmosphere packaging presented in Figure 8.13, the gas content inside the package depends on the diffusion of gases through the film and the respiratory activity of the plant product. After a transition phase where the oxygen content decreases and the carbon dioxide content increases, a stationary phase is established during which diffusive exchanges through the film compensate exactly for the production of carbon dioxide and the consumption of oxygen by the product (Figure 8.15).

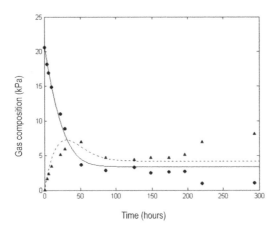

Figure 8.15. *Change in levels of oxygen (●) and carbon dioxide (▲) in a LDPE package containing 500 g of endives (20°C) [CHA 05]*

Provided that the respiratory ratio is 1 (i.e. $IR_{O2}=IR_{CO2}$), the composition of the atmosphere at equilibrium can be easily determined since it follows the following law (Ozdemir *et al.* [OZD 05]):

$$[O_2]_{eq} + S_e[CO_2]_{eq} = 21 \qquad [8.3]$$

with $[O_2]_{eq}$ and $[CO_2]_{eq}$ the gas concentrations at equilibrium (kPa) and S_e the selectivity of the film (P_{eCO2}/P_{eO2}). Knowing the selectivity of a film therefore makes it possible to identify the gas content in the package and at equilibrium. However, it is also important to know the change in gas content: the positive effect of an atmosphere in the steady state may be lost if the product is exposed to unsuitable gas levels during the transition phase. This is the case with mushrooms that undergo discoloration if subjected to peaks in carbon dioxide during the transition phase. Lettuce is so sensitive to ambient air that oxygen levels must be quickly reduced during packaging. Either pure nitrogen or a mixture of O_2 and CO_2 are injected into the bags before sealing so that the internal atmosphere contains 1–3 kPa of O_2 and 8–14 kPa of CO_2. These active modified atmosphere packaging techniques protect the injured tissue from premature browning. In this case, the gas levels at equilibrium are not modified and a certain film should be chosen so as to obtain the optimal atmospheric equilibrium for product preservation.

Models based on the mass balance described by simple mathematical equations have been proposed to predict changes in the gas content of packages. The following equation shows the oxygen level:

$$\frac{dO_2}{dt} = \frac{P_{eO_2} A}{e} (O_2 \text{ ext} - O_2 \text{ int}) - \frac{IR_{O2}}{3600} m \qquad [8.4]$$

where P_{eO_2} is the permeability coefficient of the film to oxygen (mmol m^{-1} s^{-1} Pa^{-1}), A and e respectively are the surface area (m^2) and thickness (m) of the packaging film, O_{2ext} and O_{2int} are the partial pressure of O_2 outside and inside the packaging, respectively, IR_{O2} is the O_2 respiration rate (depending on the partial pressure of O_2 and CO_2, in mmol kg^{-1} h^{-1}) and m is the weight of the product (kg).

Using the obtained simulations, it is possible to identify the optimum film permeability based on the respiratory parameters of the plant product and the desired packaging conditions (gas content and temperature).

8.5. Conclusion

Fresh-cut products are particularly suited to modern lifestyles and consumption habits due to their freshness, ease of use and nutritional properties. Progress in the fresh-cut sector depends on technological and scientific advances that optimize unit production operations and improve the control of post-harvest plant physiology. Research is necessary to further improve the shelf life of such products.

It is important to select suitable plant varieties for this sector. Since it is possible to identify groups of susceptible or resistant strains to certain storage conditions such as relatively high levels of carbon dioxide, plant breeders could use genomic tools for plant selection. In addition, systematic studies could be conducted on the quality of processed salad plants depending on exact harvest conditions. The quality (browning, physiological disease due to carbon dioxide, etc.) of bagged salad greens could thus be evaluated according to the different harvesting conditions of the raw materials.

Injury remains a critical point in the production process on a technological level. Washing must also be optimized by exploring alternatives to chlorine. Research should focus on plastic polymers, biopolymers and techniques for modifying packaging film permeability to obtain optimal storage atmospheres. In addition, the behavior of films depending on environmental conditions must be adjusted to the plant product to avoid, for example, anoxia that may arise if the cold chain is broken. The ability to actively modify the atmosphere can also improve the quality of products by significantly reducing the length of the transition phase. Currently, gas flushing is carried out for fresh-cut salad products but other techniques could be used like the use of gas absorbing bags. Finally, preservation treatments may be combined with modified atmosphere storage. Heat treatment is a potential solution provided that the treatment is used correctly to inhibit browning and simultaneously prevent necrosis caused by burns.

The mechanisms leading to the degradation of fruit and vegetables are still not fully clear. During browning, for example, specific enzymes have been identified but the effect of gas levels on their activity is not understood. The mechanism explaining the effect of carbon dioxide on preservation is still vague: effect on pH and enzyme activity, etc. It is therefore necessary to improve the knowledge of post-harvest plant physiology and the effects of preservation treatments at both cellular and gene expression level.

The growth in fresh-cut products therefore requires multidisciplinary research in technology, physiology, microbiology and genetics. This research should be carried out in the near future to meet the needs of a rapidly developing market with the introduction of innovative products (ready-to-cook meals, various salad mixes, mixed fruit, etc.).

PART 3

Food Ingredients

9

Functional Properties of Ingredients

Food is a heterogeneous system both in its composition and structure; it may consist of several phases with differences in density and chemical potential, at the roots of its thermodynamic instability. The challenge facing food producers is to stabilize such systems during production and storage. Mechanical (transport) and thermal stress, with or without a change in state (refrigeration, freezing, heating), should not affect the quality of the food. Knowing the physicochemical properties affecting the structure of foods and their development over time helps to identify ingredients manufacturers have to use in order to control food quality and stability. Ingredients can contribute to texture development through their thickening or gelling properties, to air or fat dispersion (foam, emulsions) and to the stability of multi-phase systems. Since the functional properties of ingredients are an expression of their physico-chemical properties, functionalities can be traced to molecular interactions (interactions between constituents and between constituents and water) in solution or at interfaces; these interactions are dependent on molecular or supramolecular structures in the case of proteins and polysaccharides compounds as well as the ionic environment (pH, ionic strength, type of ions).

Chapter written by Gérard BRULÉ and Thomas CROGUENNEC.

9.1. Interactions with water: hydration and thickening properties

9.1.1. *Types of interaction*

Interactions between water and constituents occur in ionizable groups capable of solvation such as acid ($-COO^-$, $-O-PO_3^{2-}$, $-O-SO_3^-$) and amino groups ($-NH_3^+$, $-NRH_2^+$) present in proteins and polysaccharides, or in uncharged polar groups ($-OH$, $-COOH$, $-CONH_2$, $-NH_2$, $-SH$) capable of forming hydrogen bonds with water. Non-polar aliphatic or aromatic groups ($-(CH_2)_n-$, $-C_6H_5$) can help structure water in their immediate environment due to their hydrophobic nature. The ability of constituents to hydrate strongly depends on the physicochemical properties of the solvent, in particular pH, ionic strength, the type of ions, the dielectric constant and temperature; these factors affect the level of ionization of acidic and basic groups as they determine their apparent pK.

In the case of amphoteric components (amino acids, peptides, proteins), the level and type of ionization change throughout the pH range; the overall charge is zero at the isoelectric point (pI), positive below pI and negative above it. The hydration and solubility of amphoteric molecules such as proteins are generally low at pI. Hydration may also depend on the structure of protein and polysaccharide compounds; the secondary structure of proteins (β-sheets, α-helix) and the crystalline or semi-crystalline structure of polysaccharides involves hydrogen bonds between amide groups of peptide bonds (-CONH-) and hydroxyl groups (-OH) of carbohydrate units, which limits hydration sites; the destructuring of these compounds by rupture of the hydrogen bonds can release polar sites that become accessible to water (gelatinization of compounds such as starch by heat treatment). The tertiary structure of proteins is strongly influenced by the presence of hydrophobic groups and the properties of the solvent (temperature, dielectric constant, pH and ionic strength). In their native state, non-polar groups are generally located within the structures whereas polar groups are exposed to the solvent, facilitating their hydration; denaturation by heat treatment or modification of the dielectric constant can result in an unmasking of hydrophobic sites and a reduction of solubility.

9.1.2. Influence of hydrophilic components on water availability and mobility

Water associated with ionic organic or inorganic groups or non-ionic polar groups via hydrogen bonds exhibit varying solvent properties given the different binding energy. Hydrophilic components can be seen as a_w depressants, which is an important part of the biological stability of food. Water binding properties can be characterized by sorption isotherms such as those described in Volume 1 (Chapter 1) [JEA 16a]; sugars and salts are the most common depressants.

Another property of immobilized water is its non-freezable character. The reduction in freezable water lowers the energy cost of freezing and reduces freezing and thawing time. The presence of hydrophilic components in food also increases the glass transition temperature (T_g), which limits the crystallization of water during freezing and the vitrification of frozen products (formation of ice crystals) when the T_g value is higher than the storage temperature.

9.1.3. Influence of hydration on the solubilization, structure and mobility of compounds

The solubility of protein and carbohydrate compounds is closely linked to their hydration capacity, which depends on their hydrophilic/hydrophobic balance, their structure and the physicochemical conditions of the solvent. It has already been mentioned that the structural changes caused by heat treatment for example can promote the hydration of compounds; at the same time, hydration can change the balance of intra- and intermolecular reactions making polymer chains more flexible and mobile, which is key for the functionalities at interfaces.

Most ingredients are sold in dry form: the ability to rehydrate is an essential quality criterion for their functionality; the rate of rehydration depends on the powder composition and its physical characteristics. Water penetration is faster the larger the contact area with the dispersing medium and therefore the smaller the particle diameter and the higher their levels of porosity and capillarity. Water migration within the particle strongly depends on the hygroscopicity of its components, and is better if the powder contains hydrophilic components such as salts and sugars. In the case of protein-rich or polysaccharide-rich ingredients, a viscous layer could form on the powder particle surface by the hydration of surface molecules, which limits water mobility and slows down its transfer to the center of the particle. The formulation of ingredients before drying should be based on its dehydration and rehydration ability and its sensitivity to heat treatment to limit a reduction in functional properties.

9.1.4. *Effect of the hydration of components on rheological properties*

The hydration of macromolecules reduces the amount of solvent water resulting in a concentration of organic and inorganic solutes in the residual solvent phase, which can induce an increase in viscosity. Conformational changes of macromolecules after hydration may lead to an increase in their hydrodynamic volume or to their unfolding, which increases viscosity and viscoelasticity. When compounds are amphoteric (protein), viscosity can be controlled by pH, ionic strength and the type of counter-ions that determine electrostatic attractions and repulsions.

9.2. Intermolecular interactions: texture properties

9.2.1. *Aggregation/gelation by destabilization of macromolecules or particles*

The stability of macromolecules or particles in solution is highly dependent on the balance between attractive (Van der Waals) and repulsive forces (electrostatic, steric) shown in Figure 9.1. Electrostatic repulsion depends on the surface potential and the thickness of the hydration and counter-ions double layer surrounding charged macromolecules or particles. These physical properties are determined not only by the type of ionizable groups of macromolecules or particles but also by the physicochemical characteristics of the solvent phase (pH, ionic strength). Steric repulsion is generally determined by interactions between chains at the surface of macromolecules or particles and the solvent.

The formation of aggregates or gels is generally caused by a reduction in repulsive forces between macromolecules or particles, which is a result of changes such as: a reduction in pH for anionic polysaccharides, an adjustment of pH to the pI of proteins, the addition of ions (increase in ionic strength, screening of charges, interactions with ionisable groups) and the hydrolysis and release of ionized fragments. Some treatments also change the structure of compounds, which may result in an increase in hydrodynamic volume and surface hydrophobicity in the case of proteins; these structural changes promote hydrophobic interactions and increase attractive forces. The rheological (viscous and elastic component) and optical (opacity, transparency) properties are dependent on the level of structural changes before the creation of intermolecular bonds.

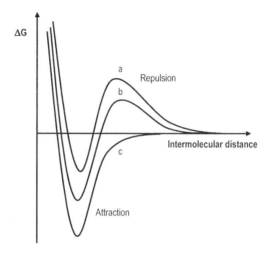

Figure 9.1. *Interactions between macromolecules or particles for increasing electrolyte concentrations (a, b, c)*

9.2.2. *Aggregation/gelation by covalent cross-linking*

Thermal aggregation and gelation may result from covalent interactions, particularly with proteins. Thiol and amine groups are highly reactive, especially in a slightly basic medium; by nucleophilic attack of thiolates (-S$^-$) on disulphide bonds (-S-S-), there is a rupture of intramolecular -S-S- bonds and the formation of new intermolecular disulfide bonds as shown in Figure 9.2. The rupture of some S-S bonds changes the secondary and tertiary structure of proteins, which can destabilize the dimers or polymers formed. Under certain conditions during the heat treatment of proteins, dehydroalanine is produced from serine or cysteine followed by a nucleophilic attack of the NH_2 group of lysine, which leads to the formation of lysinoalanine and creates covalent intermolecular bonds (Figure 9.3).

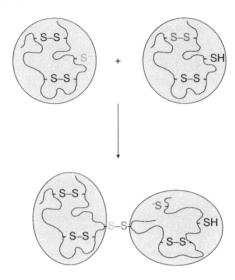

Figure 9.2. *Crosslinking of proteins by nucleophilic attack of thiolate on disulphide bonds*

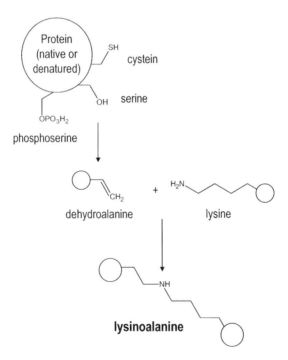

Figure 9.3. *Formation of lysinoalanine*

9.2.3. Sol-gel transitions

Some polysaccharides have loose helical chains in dilute solution at high temperature, which, during cooling, can combine to form double and triple helices by hydrogen bonding to form a three-dimensional gel network. When these chains are polyelectrolytes (alginates, pectin), divalent cation bridging may contribute to the formation of the network. At low electrolyte concentrations, anionic groups ($-CO_2^-$, $-O-PO_3^{2-}$) may limit intermolecular associations and prevent gel formation by electrostatic repulsion.

9.2.4. Influence of denaturation kinetics and molecular interactions

In the absence of an energy barrier, the rate of molecular interactions is diffusion-limited; the disappearance of molecules (concentration C) is generally a second order reaction, expressed as:

$$\frac{dC}{dt} = -k_a C^2 \qquad [9.1]$$

When the macromolecules or particles are polyelectrolytes, the aggregation rate constant k_a also depends on the surface potential, which contributes to repulsive forces. It is therefore possible to change this rate constant by modifying the characteristics of the solvent or dispersing phase (pH, ionic strength), which would raise or lower the energy barrier.

The denaturation of macromolecules by weakening hydrogen/ionic interactions is usually a first-order reaction:

$$\frac{dC}{dt} = -k_d C \qquad [9.2]$$

where k_d is the denaturation rate constant. Denaturation sometimes results in an unfolding of the macromolecules and exposure of hydrophobic sites (proteins).

When aggregation kinetics is very fast compared to denaturation kinetics (low energy barrier), slightly denatured macromolecules or particles aggregate (Figure 9.4); a concentration of macromolecules above the critical gelling concentration (path A; Figure 9.4) yields an opaque gel with a granular appearance (coagulum). Below this critical gelling concentration (path B), the result is a turbid solution of large and random aggregates. When denaturation is favored over aggregation, translucent solutions of linear aggregates or smooth and transparent gels are obtained depending on whether the concentration is above (path C) or below (path D) the critical gelling concentration (Figure 9.4).

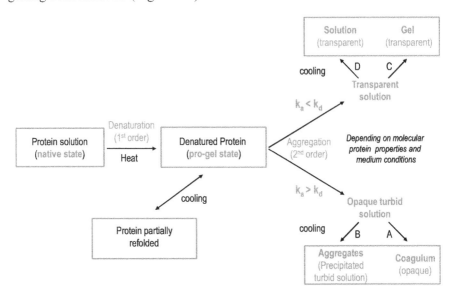

Figure 9.4. *Influence of denaturation kinetics and molecular interactions on the formation of aggregates or gels (according to [TOT 02])*

9.3. Interfacial properties: foaming and emulsification

9.3.1. *Interfacial tension*

In a pure liquid such as water, all molecules are in an equilibrated attractive force field (Figure 9.5(a)). However, at the interface between two phases (liquid/gas, liquid/immiscible liquid), the molecules are in an asymmetric environment and the attractive forces exerted by each phase on the molecules at the interface are different. The result is an interfacial energy or interfacial tension (σ; N m^{-1}) corresponding to an energy per unit area (J m^{-2}), for example the interfacial tension of pure water at 20°C is 73 mN m^{-1}.

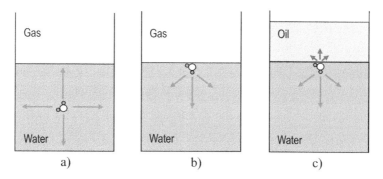

Figure 9.5. *Result of attractive forces in a), a liquid b) at liquid/gas interfaces and c) at immiscible liquid/liquid interfaces*

Organic molecules tend to concentrate at the interface and reduce interfacial tension if they have hydrophilic and hydrophobic regions (amphiphilic character; Figure 9.6). Interfacial tension decreases with the concentration of the solute in solution up to a certain level (critical micelle concentration, CMC) when amphiphilic molecules reorganize into micelles.

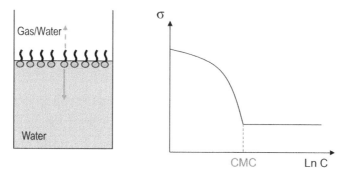

Figure 9.6. *Concentration of amphiphilic molecules at the interface and reduction in surface tension*

The diffusion and adsorption of molecules at the interface is often slow. Macromolecules in particular replace solvent molecules and subsequently undergo conformational changes that can result in intermolecular interactions and in some cases the formation of a cohesive interfacial film. The interfacial concentration of the absorbed solute C_i can be obtained by Gibb's law:

$$C_i = \frac{C}{RT}\frac{d\sigma}{dt} \qquad [9.3]$$

where C is the concentration of the solute in solution, R is the perfect gas constant, T is absolute temperature and σ is surface tension.

9.3.2. *Surfactants*

Surface-active compounds or surfactants are molecules consisting of a polar hydrophilic head (ionic or non-ionic) and a non-polar hydrophobic tail. Ionic polar groups are usually $-CO_2^-$, $-O\text{-}SO_3^-$, $-SO_3^-$, $-O\text{-}PO_3^{2-}$, $-NH_3^+$, and $-NR_3^+$ functions, and non-ionic polar groups are OH functions. Hydrophobic groups are generally aliphatic or cyclic hydrocarbon chains. Protein compounds usually have good interfacial properties as they are composed of hydrophobic regions (presence of proline, leucine, isoleucine, tryptophan and phenylalanine) and hydrophilic regions (presence of aspartic acid, glutamic acid and phosphoserine). In the case of globular proteins, the hydrophilic regions are well exposed to the aqueous solvent while the hydrophobic regions are usually located within the structure with minimum contact to water. Exposing hydrophobic sites using suitable physico-chemical treatments, especially moderate heat treatment, can improve the interfacial properties of globular proteins by promoting contact between the hydrophobic regions and the non-aqueous phase.

9.3.3. *Emulsification and foaming*

Surfactants play a key role in the formation and stability of foams and emulsions as we have seen in Volume 2 (Chapter 5) [JEA 16b].

9.3.3.1. *Emulsification*

A food emulsion is a dispersion of oil in a continuous aqueous phase (o/w) or vice-versa (w/o). These dispersions are thermodynamically unstable. Three types of instability have been defined in Volume 2 (Chapter 5): sedimentation/creaming, flocculation/droplet aggregation and coalescence.

Creaming or sedimentation is due to the movement of dispersed elements under the effect of gravity at velocity v (m s^{-1}) defined by Stokes' law in which *D* is the diameter of the dispersed particles (m), $\Delta\rho$ is the difference in density between the dispersed phase and the continuous phase (kg m^{-3}), g is gravity (m s^{-2}) and η is the viscosity of the dispersing phase (Pa s):

$$v = \frac{D^2}{18\,\eta}\,\Delta\rho\,g \qquad [9.4]$$

The most effective way to reduce the migration rate is to reduce droplet size or increase the viscosity of the continuous phase. We have seen that by using components that interact with water, it is possible to increase the viscosity of the continuous phase in the case of an oil-in-water emulsion. Reducing droplet size by shearing requires more energy the larger the desired interfacial surface. Lowering the interfacial tension by surfactants facilitates the creation of interfacial surface and the formation of droplets smaller than 1 µm in diameter, which ensures good stability. Polymers with surface-active properties form an interfacial film, the rigidity of which contributes to the stability of the emulsion. In addition, a film consisting of proteins or charged molecules can contribute to the stabilization of the emulsion due to electrostatic repulsions between the molecules. Using mechanical treatments, it is possible to create emulsions (Volume 2, Chapter 5).

Emulsifying properties are defined by three criteria:

– *emulsifying capacity*: the amount of emulsified oil per gram of emulsifier at the point of phase inversion, detected by a decrease in viscosity or electrical conductivity;

– *emulsifying activity*: the stabilized interface area per gram of emulsifier (m^2 g^{-1}) measured by optical methods;

– *emulsion stability*: the ability to maintain emulsion structure over time, determined by particle size or by evaluating its resistance to a physical treatment (centrifugation, heating).

9.3.3.2. Foaming capacity

Foam is a dispersion of gas bubbles in a continuous liquid phase; the gas volume is generally much higher than the liquid volume. Foam formation is promoted by surfactants that lower the interfacial tension. A foam produced in the presence of low molecular weight surfactants (soaps) is instable because the bubble diameter is much greater than that of the emulsion droplets and the difference in density Δρ between the air and the liquid is around 10^3 kg m^{-3}, that is ten times greater than between water and oil. In the presence of surfactant polymers such as proteins, a cohesive film is formed, which is enhanced by the concentration and denaturation of the macromolecules at the interface and stabilized by intermolecular interactions. The physicochemical properties of the aqueous phase (pH, nature of the ionic species, ionic strength) significantly influence the properties of interfacial films. Foam destabilization is described in Volume 2 (Chapter 5). One of the main processes is drainage induced by pressure gradients between the lamella and the Plateau borders. Drainage can be limited by increasing the viscosity of the lamellar liquid by adding sugar or hydrocolloids.

Foams are formed by bubbling, depression (aerosol container) or beating. Foaming properties are defined by two criteria:

– *foaming capacity*: the amount of foam formed per unit volume of solution or solute mass, which can also be determined by measuring the foam density;

– *foaming stability*: the ability of the foam to maintain its structure over time. It is usually determined by measuring the volume of liquid discharged (drainage) for a given volume of foam.

The ability of a surfactant to create an interface (foam or emulsion) and its ability to stabilize two-phase systems does not always go hand in hand. Moreover, the physicochemical conditions of the medium can promote the formation or stabilization of dispersed systems. As a result, for an ingredient and the physicochemical conditions of the food, manufacturers often need to find a compromise between the ability to create an interface and the ability to stabilize the foam or emulsion.

10

Separation Techniques

10.1. Proteins and peptides

10.1.1. *Milk proteins and peptides*

Milk proteins have nutritional, sensory and technological characteristics that make them ideal ingredients for a wide range of food applications (dairy products, ice cream, pastries, confectionary, cooked meats, soups and sauces, beverages, etc.). These proteins contain a balanced amino acid profile and do not cause flavor or color defects in food. As a whole, dairy ingredients exhibit good reconstitution properties and good solubility at neutral pH. However, some processing treatments can alter the solubility of dairy protein ingredients. In general, milk proteins are used for their ability to structure water in foods, their gelling properties (by acidification, the action of rennet or during heat treatment) or interfacial properties (emulsification or foaming).

In addition, some proteins or protein fragments possess biological or physiological activities:

– morphinomimetic activity (β–casomorphin, fragment 60-70 of β casein);

– antihypertensive activity (casokinines from αs1 and β caseins);

– antithrombotic activity (fragment 106-116 of κ casein);

Chapter written by Thomas CROGUENNEC and Valérie LECHEVALIER.

– immunomodulatory activity (β and κ casein fragments);

– bacteriostatic activity (lactoperoxidase, lactoferrin, glycomacropeptide, lactoferricin);

– transport/adsorption of minerals (α_s casein and β casein phosphopeptides);

– protection of hydrophobic molecules (β-lactoglobulin, bovine serum albumin).

The development of separation techniques and the increased knowledge of the physicochemical properties of milk proteins or their hydrolysates have led to the introduction of several purification methods based on differential solubility as a function of pH, ionic strength and/or temperature, size differences, charge or specific affinities (see Chapter 6, Volume 2 [JEA 16b]). Figure 10.1 shows a breakdown of the main milk proteins.

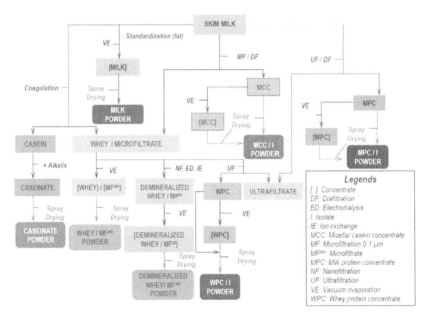

Figure 10.1. *Overview of the main protein ingredients from milk. For a color version of this figure, see www.iste.co.uk/jeantet/foodscience.zip*

10.1.1.1. Protein concentrates

Protein concentrates are mainly obtained from skim milk by the precipitation of milk proteins (co-precipitates) or by ultrafiltration (Milk protein concentrates, MPC).

Co-precipitates are usually obtained by heat treating (90°C for a few minutes) skim milk at its natural pH to denature whey proteins. Subsequent acidification and/or the addition of calcium causes aggregation and simultaneous precipitation of caseins and whey proteins (Figure 10.2). Co-precipitates may contain low (0.5 – 0.8% w/w), medium (approx. 1.5%) or high levels of calcium (2.3 – 3%) depending on the precipitation conditions (pH and amount of calcium added before or after heat treatment). The reduction in pH promotes the release of calcium from protein complexes and the solubilization of calcium phosphates that are removed in the whey. The extraction yield of total milk protein co-precipitates is generally greater than 90%. Their solubility in water is low and inversely proportional to the calcium content. However, co-precipitates can be partially solubilized at alkaline pH in the presence of calcium chelating agents.

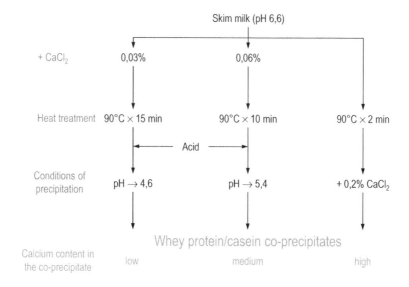

Figure 10.2. *Production processes of whey protein/casein co-precipitates (heat treatment carried out at natural pH of milk)*

The functional properties of total protein co-precipitates can be improved by modifying the pH of skim milk before heat treatment (Figure 10.3). Partially soluble co-precipitates at neutral pH are obtained by heat treating skim milk that has been adjusted to pH 7 – 7.5. Complexes formed in this way are smaller, and protein precipitation at pH 4.6 can solubilize calcium. Other production methods involve the heat treatment of skim milk that has been adjusted to more extreme pH levels (pH 3 or 10). Regardless of which process is used, co-precipitates are separated from whey by filtration or sedimentation, then washed and dried. Acid and highly mineralized by-products are difficult to further develop.

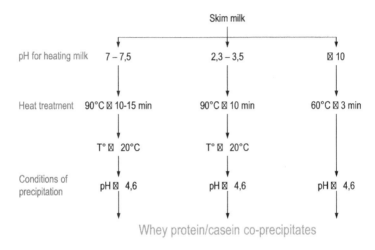

Figure 10.3. *Production processes of whey protein/casein co-precipitates (heat treatment carried out at a different pH to that of milk)*

Ultrafiltration is used to concentrate proteins from skim milk without denaturation up to a concentration factor of about 6, which corresponds to a protein concentration of approximately 18% and a protein content on a dry matter basis of 50 to 70%. Above this concentration factor, the increase in the viscosity of the concentrate will result in a considerable pressure drop and a lower permeation rate. Diafiltration (see Volume 2) is used to increase the purity of protein concentrates; it reduces the amount of lactose and soluble minerals in the concentrate and produces ingredients with up to 90% protein on a dry basis. When filtration is carried out at the natural pH of milk, a large portion of the calcium and phosphorous bound to caseins is concentrated as colloidal calcium phosphate (about 32 mg of calcium per gram of casein). The calcium content of the concentrate can be reduced by lowering the pH before filtration, which solubilizes part of the colloidal calcium phosphate. However, the reduced heat stability and the increased viscosity resulting from a drop in pH below 5.6 can be limiting factors to filtration. After filtration, the milk protein concentrates are pasteurized and spray dried. A fraction of whey protein is denatured during this process; the greater the protein content of the concentrate, the larger this fraction is. The by-product (permeate) can also be concentrated and dried. Protein concentrates obtained by ultrafiltration have better functional properties than co-precipitates because the concentration method does not affect the protein structure and the properties of the serum (pH, ionic strength) remain the same.

10.1.1.2. *Total casein (casein/caseinates)*

Total casein can be obtained from skim milk by destabilizing the casein micelle structure either by acidification (acid casein) or the action of rennet (rennet casein). Acid casein is converted to caseinates by neutralization. These methods alter the functionality of native casein micelles in milk, which can be regarded as a disadvantage (in cheese-making for example). Microfiltration (0.1 μm) can be used to obtain native phosphocaseins (micellar casein) with the same properties as milk casein micelles.

Acid casein is obtained by precipitation of casein at its isoelectric point (pH 4.6) using biological agents (lactic acid bacteria), chemical agents (hydrochloric acid, sulfuric acid) or ion exchange resins. The final pH of precipitation and the acidification temperature affect the extraction yield and properties of acid casein. The production of acid casein using biological agents (lactic casein) is performed in batch and requires the use of large capacity tanks. The pasteurized skim milk is heated to around 25°C before being inoculated with lactic acid bacteria. When the pH is close to 4.5 (after approximately 15 hours of fermentation), the curd is heated to 50–65°C to promote the agglomeration of caseins and inactivate the lactic acid bacteria. During drainage of the tank, the pH of the curd can change slightly and consequently affect the quality of the lactic acid casein (Figure 10.4).

The use of mineral acids is easier than biological agents, especially in terms of the choice of acidification temperature (which affects the properties of acid casein) and the continuity of the process for casein precipitation. Acidification temperatures below 10°C promote fine aggregates and a low mineral content whereas higher temperatures (40–45°C) result in coarse aggregates and a higher mineral content. Mineral acid is added to skim milk at a pre-defined temperature. When the pH reaches a value close to 4.3–4.4, the mixture is heated to 45–55°C to induce the flocculation of casein micelles (Figure 10.4).

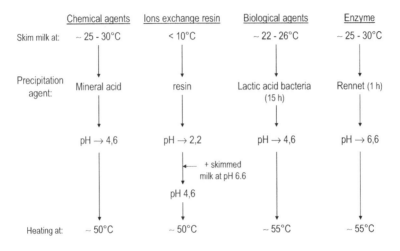

Figure 10.4. *Production processes of rennet and acid casein*

The main disadvantage of destabilizing casein micelles by chemical or biological acidification is that acid whey is produced, which is rich in minerals and difficult to process. One alternative is to use ion exchange resins, which release protons into the medium, ensuring the destabilization of casein micelles without any soluble anions. At the same time, they bind some of the cations and consequently considerably reduce the mineral content of whey. Resins are used either to partially acidify skim milk (pH 5.2) followed by chemical acidification up to pH 4.6, or to lower the pH of milk to 2.5, which is then blended with skim milk (pH 6.6) to obtain the precipitation pH of casein (Figure 10.4); the acidified milk should be separated from the resin before the precipitation of casein. The acid whey obtained is low in minerals, which facilitates further processing.

The production method of rennet casein derives from the manufacture of hard cheeses. The casein micelle is destabilized in the presence of chymosin at around 30–35°C. After coagulation, the gel is cut and heated to around 60°C by direct steam injection to inactivate the enzyme and allow whey expulsion (Figure 1.4). The properties of rennet casein differ from those obtained by acid precipitation because the latter contain colloidal calcium phosphate. Rennet casein is insoluble at neutral pH unless it is dispersed in the presence of a calcium chelating agent.

Regardless of the production method, acid, lactic acid or rennet caseins are separated from whey by sieve filtration or sedimentation. They are washed in water at 50 to 60°C to remove whey proteins, lactose and residual soluble minerals, air-dried at around 100°C in a fluid bed unit and finally ground. A partial solubilization of calcium phosphate during acidification results in an increase in pH during washing and a lower overall yield.

Acid casein is insoluble in water but can be partially or totally solubilized by conversion to sodium, potassium or calcium caseinate. Acid casein is neutralized by the addition of a base (sodium, potassium or calcium hydroxide) to a reconstituted suspension of casein with a protein content of 25% (Figure 10.5). Due to the rapid conversion of acid casein to caseinate and the high affinity of caseinate for water, a thin film of gelled proteins forms on the surface of the acid precipitate or powder particles, which slows down the penetration of the alkaline solution. Thus, fine grinding is required to increase the surface area in contact with the solution. The conversion of casein to caseinate is accompanied by a considerable increase in viscosity, especially in the case of sodium caseinate; this increase can be limited by raising the temperature to 70–80°C, which also facilitates the preparation of sodium caseinate for spray drying. For the preparation of calcium caseinate, the temperature should not exceed 40°C throughout the treatment in order to avoid gelation. Any water added to suspend casein before being converted to caseinate must be removed, which increases the production cost. To reduce this cost, an alternative is the attrition drying of a mixture of acid casein (protein content close to 60%) and sodium carbonate (Na_2CO_3). The nutritional and functional qualities of caseinates obtained by attrition drying are different to those obtained by spray drying.

The microfiltration of skim milk (0.1 μm membrane pore size) coupled with diafiltration produces concentrates containing up to 95% native phosphocasein (micellar casein) on a protein basis. Due to the pH (neutral) and low mineral content, the by-products obtained from the microfiltration of milk or enzymatic coagulation are easier to process than those from the acid coagulation of casein.

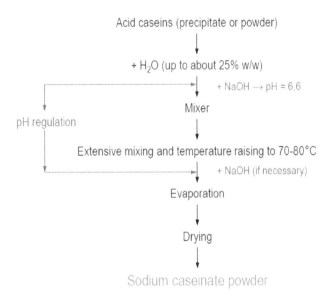

Figure 10.5. *Production process of sodium caseinate from acid casein*

10.1.1.3. *Soluble proteins*

Whey proteins are mainly derived from casein or cheese whey and may also be obtained from the microfiltration of skim milk. They are concentrated by ultrafiltration and extracted by ion exchange resins. They can also be recovered after heat denaturation.

When whey proteins are denatured, they are insoluble at their isoelectric point (pH 4.5 to 5.0): thus, an intense heat treatment (90°C for 3 to 5 minutes) combined with an adjustment in pH between 4.5 and 5.0 is used to precipitate 90–95% of whey proteins that are recovered by sedimentation or centrifugation. The precipitate is then washed with water to remove salts and lactose and dried in a fluid bed unit. Whey proteins obtained by thermal denaturation are generally difficult to resolubilise, especially if the pH during heat treatment is greater than 6, as this promotes the formation of aggregates stabilized by covalent disulphide bonds. The solubility of whey proteins can be improved by keeping the pH between 2.5 and 3.5 during heat treatment, followed by an adjustment in pH between 4.5 and 5, resulting in finer aggregates. However, their recovery is more difficult, which reduces extraction yields.

Unlike the previous treatment, ultrafiltration preserves the native state and the original techno-functional and nutritional properties of whey proteins. The whey protein content based on dry matter varies between 35 and 85% depending on the volume reduction factor applied and whether diafiltration is carried out. The concentration limit imposed by the increase in viscosity occurs when the protein concentration reaches 15% in liquid; this is why it is difficult to obtain whey protein concentrates with a dry matter protein content above 65% by ultrafiltration alone. To achieve this, diafiltration must be carried out, which reduces the amount of soluble minerals and lactose in the retentate. However, the residual fat from the whey, mainly from the fat globule membrane, is subject to the same concentration factor as whey proteins, which limits the purity of the whey protein concentrates to around 85%. In addition, residual fat reduces the mass transfer through the membrane and degrades the foaming properties of the ingredient by competitive adsorption at gas/water interface with proteins. It is also more sensitive to oxidation and can cause undesirable flavors during the storage of powders. The fat content can be reduced by applying thermo-calcium treatment to the whey adjusted at pH 6.5–7.5. In the presence of calcium and at 50°C, membrane lipids of fat globules (phosphoglycoproteins and phospholipids) form aggregates that are then removed by microfiltration or centrifugation.

Ion exchange chromatography is used to obtain whey protein isolates with a protein content above 90%. In the Spherosil® method, negatively charged whey proteins at pH 6.5 (β-lactoglobulin, α-lactalbumin) are retained on an anion exchange resin. Positively charged whey proteins at pH 6.5 (around 10% of whey proteins) and lactose pass through the resin without being retained. By changing solvent conditions (lowering pH, increasing ionic strength), proteins previously attached to the anion exchange resin are eluted. They are then concentrated by ultrafiltration before being dried. Using an anion and cation exchange resins in series allows the recovery of positively charged whey proteins at pH 6.6 on the cation exchange resin (lactoferrin, immunoglobulin and lactoperoxidase).

10.1.1.4. Individual proteins

Some milk proteins or protein fragments have certain useful nutritional, biological and technological properties. Purification processes have been developed to prepare these proteins as ingredients intended for high value-added applications. The separation of different milk caseins (α_{s1}, α_{s2}, β, κ) is the basis for the preparation of certain peptides. Purification processes are also used to fractionate major whey proteins (β-lactoglobulin and α-lactalbumin) and isolate minor proteins such as lactoferrin, lactoperoxidase or immunoglobulins, which have biological activities.

At low temperatures, β casein tends to dissociate from the casein micelle due to lower hydrophobic interactions. By carrying out microfiltration on a solution of casein micelles adjusted to 2°C, casein micelles low in β casein are concentrated in the retentate, while part of the β casein passes into the microfiltrate. It is then concentrated by ultrafiltration at 40°C (Figure 10.6). Another method involves cold precipitation of α_s and κ caseins at pH 4.6 and low ionic strength. Under these conditions, β casein remains in solution and is easily isolated.

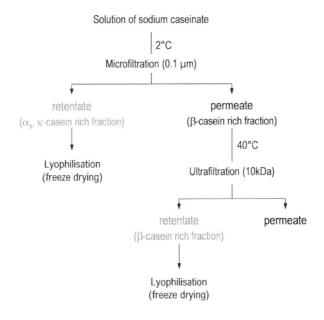

Figure 10.6. *Separation of caseins*

The separation of major whey proteins (β–lactoglobulin and α–lactalbumin) is achieved either by ion exchange chromatography or the selective precipitation of α–lactalbumin. In general, β–lactoglobulin adsorbs more strongly on anion exchange resins than other whey proteins. β–lactoglobulin replaces α-lactalbumin on the resin by competitive adsorption; α-lactalbumin is then eluted while β–lactoglobulin is concentrated on the resin. After washing the resin, β–lactoglobulin is eluted by changing the solvent properties (change in pH or addition of salts).

Another method is based on the aggregation properties of α–lactalbumin at acidic pH (below 4) and 55°C (Figure 10.7). Under physiological conditions, α–lactalbumin binds one calcium ion that is released from protein at pH values below 4. Apo-lactalbumin (calcium-free) undergoes reversible denaturation and aggregation when temperature is increased to 50–60°C. These conditions do not affect β–lactoglobulin, which remains soluble. Aggregates of α–lactalbumin are recovered by centrifugation or microfiltration. By increasing pH, adding calcium and lowering the temperature, α –lactalbumin returns to its original conformation.

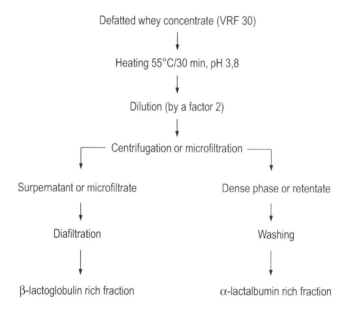

Figure 10.7. *Fractionation of α-lactalbumin*

Unlike major whey proteins, lactoferrin and lactoperoxidase have a positive charge at neutral pH. They are extracted by cation exchange chromatography; both proteins are recovered by sequential elution.

Immunoglobulin-enriched fractions are also obtained by extraction from colostrum.

10.1.1.5. Bioactive peptides

Bioactive peptides are defined as protein fragments that have a positive impact on the health and functions of an organism (affects cardiovascular, digestive, immune, or nervous systems, reduces the risk of chronic diseases, etc.). Many bioactive peptides have been identified in milk protein hydrolysates or fermented dairy products. On an industrial level, they are generated from milk proteins (either individually or in a mixture) by enzymatic hydrolysis using digestive enzymes (pepsin, trypsin) or proteases derived from microorganisms and plants, or by proteolytic microorganisms (rarely used). Enzymatic hydrolysis may be carried out continuously in reactors coupled with an ultrafiltration membrane, which removes hydrolysis products from the enzymatic reactor during the reaction while the enzyme and the substrate remain confined to the reaction vessel (see Chapter 7, Volume 2). The enzyme/substrate ratio, pH, temperature and hydrolysis time are the main parameters controlling the rate of hydrolysis. Hydrolysates are then fractionated based on the physicochemical properties of the peptides (size, charge, hydrophobicity, specific affinity). For example, the separation and purification of phosphopeptides is based on their ability to chelate calcium and therefore to form large soluble aggregates, making possible their separate from non-phosphorylated peptides by ultrafiltration.

10.1.2. Extraction of lysozyme from egg white

Of the many egg proteins, only lysozyme is extracted for applications in the agri-food and pharmaceutical industries where it is used for its antibacterial properties. Its isoelectric point of 11 gives it a positive charge at the pH of egg white and different behavior compared to other egg white proteins (mainly acid and negatively charged proteins).

Lysozyme can be removed from egg white by selective precipitation by combining neutralization at pH 10 (close to the pI of lysozyme) and the addition of NaCl (5%). The main drawback of such a method is that it generates a by-product, "lysozyme-free" egg white, which is substantially lower in value due to its salt concentration.

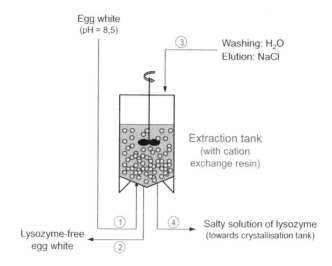

Figure 10.8. *Extraction of lysozyme from egg white*

Lysozyme can also be extracted from egg white by ion exchange chromatography (Figure 10.8). Egg white adjusted to pH 8.5 using citric acid is placed in an extraction vessel containing cation exchange resin. The agitation of the extraction vessel keeps the beads suspended in the egg white and improves the binding of lysozyme. After a few minutes of agitation, the lysozyme-free egg white is removed from the extraction vessel and the resin beads are rinsed in water. The extraction vessel is then filled with a salt solution to dissociate any lysozyme from the resin beads and thus ensure its elution. Yields can be improved by carrying out two or three successive batches (see Volume 2). The eluted lysozyme solution also containing some minor contaminating proteins is then fed to a tank that is used to crystallize lysozyme and increase its purity level. For lysozyme crystallization, the pH of the solution is adjusted to around 10 with sodium hydroxide. Crystallization is triggered by the addition of lysozyme seed crystals from a previous crystallization and a gradual cooling of the lysozyme solution. After approximately 10 hours, the lysozyme crystals are separated from the solution by filtration. To improve purity and remove residual contaminating proteins, several re-crystallization cycles are carried out whereby the lysozyme crystal cake is solubilized at 20°C at slightly acidic pH (around 6) and re-crystallized at pH 10. The lysozyme cake is then solubilized at acidic pH (around 3) using HCl, concentrated by ultrafiltration and dried as lysozyme hydrochloride.

10.1.3. *Extraction of gelatin*

Gelatin is derived from collagen in the skin and bones of animals such as pigs, sheep, chickens, cattle and fish. Skin is usually processed directly after cutting and washing whereas fatty bones must first be cleaned of fat and meat residue by washing in hot water. After drying, the mineral fraction of fat-free bones (mainly calcium phosphate) is removed. Demineralization is carried out in counter-current flow using hydrochloric acid. The demineralized bone, or ossein, is flexible and consists only of organic matter (collagen); it is the raw material for the extraction of gelatin. Calcium phosphate is recovered by lime precipitation.

There are many extraction methods for commercial gelatin. The two main methods differ in the pre-treatment of raw materials resulting in gelatins with different functionalities (Figure 10.9). An acid process for pre-treating raw materials produces type A gelatin while an alkaline process produces type B gelatin. Some raw materials are more suited to one particular process while others are suitable for both:

– the acid process is mainly used for the pre-treatment of pig skins and to a lesser extent ossein. It involves immersing the raw material in an acid solution (HCl, H_2SO_4, etc.) for several hours at room temperature. It is then washed in water and its pH is adjusted to around 4. Gelatin is extracted from this substrate by cooking in water;

– the alkaline process involves pre-treating the raw material (ossein or trimmings) with an alkali, usually lime, for several weeks at room temperature to solubilize components other than collagen (keratin, globulins, pigments, mucopolysaccharides, etc.). After liming, the excess lime is washed off and the pH is adjusted to close to neutral before being cooked to solubilize the gelatin.

Gelatin undergoes the same processing operations after cooking regardless of whether it is generated from an alkaline or acid pre-treatment. During cooking, the gelatin is solubilized, whereas the other proteins or mucoids become insoluble. The soluble fraction is then separated by filtration and cleared of mineral matter by passing through ion exchange resins before being sterilized at 140°C. The sterilized solution is then rapidly cooled to ensure gelation of the gelatin, which is extruded through a die and placed on a steel belt where it undergoes continuous hot air drying without causing the gel to melt.

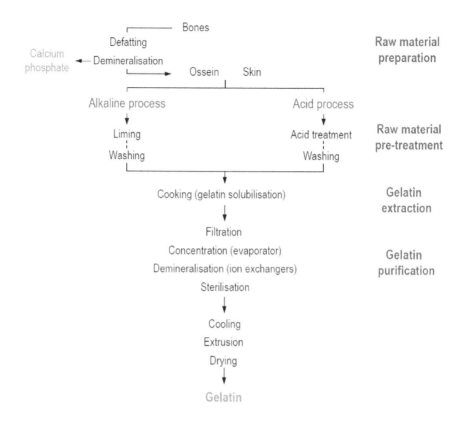

Figure 10.9. *Extraction of gelatin*

10.1.4. *Plant proteins*

Plant proteins, sold as flours, concentrates or isolates are derived from a wide variety of raw materials: grains or products derived from cereal grains, legumes, oilseeds or other plant organs (roots, stems, leaves). Flours are obtained by hulling, milling and defatting grains. They are depleted of cellulose, starch and oil and have a protein content ranging from 50 to 60%. Enriching flour with protein is achieved by removing non-protein compounds soluble in water or in certain hydro-alcoholic solvents. The protein content of the concentrates varies between 60 and 80%. Isolates (protein content higher than 80%) are obtained by the direct selective extraction (in the case of gluten) or selective extraction after solubilization (most plant proteins) of protein material from the flour (Figure 10.10).

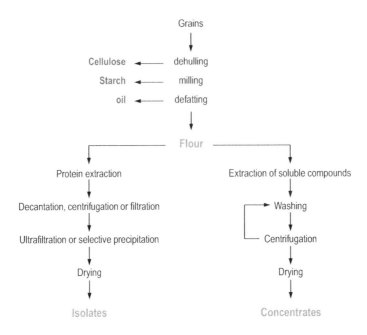

Figure 10.10. *General outline of the preparation of flours, concentrates and isolates (based on [BER 85])*

The production of flour involves removing the outer parts of the grain, which are generally low in protein and rich in cellulose. Cereal grains are ground to separate the endosperm from the bran and germ by passing over grooved steel rollers. The kernel is then ground and the resulting grist is divided into a starch-rich coarse fraction and a protein-rich finer fraction using a cyclone separator. In the case of oilseed, oil is removed by mechanical extraction followed by solvent extraction (see Chapter 6, Volume 2). The desolventized meal is then processed into flour.

The production of protein-rich flour (concentrate) involves removing water-soluble compounds or compounds soluble in certain hydro-alcoholic solvents while retaining the proteins in the insoluble fraction. The extracted compounds are simple sugars, amino acids, minerals, in particular those bound to phytates, but also anti-nutritional elements and colored compounds or compounds responsible for undesirable taste or flavor. Hemicellulose and pectin components remain associated with the protein fraction in the concentrates. To minimize the loss of protein in water, soluble compounds are usually extracted at a pH close to the average pI of proteins (e.g. acidic pH for the production of soy protein concentrates) or by decreasing the dielectric constant of the medium by adding alcohol. After several acid or hydro-alcoholic baths to remove the soluble compounds from the flour, the suspension is centrifuged, neutralized and dried.

The production of isolates involves specifically isolating the protein fraction from the complex raw material. The plant proteins (flour, tubers, leaves) are initially solubilized at neutral or slightly alkaline pH and separated from the insoluble fraction by decantation, centrifugation or filtration before being recovered according to their specific physicochemical properties. Proteins are usually precipitated by changing the physicochemical conditions of the medium (pH, ionic strength, dielectric constant, temperature) without affecting the other soluble compounds. It is also possible to vary the size (ultrafiltration) or charge (ion exchange) of the soluble elements to be separated. In the case of soy, proteins are solubilized at alkaline pH (8–9). The insoluble and fibrous compounds are removed by centrifugation or filtration. By lowering the pH of the solution containing solubilized proteins to around 4.5, the main soy proteins (glycinin and b-conglycinin) precipitate. The precipitate is washed, neutralized and dried. When separating gluten and starch from wheat, flour is mixed with water to facilitate the formation of large gluten aggregates, which physically separate from starch. Gluten is the exception because proteins are poorly soluble in water and tend to agglomerate independently of other compounds. Starch is separated from gluten using a decanter or a hydrocyclone. At this stage, the gluten forms a cohesive mass that is then washed and dried.

10.2. Carbohydrates

10.2.1. *Sucrose*

Sucrose, which consists of one glucose molecule and one fructose molecule linked together (Figure 10.11), is the most common carbohydrate, which is why it has become synonymous with sugar in everyday language. It is produced by a number of sugar crops including cane sugar and sugar beet through photosynthesis, and can be extracted industrially. Sugar beet accumulates sucrose in its roots. It can contain 15 to 20% sugar depending on the variety. In sugar cane, sucrose accumulates in the stem. Its sugar content ranges from 11 to 18% depending on the species, climate, terrain and maturity. In both cases, the plant must be processed immediately after harvest because the sugar content drops rapidly as it is metabolized by the plant.

Sucrose is one of the components that can be obtained industrially in high purity and at low cost; its separation from the other constituents of beet or cane involves a series of unit operations. The sugar industry generates many by-products that are converted into alcohol, fertilizer, animal feed or an energy source. The latter allows the sugar industry to minimize its energy consumption.

The main stages in obtaining sugar are extraction and refining. The latter is common to sugar cane and sugar beet. However, the first extraction stages differ depending on the plant.

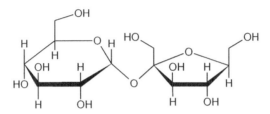

Figure 10.11. *Chemical formula of sucrose*

10.2.1.1. *Raw juice extraction*

Extraction by diffusion

A root of sugar beet contains 25% dry matter consisting of 17% insoluble cell wall, 73% sugar and 10% soluble material. The sugar in the cells of the beet root is extracted by diffusion and dissolution in hot water (Figure 10.12).

Diffusion is described by Fick's law (Volume 2), which expresses the amount of diffusing material per unit of time as a function of the diffusion coefficient D_m, the exchange surface and the concentration gradient of the solute. The last two parameters can be adjusted to optimize sugar extraction. The first stage involves cutting sugar beet into thin strips 5 to 6 cm long called cossettes; this increases the extraction rate as the contact surface between the raw material and the solvent is greater. In addition, the water and cossettes flow counter-currently in the diffuser to improve the extraction yield by maintaining the concentration gradient between the juice and the cossettes. The juice obtained at the top of the diffuser is an opalescent grayish-brown solution with around 15% dry matter. It is slightly acidic (pH 6.0) and contains 13 to 14% sugar and 1 to 2% organic impurities (protein, pectin, other sugars, organic acids) and inorganic salts (sodium, potassium, calcium, magnesium and other salts). At the bottom of the diffuser, the sugar-depleted cossettes are recovered as pulp and used for animal feed.

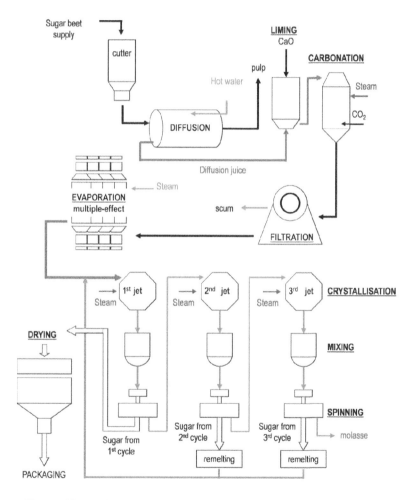

Figure 10.12. *Extraction of sugar from sugar beet. For a color version of this figure, see www.iste.co.uk/jeantet/foodscience.zip*

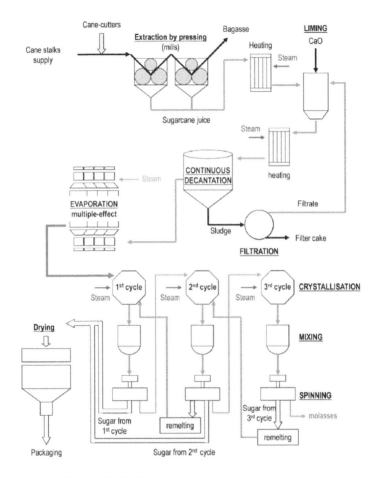

Figure 10.13. *Extraction of sugar from sugar cane*

Extraction by milling/pressing

The diffusion process is not suitable for sugar cane given its high fiber content. Sugar is extracted from cane by milling and pressing the cane stalks (Figure 10.13). The washed cane stalks are fed into "cane-cutters" that slice them into pieces 10 cm in length and 4 mm in diameter; these pieces are crushed and pulped by a mill consisting of two grooved rollers, then chopped in a shredder to facilitate juice extraction. The cane pieces then pass through a series of 4 to 6 mills, each consisting of three slowly-rotating (4–6 rpm) horizontal rollers arranged in the shape of a triangle. The cane is firstly pressed between the feed roller and the top roller and then between the top roller and the discharge roller; it is therefore pressed twice. The crushed cane, known as "bagasse", is fed through the subsequent mills. After passing through the first mill, juice extraction is facilitated by a process called imbibition. This involves adding water to the bagasse, which dilutes the juice in the latter and improves the efficiency of pressing. After passing through the final mill, the bagasse represents 20 to 30% of the cane weight and contains 48 to 50% moisture. It is usually the primary fuel source for sugar mills due to its high calorific value (17,000 kJ kg^{-1} in a dry state).

The sugarcane juice collected at the bottom of the mill is cloudy, ranging in color from greenish yellow to dark brown. Its composition and quality vary depending on the variety and quality of the cane but on average it contains 80 to 85% water, 10 to 18% sucrose, 0.3 to 3% reducing sugars and 0.7 to 3% other organic (soil and wood particles) and inorganic compounds.

10.2.1.2. Juice purification

The purification of sugar beet juice involves two stages: liming, whereby some of the impurities are precipitated by forming insoluble calcium salts, and carbonation, whereby excess lime is precipitated. Liming is carried out in two stages to minimize the amount of lime used; pre-liming, which represents about 20% of the total lime added, is a progressive alkalinization of diffusion juice (final pH of 11.5) in order to selectively precipitate certain impurities (acids, inverted sugars, nitrogenous substances, bi- and trivalent cations in the form of hydroxides, sulfates or phosphates in the form of insoluble calcium salts); the juice is heated to 85°C and fed into liming tanks where large quantities of milk of lime is added, triggering the degradation of nitrogenous substances (formation of insoluble calcium salts and ammonia) and reducing sugars (formation of brown compounds and lactic acid).

Carbonation is performed in two successive stages: carbonation and filtration. The limed juice is heated and carbon dioxide is bubbled through the mixture. The first stage of carbonation involves precipitating the excess lime in the form of calcium carbonate. Since the reaction is highly exothermic, some water can at the same time evaporate from the juice. In addition, the brown compounds resulting from the decomposition of reducing sugars are adsorbed on the newly-formed carbonate crystals. In order to avoid the dissolution of impurities, the juice is maintained at pH 11.2; it is then filtered to obtain a clear juice, referred to as first carbonation. Impurities precipitated by lime and/or adsorbed on the calcium carbonate crystals (scum) are removed by filtration. The insoluble material is washed in water to recover any sugar; the wash water is then used to dissolve the quicklime (calcium oxide) and form milk of lime, while the residual insoluble substances are used for liming.

The first carbonation juice heated to 95°C undergoes a second carbonation to precipitate the remaining lime; the final pH is around 9.2. Before concentration, the juice is decalcified on a cation exchange resin to prevent scaling of the evaporator tubes, and de-colored using SO_2. At this point, around 30% of impurities have been removed and the purity of the juice is about 93% on a dry matter basis. The purified juice is a clear straw-colored sugar solution containing approximately 86% water, 13% sugar and 1% dissolved impurities. These impurities differ from those in beet juice, as they mostly include sugars other than sucrose, proteins, organic acids, dyes (chlorophyll, tannins), fats (wax), inorganic salts (iron), gums and pectin in a colloidal state, which are particularly difficult to remove. The purification of sugarcane juice does not require carbonation since all the lime is used up in the conversion of acids to insoluble salts and the coagulation of proteins.

After liming, the juice is boiled to promote the flocculation of residual impurities (sludge) that settle to the bottom of the tank (Figure 10.13). The clarified juice is then concentrated by evaporation. The settled sludge is mixed with fine bagasse (filter aid) and filtered on rotary vacuum filters. The filtrate undergoes another liming process and the desugared residue (known as filter cake or filter press mud) is used as fertilizer.

10.2.1.3. *Concentration of juice by evaporation*

Sugar beet or sugar cane juice is concentrated around five times by evaporation, achieving a sucrose concentration close to saturation of about 60 to 70% (w/w). Multiple-effect climbing film evaporators are used; the optimization parameters for the concentration process include the exchange surface and the temperature gradient between the heat transfer fluid and the product. The level of energy consumption is around 3.1 kWh for 100 kg of sugar beet.

10.2.1.4. *Crystallization of sucrose*

Crystallization is the final stage in the purification of sucrose. It isolates the remaining impurities that are concentrated in the liquid phase, known as molasses, while sucrose is extracted in the form of crystals.

Crystallization involves a first phase known as nucleation, that is the formation of crystal nuclei or seed crystals and a second phase where the seed crystals develop during the growth phase. In the case of sucrose, it takes at least 6 molecules to form the basic unit of a crystalline structure. From a thermodynamic point of view, a nucleus must cross an energy barrier corresponding to a critical size of about 100 sucrose molecules to ensure its stability in a three-dimensional state. Crystals grow by the attachment of sucrose molecules to the crystal surface. This is a heterogeneous process involving the diffusion of sucrose molecules from the solution to the surface of the crystal and the incorporation of molecules into the crystal lattice.

After concentration, the sugar syrup contains 60 to 65% sucrose, which is an insufficient concentration for crystallization to occur. The syrup must be concentrated to supersaturation, usually by isothermal evaporation, to allow the development of seed crystals and/or the growth of existing crystals. The supersaturation region can be divided into three zones (Figure 10.14):

– a metastable zone where crystallization can occur, but only after crystal seeding for triggering the crystallization process;

– an intermediate zone where heterogeneous nucleation is possible;

– a labile zone where nucleation spontaneously occurs.

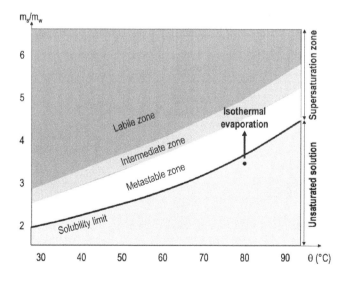

Figure 10.14. *Solubility curve of sucrose as a function of temperature*

In order to facilitate the separation of crystals and the transfer of the product (high viscosity), crystallization is carried out in three cycles (Figure 10.15) of three steps: boiling, mixing and centrifuging. After three cycles, the sugar stops crystallizing due to an increase in the impurity content.

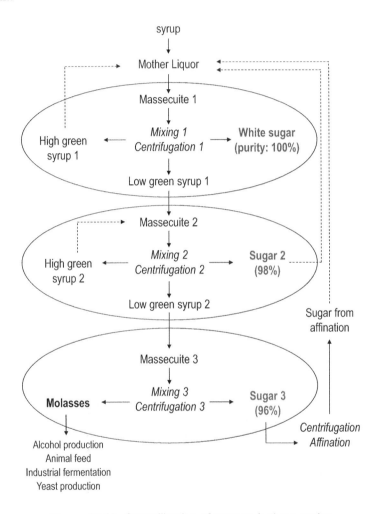

Figure 10.15. *Crystallization of sucrose in three cycles*

The syrup from the first cycle, known as mother liquor, is a mixture of different products. It is added to large boilers (called vacuum pans) that operate under partial vacuum at 80°C in which isothermal evaporation occurs bringing the liquor to the metastable zone (boiling). Fine sucrose crystals are then added to initiate crystallization: this is called graining. It is possible to control the size of the sugar crystals formed: the number of recovered crystals corresponds to the number of crystals added and the sucrose in the mother liquor is solely involved in the growth of these crystals. The syrup/crystal mixture is known as "massecuite". As the crystals grow, the syrup concentration drops. To maintain supersaturation, syrup is constantly added to the boiler while evaporating under vacuum. When the rate of crystallization decreases and the boiler is full, no more syrup is added but evaporation continues. At the end of the boiling step, all excess water is evaporated, which improves crystallization yield. In general, the yield in the first cycle is 55%.

The massecuite is then cooled from 75–85°C to 45–50°C with regular agitation (mixing) allowing the crystals to finish growing. The crystals are separated by centrifugation and the discharged liquid phase is known as "green syrup". The crystals are washed and dried by adding hot water followed by steam. The highly-pure syrup collected at this stage is called high green syrup. After being discharged from the centrifuge, the crystals contain less than 1% water.

The low green syrup from the first and second cycles is used in the third cycle, while the high green syrup is fed back into the massecuite from which it is derived.

The beet sugar recovered after the first cycle is very white and pure. It is used directly for human consumption. On the contrary, cane sugar from the first cycle is reddish-brown due to the large number of polyphenols and tannins. It can be sold as such or refined. After the second cycle, a brown sugar with 98% purity is obtained. It is fed back into the mother liquor to be remelted and re-purified. After the third cycle, the recovered sugar is only 96% pure. Before being remelted and added back into the mother liquor, it undergoes affination, which reduces the amount of impurities reintroduced into the mother liquor.

Sucrose crystallization generates a by-product that is quantitatively and qualitatively important in terms of valorization: molasses. It consists of 50%

sugar and contains potassium salts and various organic and nitrogenous substances. Molasses is used for animal feed and industrial fermentation, in particular alcohol fermentation (production of ethanol), but also as a substrate for the production of baker's yeast, antibiotics, glutamic acid, citric acid and so forth.

10.2.1.5. Refining

Refining is an additional industry to the sugar industry: it treats brown cane sugar, raw beet sugar and sugar syrups.

The purpose of refining is to remove impurities (minerals, organic matter) by remelting, adding lime, carbonation, filtration and recrystallization, which is quite similar to what is done during the purification and crystallization of beet sugar.

The first stage is affination: raw sugar is vigorously mixed with a warm, slightly under-saturated syrup, which promotes the dissolution of the more impure surface crystals. The massecuite is then centrifuged and washed resulting in "affined" sugar.

The next stage is remelting: the affined sugar is dissolved in hot water and the syrup formed is made alkaline by adding milk of lime. Any impurities precipitate out and are separated from the syrup by filtration. The syrup then passes over a bed of activated carbon, which is a decolorization process. The resulting juice gives a quality sugar in four or five crystallization cycles.

10.2.1.6. Drying

The white crystal or granulated sugar from the first crystallization cycle is removed while still warm (45 to 60°C) and moist (1%). The sugar crystals are coated with a film of saturated syrup, in equilibrium with the atmosphere depending on the sorption isotherm of the crystalline sugar (Figure 10.16). To ensure stability of the sugar, the moisture content of the film must be reduced from about 30% (a_w of around 0.8) to 0.03–0.06%. To limit heating and the risk of caramelization, sugar is dried co-currently with hot dry air (50°C); then, it is cooled counter-currently with cold dry air to obtain a stable product.

If drying is too quick, a layer of amorphous sugar forms on the surface. Its properties are very different from those of crystalline sugar. It is in a metastable state and tends to absorb water even at low a_w values, which predisposes it to caking, agglomeration and a loss of flow. After the adsorption of water (drop in T_g) or an increase in temperature, amorphous sugar crystallizes and releases water molecules. The presence of amorphous sugar on the surface of the crystals is therefore an instability factor, and thus drying parameters (θ, H_R) must be controlled.

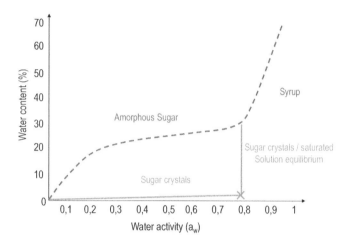

Figure 10.16. *Water vapor sorption isotherm of crystalline sugar and amorphous sugar at 25°C*

10.2.1.7. Packaging

After drying, sugar is sieved, graded and weighed; it is then either bagged, packaged as blocks or stored in silos. Sugar intended for storage in silos is dry and flows freely, but crystallization of the residual amorphous sugar may occur on the crystal surface during the first few days of storage; this is accompanied by a release of water that must be removed by ventilation (maturation).

10.2.1.8. *Use of sucrose and its derivatives in food*

The sugar industry has diversified its range of products to provide consumers or other industries with sugar in forms most suited to their needs (Figure 10.17).

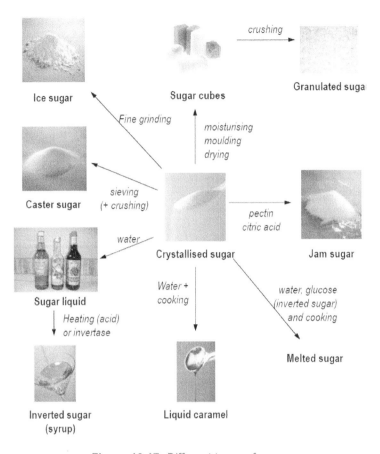

Figure 10.17. *Different types of sugar*

Sucrose

It is possible to obtain a large range of products from white crystal sugar by sieving, grinding, crushing, or molding:

– *caster sugar* (crystal size around 0.4 mm): this is used to make desserts and puddings and to sweeten dairy products and fruit. It can be lightly browned and flavored with vanilla extract or essence;

– *icing sugar* (crystal size around 0.15 mm): this is very hygroscopic and often supplemented with 3% cornstarch to avoid caking. It is used in all uncooked desserts, for decorating pastries and making glazes;

– *lump sugar* or *sugar cubes*: warm moistened crystalline sugar is moulded and dried to form different sized shapes. It is used to sweeten hot drinks or make sugar syrup or caramel;

– *granulated sugar*: round grains are obtained by crushing pieces of very pure refined sugar. The grains are then graded according to size by sieving;

– *jam sugar*: white sugar is supplemented with natural fruit pectin (0.4–1%), food-grade citric acid (0.6–0.9%) and sometimes tartaric acid. Pectin promotes jam setting and its action is enhanced by the presence of acid;

– *liquid sugar* or *sugar syrup*: this colorless sucrose solution has a minimum dry matter content of 62% (of which maximum 3% is inverted sugar); it is generally intended for the agri-food industry except for colorless or yellow sugar cane syrup, which is used in the production of desserts and cocktails.

Other forms of sugar can be obtained by modifying the extraction or refining processes:

– *rock sugar*: this is obtained by the slow crystallization of very pure, hot, concentrated sugar syrup, which is left to cool slowly in trays containing stretched linen or cotton threads around which the crystals can grow. The slower the crystallization, the larger the resulting crystals. This sugar is recommended in the production of liqueurs, and is used by some champagne producers to make expedition liqueur;

– *brown sugar*: this is a raw crystal sugar, extracted directly from sugar cane juice giving it its brown color and specific flavor;

– *soft brown sugar*: this is a soft, colored and fragrant sugar. It can be light or dark depending on the syrup used, either from the first or second centrifuging of sugar. It is widely used in baking.

Derivatives of sucrose

Sucrose may also be converted into other products, some of which are used in the food industry:

– *inverted sugar*: results from the hydrolysis of sucrose into an equimolar mixture of glucose and fructose. It is called inverted sugar because of the change in direction of optical rotation from positive (dextrorotatory) to negative (levorotatory). It is obtained industrially by acid hydrolysis or enzymatic hydrolysis (action of β-fructosidase, invertase produced by yeast).

Unlike sucrose, inverted sugar is a reducing sugar. It can therefore be the substrate of non-enzymatic browning reactions (see Chapter 5, Volume 1);

– *caramel*: this is obtained by boiling crystal sugar in water. Caramelization occurs when sucrose is heated beyond its melting point (186°C), in preference under acidic conditions. Only sugars are involved in this thermal degradation. This results in the breakdown of sucrose into glucose and fructose followed by a series of dehydration and isomerization reactions, which vary depending on the heat treatment and the level of acidity. These reactions lead to the formation of colored and aromatic compounds;

– *polyols*: these are bulk or nutritive sweeteners with a lower energy value than sucrose. They are obtained by the reduction of sucrose. Depending on the reaction conditions (temperature, pressure, catalyst), various hydrogenated products are obtained such as sorbitol, mannitol or an equimolar mixture of the two (isomalt). These components are non-fermentable, and thus do not contribute to tooth decay;

– *sucrose esters*: these are obtained by the transesterification of sucrose and fatty acid methyl esters, usually of plant origin. These molecules have many applications especially due to their amphiphilic properties. They are mainly used for their surfactant properties (emulsion stabilization), but also for their anti-adhesive and lubricating (confectionery), texturing (bread and biscuit making), dispersion and solubilization (beverages), as well as antibacterial and antifungal properties [CEC 01].

10.2.2. *Lactose*

Lactose is found exclusively in the milk of mammals. In cow's milk, the lactose concentration varies between 48 and 50 g L^{-1}. It has unique physicochemical characteristics (low sweetness, high glass transition and melting temperatures, etc.) and may be the precursor of certain food and pharmaceutical molecules. These are obtained by chemical, enzymatic or microbial transformation.

Lactose is extracted industrially from sweet whey, a by-product of the production of hard cheese, certain soft cheeses and rennet casein, or acid whey, a by-product of the production of fresh cheese and acid casein, or also from milk permeate (Table 10.1).

	Solids (g L^{-1})	Lactose (g L^{-1})	Total nitrogenous content (g L^{-1})	Minerals (g L^{-1})
Sweet whey	67	50	9.5	7.5
Acid whey (cheese)	64	44	8.0	12.0
Acid whey (casein)	67	50	8.0	9.0
Milk permeate	60	50	2.5	7.5

Table 10.1. *Average composition of the various milk liquids that are sources of lactose*

10.2.2.1. *Extraction and purification*

Industrially-produced lactose is in a crystalline state in the form of α-monohydrated crystals. It can be purified from whey by three different methods that vary in their initial stages (Figure 10.18):

– in the first method (Figure 10.18, pathway 1), whey is heated to 90°C in the presence of CaCl$_2$ (0.15%) to induce the precipitation of whey proteins. After separating denatured proteins by filtration, the filtrate containing lactose (4.8%) is concentrated by evaporation up to a lactose content of 15%. Additional heating precipitates excess calcium, which is then removed. The lactose solution (15%) is concentrated a further four times approximately. Crystallization is triggered by the addition of α-lactose crystals finely dispersed in the concentrated lactose solution at 60°C; the solution is then slowly cooled from 60°C to 15°C in about 20 hours, which allows the formation of large crystals that are more easily separated by decantation or centrifugation. The by-product of crystallization is a highly mineralized lactose solution;

– in the second method, protein denaturation is limited by keeping the temperature of the whey below 70°C during all stages (Figure 10.18, pathway 2). The whey is first concentrated 12 times by evaporation at 65°C before being seeded with finely dispersed α-lactose crystals and then cooled to 15–20°C in about 20 hours (crystallization yield of 70 to 80%). The by-product formed is a delactosed whey with low lactose content (the composition of the delactosed whey is equivalent to a milk concentrated 2.5 times), which is concentrated to obtain a solids content of about 50% before drying;

– in the third method (Figure 10.18, pathway 3), protein is removed from the whey by ultrafiltration and then the permeate is concentrated by evaporation to obtain a lactose concentration of around 60%. The saturated lactose solution is then seeded with finely dispersed α-lactose in a crystallization tank. Cooling induces the crystallization of α-lactose. The ultrafiltration retentate is used for the production of whey protein concentrates, the lactose content of which varies depending on the concentration factor applied. The by-product of crystallization, which is rich in lactose and minerals, is used in the production of animal feed.

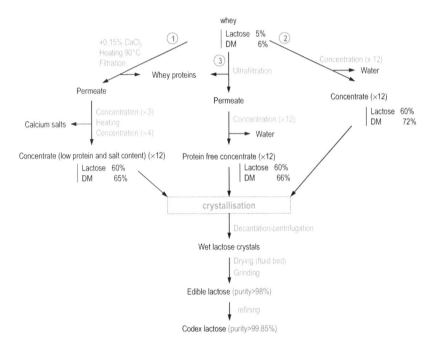

Figure 10.18. *Purification of lactose. For a color version of this figure, see www.iste.co.uk/jeantet/foodscience.zip*

Whatever the method used, α-lactose crystals (crystallization yield of about 80%) are then separated from the mother liquor by decantation or centrifugation, washed, dried in a fluid bed and then ground. The lactose obtained is 98% pure (edible lactose). Its purity can be increased to 99.8% by refining (codex lactose), which involves the hot dispersion (above 90°C) of lactose crystals at a concentration of around 60%. This temperature allows the denaturation of proteins and the precipitation of residual calcium salts, which are removed by filtration before carrying out another cycle of crystallization, centrifugation and drying. The purification yield of lactose in refining is around 75%.

10.2.2.2. Lactose derivatives

Hydrolysis, oxidation, reduction, isomerization, condensation with an amino group or fermentation are all ways of derivatizing lactose and thus increasing its field of application (Figure 10.19). In addition, lactose can be used as a substrate for the production of microbial enzymes or biomass under aerobic conditions (yeasts).

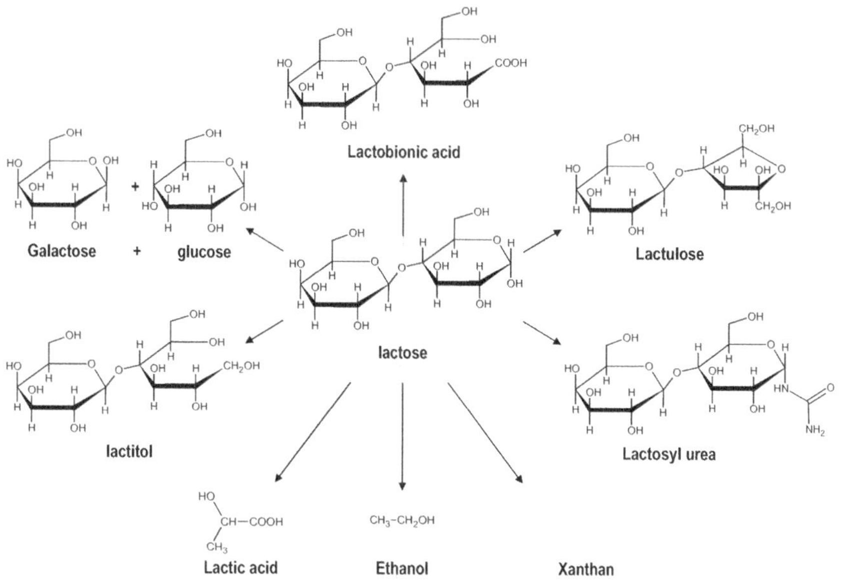

Figure 10.19. *Main lactose derivatives*

The hydrolysis of lactose into glucose and galactose can improve the nutritional quality of foods intended for lactose-intolerant customers, increase its solubility and sweetening power and double its reducing power. The hydrolysis of lactose can be done either chemically at pH 2 at around 100°C (extreme conditions) or enzymatically in the presence of β-galactosidase. Commercial β-galactosidase from moulds (Aspergillus) or yeasts (Kluyveromyces) have different properties, in particular regarding optimum activity pH, temperature stability or sensitivity to inhibition by reaction products. Hydrolysis can be carried out in batch with free enzymes that are inactivated by heat treatment or removed by filtration after the reaction. This can also be achieved in enzyme reactors equipped with an ultrafiltration membrane that retains β-galactosidase or with β-galactosidase immobilized on a support in batch or fixed bed.

Lactulose is an epimer of lactose where the glucose residue is isomerized to fructose. Epimerization does not occur spontaneously, but is facilitated by heat treatment under basic conditions. Lactulose is sweeter than lactose and non-cariogenic. It is not hydrolyzed by intestinal β-galactosidase, which allows it to reach the large intestine and be degraded by lactic acid bacteria of the *Bifidobacterium* genus. Lactulose therefore acts on the intestinal flora by decreasing the pH of the intestinal contents, thereby preventing the growth of undesirable spoilage bacteria (coliforms); its pharmaceutical applications are based on this property.

Lactitol is produced by the reduction of the reducing function of lactose in the presence of a catalyst (Raney nickel). It is a sweeter polyol than lactose with solubility similar to that of sucrose. It cannot be assimilated as such (non-calorific), but is fermented in the large intestine, which pH is lowered and thus limits the growth of ammonia-producing flora. It can be esterified by one or more fatty acids to produce food or non-food (toiletries, dentrifices) emulsifiers.

Lactobionic acid is produced by the chemical, enzymatic or biological oxidation of the reducing function of lactose. It can chelate minerals and heavy metals. It has applications in the food and pharmaceutical industries, and in the preservation of organs before transplant. Its lactone, obtained enzymatically or microbially, can be used as an acidifier of the glucono-δ-lactone family (see Chapter 1).

Lactosyl urea is produced by the nucleophilic reaction of the amine group of urea with the reducing function of lactose (first step of the Maillard reaction). It is used as a source of nitrogen in animal feed.

Lactose is quickly transformed under anaerobic conditions to lactic acid and ethanol by lactic acid bacteria and yeast, respectively. After fermentation, microorganisms are removed from the reaction medium by centrifugation. Lactic acid is used as a food acidulant, a preservative, an anti-crystallizing agent and a flavoring agent. After distillation, 96.5% (w/w) ethanol can be used in the production of spirits (whiskey, vodka). Lactose is generally used as a substrate in the production of xanthan gum by *Xanthomonas campestris*.

10.2.3. Polysaccharides

The methods used to purify polysaccharides depend primarily on their solubilizing properties. Poorly soluble polysaccharides such as starch or cellulose are purified by successive fractionation of the various contaminants. In the case of highly soluble polysaccharides (carrageenan, pectin, alginates, etc.), purification methods are essentially based on selective precipitation by modifying the temperature, pH and polarity of the solvent.

10.2.3.1. Starch

Starch is a mixture of two polysaccharides: amylose, which is a linear chain of α(1-4)-linked D-glucose and amylopectin (Figure 10.20), which is a chain of α(1–4)-linked D-glucose with α(1–6)-linked side chains.

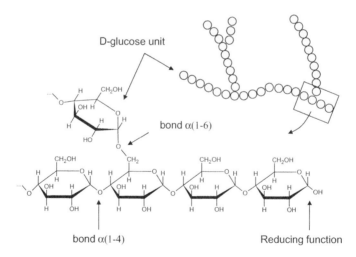

Figure 10.20. *Structure of amylopectin*

The main sources of starch used in the food industry are cereals (corn, wheat, rice, sorghum) and tubers (potatoes, cassava). Starches differ from each other by the shape and size of the granule, the respective proportions of the amylose and amylopectin chains, which determine their physical properties, and the extraction method used. In the case of corn, the mechanical extraction method involves a series of operations such as soaking, milling and separation of starch by centrifugation or decantation (Figure 10.21).

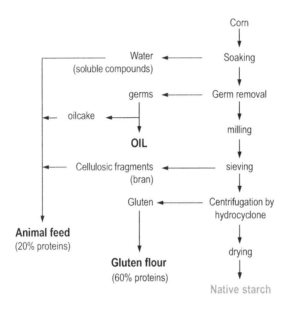

Figure 10.21. *Diagram of the purification of starch*

Soaking in water at around 50°C for 30 to 40 hours softens the corn and weakens the protein membrane around the starch granules. At the same time, the water is enriched with soluble molecules like simple sugars, amino acids, salts or phytates that are recovered for animal feed. After soaking, the corn kernels are degermed by coarse grinding and the germs containing the lipid fraction of the grain are removed using a hydrocyclone. The extracted oil is used to make cooking oil or margarine due to its high level of linoleic acid. The by-product of oil extraction (oil cake) is dried for further processing. The grain consisting of a mixture of starch, protein (gluten) and cellulosic material is then finely ground and sieved to separate the cellulosic fragments (bran) from the finer grind, containing starch and protein. The protein particles can then be separated from the starch granules, which have a higher density, by decantation or centrifugation in hydrocyclones. This separation can also be done by sieving because of their different sizes.

Starch is the basis for many derivatives (Figure 10.22 lists the main ones). They are obtained by chemical, physical or enzymatic modification.

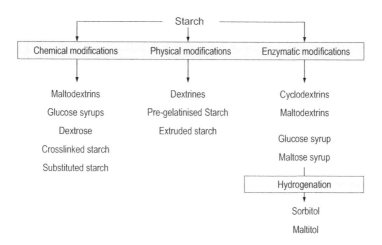

Figure 10.22. *Derivatives of starch*

Chemical modification

There are three types of chemical modification: hydrolysis, cross-linking and the grafting of functional groups. In the presence of a mineral acid (0.1–1% HCl) and under certain conditions, the hydrolysis of amylopectin and amylose in solution (30 to 40%) generates the following compounds:

– maltodextrins resulting from partial hydrolysis (3-20 dextrose equivalent);

– glucose syrups (20–70 dextrose equivalent);

– hydrolysates from extensive hydrolysis (dextrose equivalent close to 95);

– dextrose (pure glucose).

Chemical cross-linking involves creating bridges between amylose and amylopectin molecules, which increases the degree of polymerization of the starch. Cross-linking reinforces the macromolecular network and the internal cohesion of the starch granule, and stabilizes the viscosity of the starch paste. Cross-linked starch is more resistant to high temperatures, acidic pH or mechanical treatments such as pumping, shear, etc. The cross-linking reaction takes place below the gelatinization temperature with a cross-linking agent (adipic acid, acetic anhydride, sodium trimetaphosphate, etc.).

The grafting of esters (succinylated starch, starch acetate) or ethers (hydroxypropyl starch) onto hydroxyl groups of amylose and amylopectin chains can lower the gelatinization temperature, reduce the risk of interaction between amylose and amylopectin chains and consequently limit the retrogradation of starch. The reaction conditions are controlled in such a way as to ensure sufficient swelling of the granules (accessibility of hydroxyl groups) while keeping them whole.

Physical modification

Physical treatments allow the formation of dextrin, pre-gelatinized starch or extruded starch. Dextrins are generated by applying heat treatment to dry starch granules. Pre-gelatinized starch is obtained by cooking/drying a starch suspension on hot rollers; it swells in cold water, and has a high thickening capacity and high digestibility. Extruded starch differs from pre-gelatinized starch by its higher solubility.

Enzymatic modification

Enzymatic treatments are increasingly used in the processing of starch. α–amylase and pullulanase (debranching enzyme) are used to produce maltodextrins and glucose syrups. β-amylase is used to generate maltose syrups. Glucose isomerase converts glucose to fructose. Cyclodextrin-glycosyltransferase catalyses cyclization reactions of α–, β– and γ–cyclodextrins consisting of 6, 7 or 8 glucose units linked by α–(1, 4) bonds. The products of chemical and enzymatic hydrolysis (glucose and maltose) can be reduced to sorbitol and maltitol. These polyols are obtained by the hydrogenation of dextrose and maltose syrups obtained from starch in an aqueous medium, under pressure, at high temperatures and in the presence of a catalyst.

10.2.3.2. Carrageenan

Carrageenan is a linear chain of carrabiose units, consisting of two $\alpha(1-4)$ bonded galatose derivatives (sulphated galactose or anhydrogalactose) linked together by $\alpha(1-3)$ bonds (Figure 10.23).

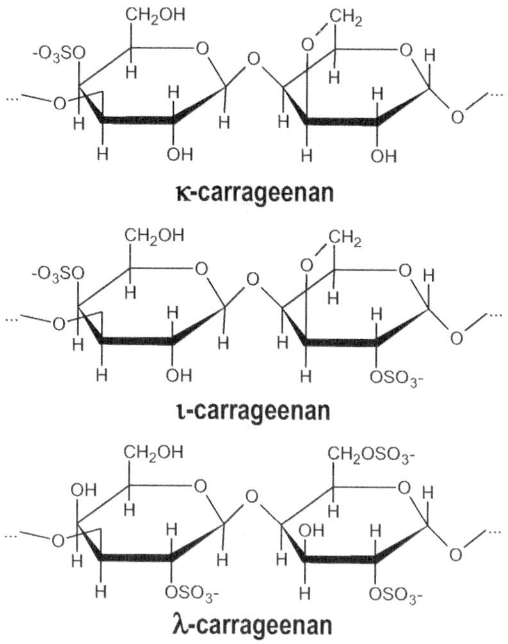

κ-carrageenan

ι-carrageenan

λ-carrageenan

Figure 10.23. *Chemical structure of carrageenan*

Extraction methods are based on the fact that carrageenan is soluble in hot water and insoluble in organic solvents. Carrageenan is extracted from dried and ground red algae (*Rhodophyceae*) of the *Chondrus* or *Gigartina* genus (Figure 10.24). Rehydrated red algae soaked in hot water at around 80°C in the presence of an alkaline (lime, caustic soda or potash) facilitates the extraction and solubilization of carrageenan. The soluble fraction containing carrageenan (syrup) is separated from insoluble impurities at high temperature by filtration (filter press) in the presence of filter aid. After concentration by evaporation or filtration, the carrageenan solution is precipitated by the addition of isopropanol. The precipitate is then washed with water, pressed and dried.

10.2.3.3. *Alginates*

Alginates consist of polysaccharide chains of homogeneous or heterogeneous blocks of β–D-mannuronic acid and α–L-glucuronic acid (Figure 10.25).

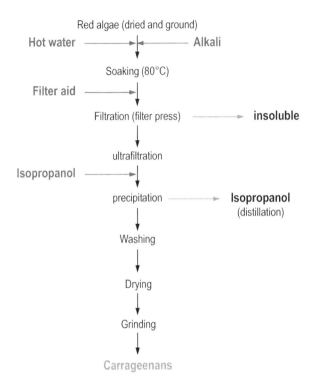

Figure 10.24. *Diagram of the purification of carrageenan*

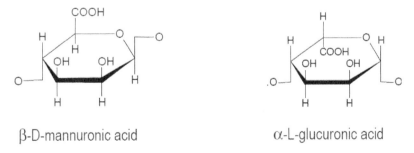

Figure 10.25. *Chemical formula of β-D-mannuronic acid and α-L-glucuronic acid*

Alginates are extracted from brown marine algae (*Phaeophyceae*) based on their high solubility in water at neutral or alkaline pH and their insolubility in acidic media (Figure 10.26). In algae, alginates are mainly present as insoluble calcium salts. The first stage of extraction involves pre-treating the algae by leaching it with a dilute mineral acid to protonate the carboxylic acid group of the alginates and convert calcium alginate to alginic acid. Demineralized algae is then treated with an alkali (caustic soda) to solubilize the alginic acid in the form of sodium alginate. After insoluble compounds have been removed by filtration or decantation, sodium alginate is again precipitated by adding a mineral acid. The precipitate is washed, centrifuged and dried. The food industry uses a wide range of alginates (sodium alginate, potassium alginate, etc.) that are produced from alginic acid by neutralization with a given alkali base. The alginate is then dried.

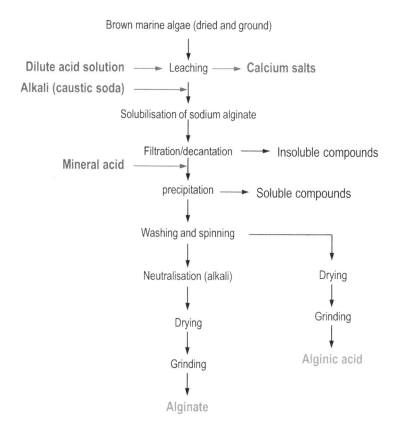

Figure 10.26. *Diagram of the purification of alginic acid and alginates*

10.2.3.4. Pectin

Pectin consists of D–galacturonic acid units linked by $\alpha(1-4)$ bonds (Figure 10.27).

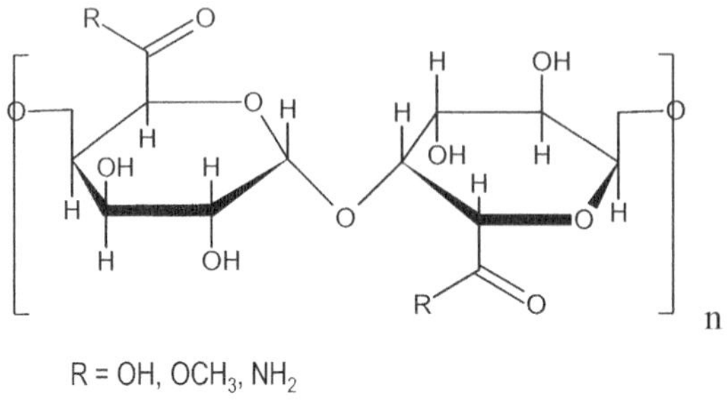

R = OH, OCH$_3$, NH$_2$

Figure 10.27. *Chemical structure of pectin*

Pectin is mainly extracted from apple pomace, citrus peel (lemons, oranges) or sugar beet pulp after the extraction of sugar (Figure 10.28). The extraction process involves the hydrolysis of protopectin in a hot dilute acid solution; pectin and other soluble substances are released under these conditions. After the insoluble fraction has been removed by filtration, the filtrate is concentrated and the pectin is precipitated by adding alcohol. The pectin obtained is generally highly methylated (HM pectin). It may be partially demethylated (LM pectin) in a hot dilute acid solution. The rate of demethylation or the final degree of esterification depends on temperature, pH and the length of the acid treatment. Amidated pectin is obtained from ammonia solutions.

10.2.3.5. Xanthan gum

Xanthan gum (Figure 10.29) is an exopolysaccharide produced by the bacterium *Xanthomonas campestris* during the aerobic fermentation of simple sugars (sucrose, lactose, etc.). Fermentation takes place under standardized pH, temperature and oxygen conditions. Xanthan gum is produced during the exponential and stationary growth phases, when *Xanthomonas* campestris is deficient in nitrogen. After sterilization, xanthan gum is precipitated by the addition of alcohol (isopropanol or ethanol) and recovered by centrifugation. It is then washed, dried and ground (Figure 10.30).

Separation Techniques 377

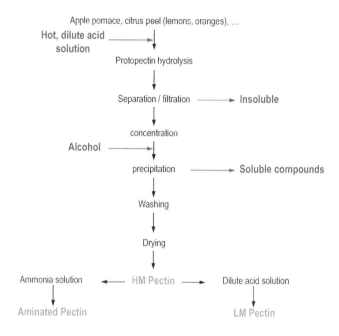

Figure 10.28. *Diagram of the purification of pectin*

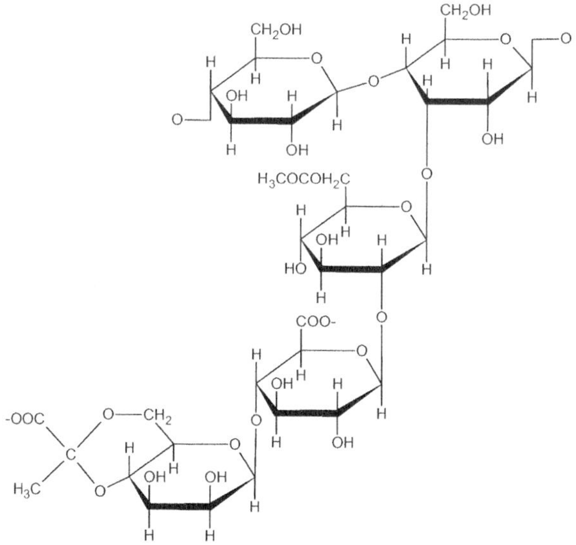

Figure 10.29. *Chemical structure of xanthan gum*

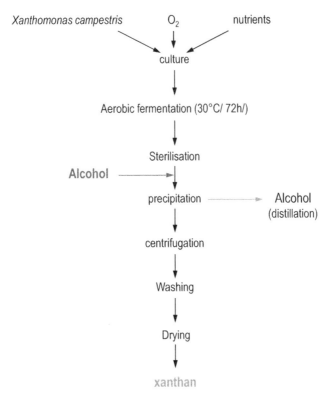

Figure 10.30. *Diagram of the purification of xanthan gum*

10.3. Lipids

The lipid fraction is an important part of the diet as it is a major source of energy as well as essential fatty acids and fat-soluble vitamins (A, D, E and K). It also contributes to the sensory properties of products in terms of texture and flavor. Oils or fats are extracted from oilseeds, drupes and/or the stone and/or the kernel of certain fruit as well as animal and fish tissue. The choice of extraction method of oils and fats depends on the raw material. For oils, methods often include pressure treatment followed by solvent extraction. The crude oil obtained cannot be sold as such and is generally refined, that is removal of its non-triglyceride fraction. Animal fats are obtained through the melting of by-products from slaughtering, butchering and rendering.

The composition and nutritional and technological functions of the lipid fraction of agricultural raw materials vary greatly. To meet the constraints of the food industry and/or consumer demands, they can be adapted by mixing oil of various origins or by physicochemical modifications (hydrogenation, transesterification, fractionation).

10.3.1. *Production of vegetable oils*

10.3.1.1. *Crude oil*

The pressure extraction and in some cases the solvent extraction of crude oil isolates the liquid lipid fraction from the solid fraction (cake), which consists mainly of protein and cellulose. It is usually preceded by various mechanical processes such as the removal of foreign matter (earth, leaves, metal catalysts of lipid oxidation), hulling, fragmentation by abrasion and grinding of oilseeds or fruit. Steam treatment can also be carried out to increase the extraction yield by causing the rupture of oil-containing cells, thereby allowing oil flow. It also destroys microorganisms and thermolabile toxic substances, and deactivates certain enzymes like lipoxygenase.

Pressing seeds or fruit in hydraulic or continuous screw presses separates the crude oil from the oil cake. Pressure and temperature both affect the extraction yield and the quality of the oil obtained. Increasing pressure and temperature improves the extraction yield but may impair oil quality. It is therefore better to apply lower pressure and temperature and carry out cold extraction using an organic solvent. In some cases, only pressure treatment is applied (peanuts, sunflower seeds, rapeseed).

After it exits the press, oil contains solid debris that is removed either by centrifugation, cloth filtration or filtration using steel meshes with metal scrapers. The oil obtained is usually dried to reduce its moisture content to below 0.1% to avoid the risk of glyceride hydrolysis. Oil can be spray-dried at 80–90°C in a vacuum chamber.

"Virgin olive oil" is oil obtained from crushed olive pulp by mechanical processes only (pressure, decantation, centrifugation). The oil is immediately ready for consumption if the olives used are of good quality. The oil from green olives is intense and fruity whereas oil from late-harvest olives is yellow and mild. Pressing separates the liquid fraction (composed of around 20 kg of oil and 40 kg of olive water per 100 kg of olives) from the solid fraction (around 40 kg of olive cake containing 10–15% oil per 100 kg of olives). An alternative is to replace the pressing operation by the continuous centrifugation of crushed olive pulp. This separates the moist solid phase (olive cake) containing 10–15% residual oil, the aqueous phase (olive water) low in oil, mainly in the form of an emulsion, and the oil phase, which is the virgin oil.

Sometimes oil is recovered from the oil cake after the pressed crude oil has been removed. The oil is extracted from the cake by solvent percolation; the solvent (hexane) and the solid material move counter-currently to one another (Figure 10.31). Solvent extraction increases the amount of oil extracted and at the same time improves the stability of the cake. Once the oil has been removed from the cake, the solvent is removed by distillation and can be re-used after condensation. The cakes exiting the extractor contain less than 1% residual oil, but have high levels of solvent (around 30% solvent by weight). Removing the solvent from cakes, using heat or vacuum, means they can be used for animal or human consumption. They are finally dried to remove any traces of water and residual solvents. Some oils can be directly extracted by solvent extraction without prior pressure extraction.

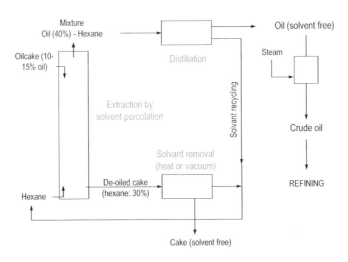

Figure 10.31. *Solvent extraction of oil from cakes*

The temperature and water content of oil cakes affect solvent extraction yield. Increasing the temperature improves the diffusion and extraction of oil from the cakes but also increases solvent loss. Generally, temperatures are slightly above 60°C, and solvent loss is constantly compensated for. The high water content of the de-oiled cakes increases the risk of them sticking to the extraction unit. However, the presence of water facilitates the removal of the solvent after the oil has been extracted.

10.3.1.2. Refining of crude oil

Crude oil is a mixture of triglycerides, diglycerides, monoglycerides, free fatty acids, phospholipids, waxes, pigments, sterols, vitamins, flavonoids, tannins, trace metals and solvents. It is generally refined (to reach appearance, sensory qualities and stability standards) before retail. Refining removes non-triglyceride lipid matter. It involves degumming or mucilage removal, deacidification, decolorization, deodorization and drying.

Degumming

In the presence of water, phospholipids form precipitates known as "mucilage". Degumming or mucilage removal involves precipitating and removing phospholipids from crude oil. It is carried out with water (2 to 4 %) and acid, usually phosphoric acid (0.1 to 0.3%), on oil at 80°C (Figure 10.32). The water and gum or mucilage are separated by natural sedimentation or centrifugation. Degumming also removes proteins and other impurities, as colloids or suspensions in the oil. The mucilage of certain oils can be used as a surfactant (e.g. soy lecithin).

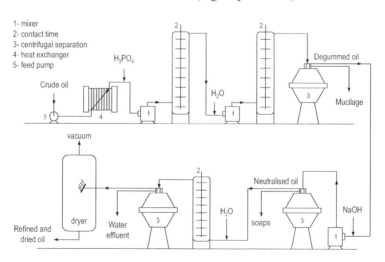

Figure 10.32. *Main stages of oil refining*

Deacidification

Deacidification or neutralization involves removing free fatty acids from crude oil, which are normally absent from the lipid fraction of living cells. They cause undesirable flavors and are formed after harvest in oilseeds and fruit, after the slaughter of animals by the action of lipase or during processing (e.g. heating). Two methods (chemical or physical) can be used to deacidify oil.

The first and the most commonly used method involves adding a base (sodium hydroxide) to degummed oil to avoid emulsification. Neutralizing oil induces either solubilization in the aqueous phase or the precipitation of free fatty acids as soaps that are removed by continuous centrifugation (Figure 10.32). The latter can hold up to 50% of their weight in oil. In addition, neutralised oil should be immediately washed after the removal of soaps to avoid the saponification of triglycerides by excess sodium hydroxide. Chemical neutralization incurs most of the losses during oil refining and generates large amounts of effluent. It also reduces the amount of sterols, tocopherols and vitamins in oil, which should be preserved for their nutritional and antioxidant role. However, chemical neutralization removes residual phospholipids and traces metals by precipitation and destroys certain colored pigments.

Physical deacidification removes free fatty acids by vacuum distillation and steam entrainment. This method has many advantages compared to chemical deacidification:

– reduction in oil loss by limiting saponification;

– no environmentally harmful by-products;

– lower energy costs;

– reduction in the amount of volatile compounds such as molecules responsible for unpleasant odors;

– suitable method for oils containing large amounts of free fatty acids.

However, physical deacidification occurs at high temperatures and as a result is not suitable for heat-sensitive oils (oils rich in unsaturated fatty acids). It causes isomerization of double bonds in unsaturated fatty acids and polymerization. In addition, trace metals such as iron in crude oil induce browning during distillation. And finally, the amounts of tocopherol and carotenoid (antioxidants) in oil are reduced after physical deacidification.

Decolorization and deodorization

After neutralization, oil is decolorized and deodorized. Decolorization not only removes chlorophyll, but also carotenoid pigments from the oil. It is therefore a compromise between eliminating compounds that reduce oil stability (chlorophyll) and those that protect against oxidation (carotenoids). For decolorization, oil is heated to around 100°C and treated with active carbon and other adsorbents before being filtered.

The last step of the oil refining process is deodorization to remove certain volatile compounds such as aldehydes and ketones, often responsible for unpleasant odors, or residual free fatty acids that are more sensitive to oxidation than triglycerides. Deodorization is carried out by vacuum distillation at about 200°C. The absence of air is essential and prolonged heating at high temperatures should be limited to avoid polymerization reactions. Antioxidants are sometimes added to oil before deodorization to prevent the risk of oxidation. Deodorization is particularly necessary for fish oils and certain oilseeds. During the process, some of the phytosterols are eliminated in the distillate. They can be purified from distillates to be reincorporated as such, as phytosterol esters (obtained by esterification of phytosterols with fatty acids) or as phytostanols (obtained by hydrogenation of phytosterols) into various fat products such as margarine.

The oil is finally dried to remove any moisture that could cause triglyceride hydrolysis, and packaged oxygen-free, using nitrogen packaging for example.

10.3.2. Lipid modification

The lipid fraction derived from plants or animals is not necessarily satisfactory from a nutritional (composition), technological (spreadability, melting point, resistance to heat treatment) or chemical perspective (sensitivity to oxidation). It may therefore be necessary to modify lipids by technological or biological treatments to meet certain specifications (margarine, frying oil, infant formula, etc.). These treatments are used to obtain lipid fractions that have a wide range of use compared to the original raw materials. Hydrogenation, transesterification and fractionation are the main treatments used. They are applied to refined oils and can be used on their own or in combination.

10.3.2.1. *Hydrogenation*

Hydrogenation is a chemical reaction that reduces the degree of unsaturation of an oil. The result of the hydrogenation reaction is expressed by a lower iodine index. Hydrogenation can be partial or total. It is used to:

– increase the melting point of oil by increasing the proportion of saturated fatty acids;

– improve the stability of the oil against oxidation.

However, hydrogenation reduces the proportion of polyunsaturated fatty acids and therefore the nutritional value of the oil. Hydrogenation is generally effective in increasing the melting point or stabilizing highly unsaturated oils such as rapeseed, soybean or fish oils. It is also used to reduce the degree of unsaturation in oils with a low level of unsaturated fats such as coconut oil.

Changes in the properties of the oil depend on the level of hydrogenation applied (percentage of hydrogenated double bonds) and the selectivity of the reaction. Selectivity corresponds to the preferred hydrogenation of certain double bonds in polyunsaturated fatty acids (e.g. linolenic acid) based on the ratios of the rate constant k_1/k_2 and k_2/k_3:

$$18:3 \xrightarrow{k_1} 18:2 \xrightarrow{k_2} 18:1 \xrightarrow{k_3} 18:0 \qquad [10.1]$$

For example, when the ratio of the rate constants k_2/k_3 is high (around 50 to 100), partial hydrogenation leads to the formation of monounsaturated fatty acids without affecting the content of saturated fatty acids.

Partial hydrogenation lowers the degree of fatty acid unsaturation, but also generates isomers, namely conjugated fatty acids and *trans* fatty acids. For the same chain length, *trans* fatty acids have a higher melting point than *cis* fatty acids with no change in the iodine index. The presence of *trans* fatty acids in foods is controversial as it could contribute to the occurrence of cardiovascular diseases by affecting atherogenic parameters. The partial hydrogenation of triglycerides consisting of three oleic acid chains (triolein) generates a wide variety of triglycerides containing oleic, elaidic (*trans* isomers of oleic acid) and/or stearic acid. Complete hydrogenation only forms tristearins. Double bond isomerization is favored when there is a shortage of hydrogen molecules on the surface of the catalyst, which limits the rate of hydrogenation. Thus, isomerization is favored by low hydrogen pressure, high temperatures, gentle agitation, highly unsaturated oil or an excessively high catalyst concentration and/or activity.

Hydrogenation takes place in a heterogeneous medium containing a gas phase (hydrogen), a liquid phase (oil) and a solid phase (catalyst). It occurs on the surface of the catalyst (usually nickel), and is exothermic. It requires intense agitation to improve the diffusion between the three phases and allow a homogeneous distribution of hydrogen and the catalyst in the oil. Temperature, hydrogen pressure, catalytic activity and concentration, agitation and reaction time are the main factors affecting oil hydrogenation.

The hydrogenation reaction is performed in a hermetically sealed tank, usually between 150 and 200°C at a pressure of 0.3 MPa. The tank is filled two-thirds full with oil and a catalyst (Figure 10.33). The catalyst should have high activity and selectivity and be easily removable from the hydrogenated oil. The activity, selectivity and dispersibility of catalysts in oil increase as particle size decreases; however, recovery is more difficult once the reaction is finished. Before hydrogenation, oil must be refined and dried as catalysts are quickly deactivated by the presence of free fatty acids, lipid oxidation products, water or sulfur compounds. Rapeseed oil contains organic sulfur compounds (from the thioglycosides in the seeds), which are not removed by chemical refining. During the hydrogenation reaction, they poison the catalysts, which must therefore be overdosed. Finally, hydrogen is circulated in the oil by bubbling it through a recirculation system (hydrogen injected into the oil is drawn from the headspace of the tank before being reinjected).

When the desired degree of hydrogenation or iodine index has been reached, the catalyst is removed from the reaction mixture by filtration. The filtration temperature is lowered to 90°C to minimize the risk of oxidation, and maintained at this temperature to avoid an increase in viscosity. The recovered catalyst is purified before being re-used. Once hydrogenated, the oil is refined to remove any remaining traces of the catalyst or color defects that may have appeared during the reaction.

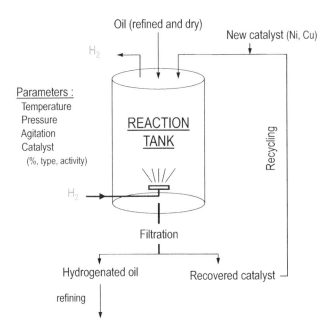

Figure 10.33. *Diagram of the hydrogenation of oils and fats*

10.3.2.2. *Transesterification*

Transesterification is a modification of the triglyceride structure by the intra- or intermolecular rearrangement of fatty acid chains on the glycerol molecules. Transesterification can be catalyzed chemically (using alkali metal alcoholates such as sodium methylate for example) or enzymatically. It changes the physical characteristics of the fat (melting and crystallization properties) without altering its fatty acid composition. Pork fat (lard) has a high proportion of palmitic acid in the sn-2 position of the triglyceride. It crystallizes in the form of β–crystals. After transesterification, the random distribution of fatty acid chains on the glycerol molecules modifies the functional and crystallization (crystallization as β'–crystals) properties of the fat. It is therefore a good alternative to the partial hydrogenation of oils as it does not generate *trans* fatty acids. Margarines with no *trans* fatty acids can be obtained by subjecting a mixture of fully hydrogenated palm oil and sunflower oil to transesterification. A solid fat (e.g. margarine) containing more than 60% essential fatty acids is obtained from a transesterified mixture of sunflower oil and 5% fat with a high melting point.

Transesterification catalyzed by chemical agents is carried out on dried refined oil, and heated to around 100°C for 30–60 minutes. The redistribution of fatty acid chains on the three hydroxyl functions of the glycerol molecules occurs randomly. The proportion of different triglycerides obtained follows a statistical distribution (Figure 10.34), which is generally not the case with natural oils and fats.

By operating at a lower temperature, for example below the crystallization temperature of high melting point triglycerides, high melting point triglycerides, produced by transesterification, crystallize out and thereby change the statistical distribution of fatty acids in the triglycerides. The shift in equilibrium progressively occurs towards the formation of higher melting point triglycerides (Figure 10.34).

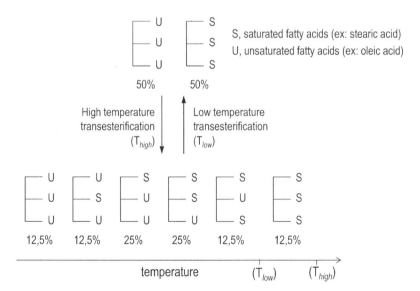

Figure 10.34. *Transesterification at high and low temperatures*

After transesterification, the catalyst is deactivated by adding water to the reaction medium. The transesterified fat is cleared of soaps that formed upon contact with water, and dried under vacuum to remove any moisture. This last operation removes methyl esters that have formed during esterification. In transesterification reactions catalyzed by chemical agents, free tocopherols can be esterified, thereby losing their antioxidant activity. However, esterification hardly affects their vitamin activity.

Transesterification can also use enzyme catalysts such as microbial lipases. Due to their specificity, products can be obtained that could otherwise not be produced by chemical esterification methods. They are used to carry out exchanges between fatty acids in the sn1 and sn3 positions of triglycerides, without affecting the fatty acids in the sn2 position. This reaction has the advantage of preserving the nutritional quality of transesterified fats because the fatty acids in the sn2 position pass through the intestinal wall as monoglycerides whereas fatty acids in sn1 and sn3 positions are sometimes excreted in the feces as soaps. In addition, the pH and temperature conditions applied in enzymatic transesterification are not extreme and therefore changes such as the degree of unsaturation of fatty acids are quite limited. However, the relatively high cost of the process means it is limited to certain value-added applications (infant formula, health products, etc.).

10.3.2.3. Fractionation

The separation of oils or fats into fractions is based on differences in the solubility and melting point of the triglycerides. The less soluble, higher-melting portion is called the stearin fraction while the more soluble, lower-melting portion is called the olein fraction. Fractionation has a number of purposes:

– to remove the small amount of triglycerides or other higher-melting lipid materials that tend to cloud the oil during storage at low temperature. This is, for example, the case with the dewaxing of sunflower oil at 5°C;

– to obtain two or more fractions, which together offer a wider range of uses than the original raw material or meet specific applications, for example tallow, anhydrous milk fat (AMF) or palm oil can be separated into olein and stearin fractions. The olein fraction of palm oil is an excellent frying oil whereas the stearin fraction is used for margarine. Medium-melting palm oil can be used as a substitute for cocoa butter. The olein fraction of AMF is used in the production of spreadable butters. Controlled-melting stearin fractions are used in the baking industry;

– to establish an alternative to other lipid modifying treatments such as hydrogenation.

The fractionation of a lipid raw material can, in theory, be repeated again and again. However, increasingly targeted fractionations (series of fractionations) can generate additional fractions that are difficult to use. Thus, these methods are only commercially viable if the target lipid fraction is used in high value-added applications.

Fractionation is carried out on an industrial scale mainly by fractional crystallization (Figure 10.35). The melted lipid (raw material) is subjected to partial crystallization by controlled cooling. The aim is to obtain high-melting lipid crystals that are sufficiently large to allow easy separation. Controlled agitation facilitates heat and mass transfer. The temperature is higher in the immediate vicinity of the growing crystals due to the exothermic nature of triglyceride crystallization and a triglyceride concentration gradient surrounds the crystals. However, excessive agitation can reduce the size of growing crystals, which makes recovery more difficult.

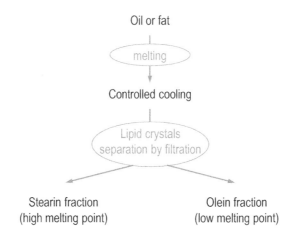

Figure 10.35. *Principle of fractionation of oils and fats*

When crystallization is complete (after a few hours), the stearin and olein fractions are usually separated by continuous or filter press filtration (Figure 10.36). The olein fraction is removed through the pores of a filtration membrane while the high-melting crystals are retained in the retentate. Occluded olein (oil trapped between the stearin crystals) can be removed by applying pressure to the stearin cake.

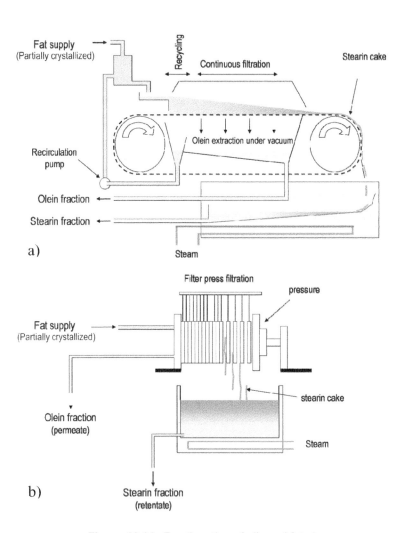

Figure 10.36. *Fractionation of oils and fats by a) continuous filtration and b) filter press filtration*

10.4. Pigments and flavorings

Agricultural raw materials, especially of plant origin, contain a wide variety of sensory components that are extracted for the production of natural colorings and flavorings. These operations are an excellent way of adding value to by-products from primary processing.

10.4.1. *Types of pigments and flavorings*

The main natural pigments can be classified into four categories:

– carotenoids including β–carotene, lycopene, xanthophyll and so on;

– porphyrin pigments: chlorophyll, heme pigments;

– flavonoids and derivatives;

– betalains.

Carotenoids are pigments extracted from fruit (peaches, cherries, oranges, strawberries, etc.), vegetables (carrots, tomatoes, etc.) and flowers. These compounds, which vary greatly in color, consist of chains of isoprene units, some of which are described in Figure 10.37. In addition to their coloring properties, these pigments have antioxidant properties capable of limiting the oxidative stress involved in cell aging; however, they are highly sensitive to light and oxygen.

Porphyrins are heterocyclic compounds composed of four pyrrole subunits (Figure 10.38). This heterocycle can chelate a certain number of metals in its center (Fe, Mg, Mn, Cu, Zn, etc.): porphyrins containing iron are called hemes (haemoglobin) and those incorporating magnesium are known as chlorophylls.

Flavonoids are a group of natural polyphenolic compounds. They can be found in all parts of higher plants (roots, stems, leaves, flowers, fruit, etc.) and have a wide variety of colors from yellow, orange, red and purple to blue. They can exhibit many biological properties: anti-viral, anti-tumor, anti-inflammatory, anti-cancerous and so on. Their basic structure is composed of 15 carbon atoms comprising two rings of 6 carbons (A and B) linked by a chain of 3 carbons. The compounds in this family differ in number, position and type of substitutions (hydroxy, methoxy and others) on the A and B rings and the C3 chain. In their natural state, one or more hydroxyl groups are glycosylated; the part of the flavonoid other than the carbohydrate is called the aglycone. The three main subgroups of this family are anthocyanins, flavones and flavonols; in this type of molecule, the C3 chain forms a ring by interacting with a hydroxyl group in ring A. Anthocyanins have a hydroxyl group in position 3, flavones a carbonyl group in position 4 and flavanols a carbonyl group in position 4 and a hydroxyl group in position 3 (Figure 10.39).

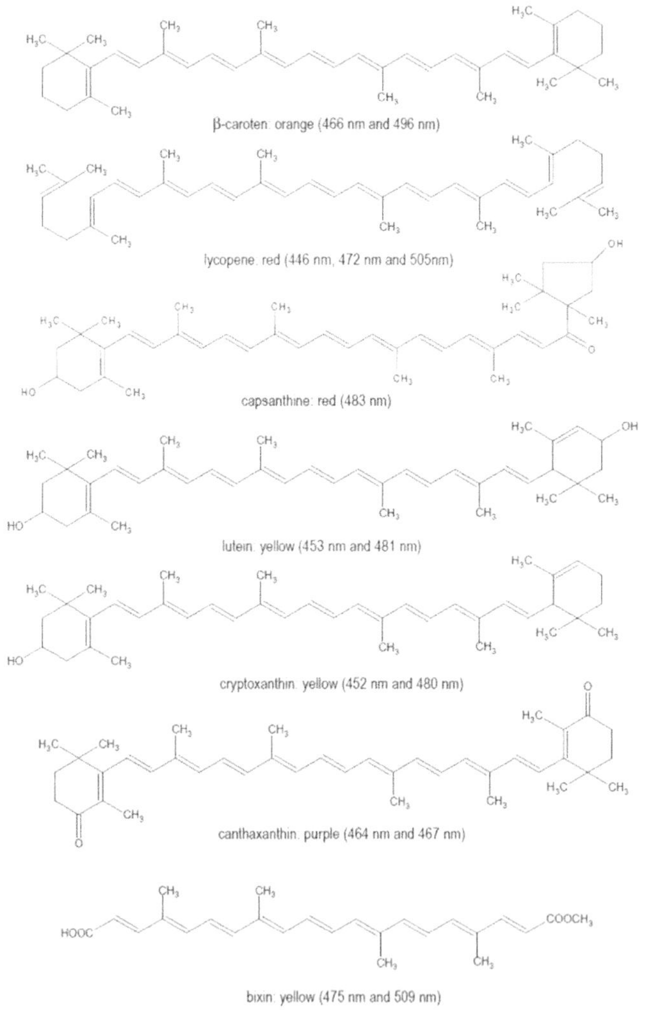

Figure 10.37. *Structure of the main carotenoids and xanthophylls (according to [LIN 94])*

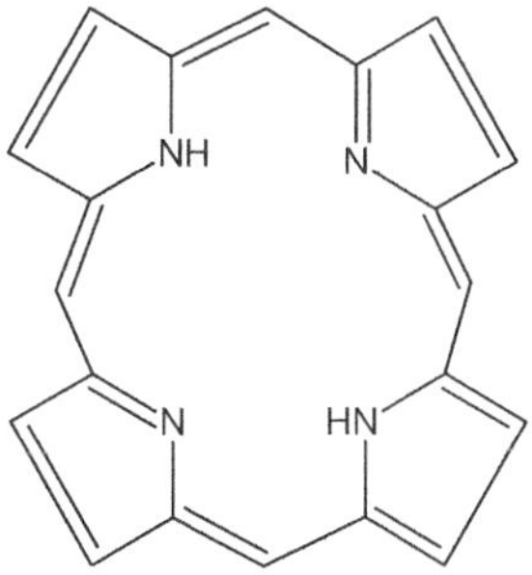

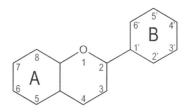

Figure 10.38. *Structure of porphyrin*

Figure 10.39. *General structure of flavonoids*

Anthocyanins that have a wide range of colors are often glycosylated in position 3 and 5, and the most common sugars are glucose, galactose, rhamnose and arabinose, which can be acylated with different acids (coumaric, cinnamic, caffeic, ferulic, etc.). Despite the presence of these sugars, which increases the solubility and stability of pigments, the use of anthocyanins remains limited to certain types of products due to their high sensitivity to physicochemical conditions and their ability to precipitate proteins. Anthocyanins are very stable at low pH, which explains their use in fruit juice and lactic fermentation products such as yoghurts; the spectral properties of anthocyanins are highly dependent on pH (Figure 10.40).

Betalains are the red and yellow pigments in beetroot (*beta vulgaricus*). These pigments have a piperidine ring in common with two carboxyl groups and an iminium group (Figure 10.41), corresponding to the nitrogen in linear α-amino acid (vulgaxanthin) or cyclic α-amino acid (betanin). Their amphoteric character makes them highly soluble in water over a broad pH range, but sensitive to heat treatment, making it difficult to produce sterile concentrates.

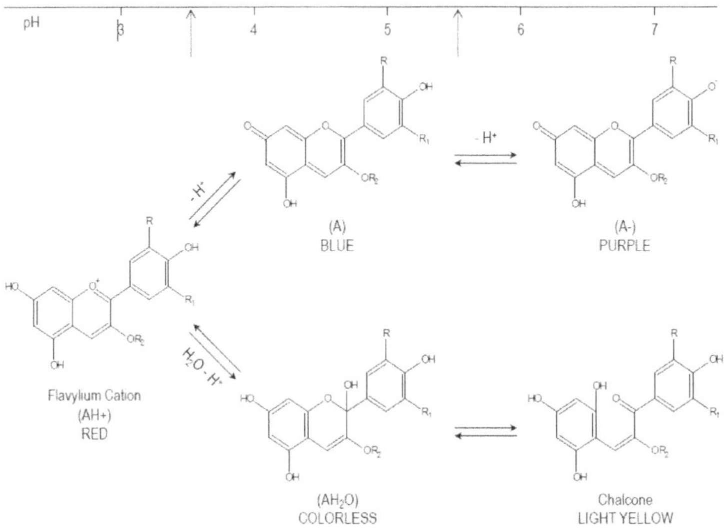

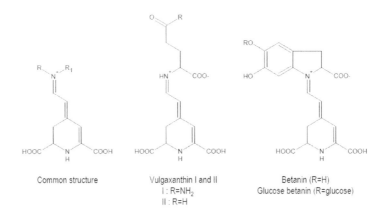

Figure 10.40. *Influence of pH on the structure and color properties of anthocyanins (according to [LIN 94])*

Figure 10.41. *Structure of betalains*

Aromas are a set of molecules that are carried in a gaseous state to the olfactory cells when food is in the mouth; the aromatic fraction includes hundreds of molecules with concentrations often less than a nanogram per kilogram. Aromatic molecules are of low molecular weight and have a sufficiently high vapor pressure at atmospheric pressure and room temperature so that a fraction can reach the olfactory mucosa retronasally (via the mouth).

Knowledge of the aromatic constituents of raw materials and foods has increased rapidly since the 1970s due to the combination of gas chromatography and mass spectrometry; thousands of compounds belonging to different classes of organic compounds have been identified. They include:

– hydrocarbons, many of which are terpenes resulting from the condensation of isoprenes;

– aldehydes, ketones, alcohols and thiols from the metabolism of amino acids and fatty acids;

– volatile acids, fermentation esters;

– volatile amines from the decarboxylation of amino acids;

– heterocycles (furan, pyrazine, etc.) of pyrolysis products of carbohydrate derivatives in the presence of proteins;

– phenols from the degradation of lignin.

The aromatic characteristics of a food often result from the olfactory properties of one or more molecules (Figure 10.42).

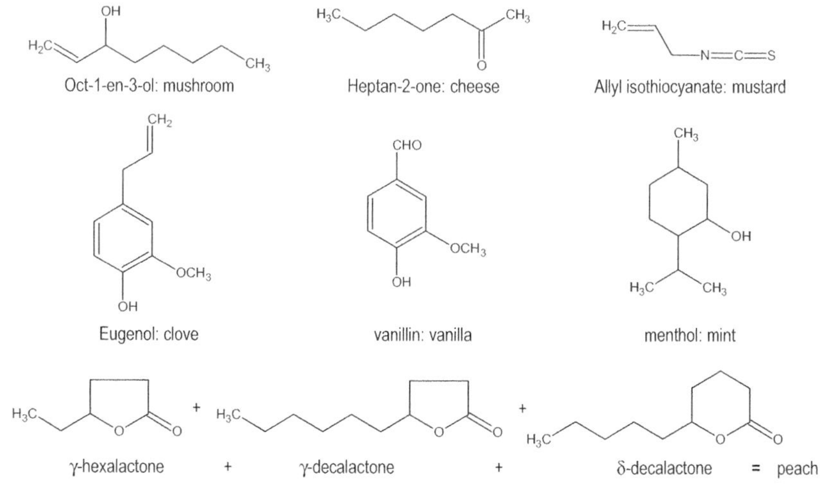

Figure 10.42. *Examples of aromatic molecules*

In most foods, flavor is caused by a large number of components, some of which in isolation have no sensory properties but still contribute to the aromatic profile of the product. This complexity makes the production of flavorings from synthetic molecules difficult. Moreover, many molecules involved in flavor have asymmetric carbons and the olfactory properties are often stereospecific, making chemical synthesis more difficult, which usually results in the formation of racemic mixtures. In general, food flavorings are classified into four categories:

– *natural flavorings* are fractions or extracts of natural products or formed by biological processes (enzymatic and/or microbial) from natural substrates;

– *nature identical flavorings* are synthetic molecules identical to those present in the natural environment (e.g. 4-hydroxy-3-methoxybenzaldehyde corresponds to vanillin or 1-octen-3-ol corresponds to mushroom);

– *artificial flavorings* are molecules that do not exist naturally (e.g. 3-ethoxy-4-hydroxybenzaldehyde corresponds to ethylvanillin, the olfactory intensity of which is 10 times greater than that of vanillin);

– *process flavorings* such as the degradation products of sugars in the presence of amino acids (Maillard reaction and Strecker degradation) or the degradation of lignin (smoke flavor).

Molecules responsible for food flavor have a number of origins:

– the raw materials;

– the biological, enzymatic and microbiological processes that occur in raw materials during storage or processing;

– the physicochemical treatments (cooking, smoking, salting, etc.).

The knowledge gained with regard to flavorings has made it possible to characterize aromatic substrates and elucidate the main biosynthetic pathways; from this work, it has also been possible to identify biological and physicochemical parameters that enhance aromatic intensity and make better use of the aromatic potential of raw materials. Many aromatic substrates in plants are glycosylated and are consequently non-volatile; enzymatic or chemical hydrolysis releases aglycone and increases the aromatic potential of the product, for example the release of vanillin from glucovanillin in the fermentation of pods. Many biotechnological processes using plant cells, microorganisms and enzymes were developed to enhance the flavor of wine, fruit juice and so forth or to produce flavorings from substrates from agricultural or industrial by-products.

These scientific findings have contributed to the development of the flavoring industry that produces an extensive range of aromatic bases to meet the needs of the food industry as well as consumer expectations.

10.4.2. *Extraction/concentration of colorings and flavors*

Some raw materials with coloring and/or aromatic properties can be used as such; they can be stored in a frozen or dried state, for example herbs.

Industry focuses more on concentrates, essences or extracts for reasons of practicality and quality consistency.

Solubility in water or organic solvents and the vapor pressure of pigments and aromatic molecules are the main factors influencing the type of extraction and/or concentration methods used.

10.4.2.1. *Solid/liquid extraction*

The extraction of pigments and organic acids is often performed directly on raw materials (beet) or by-products (grape pomace from winemaking, etc.). The technique involves placing the product in contact with a solvent that is selected based on the type of raw material, the extractable compounds and the uses of the extract and the residual by-product. The solvent can be aqueous or organic (alcohol, hexane, acetone, petroleum ether, ethyl acetate, dichloromethane, supercritical CO_2). In the case of an aqueous solvent, the diffusion transfer rate is determined by Fick's law; it therefore depends on the liquid/solid interface area, the diffusion coefficient and the concentration gradient of the components to be extracted at the liquid/solid interface. In the case of an organic solvent (flavor extraction), the extraction rate depends on the partition coefficient of the components between the aqueous phase of the product and the organic phase.

Extraction can be performed in batch (single or successive) or continuous counter-current mode. Batch extractors consist of temperature-controlled tanks, a container for the extracted product, an agitation system and in some cases a filter at the bottom of the tank or floating on the top to clarify the liquid phase. Continuous diffusers all operate counter-currently either by immersing or spraying the product; counter-current mode provides a greater level of extraction since the outgoing product is in contact with the pure solvent. Continuous diffusers are less used for the extraction of flavor and colorings because they are less flexible than batch systems where products can be more finely crushed, enzymes can be added and physicochemical parameters can be controlled (temperature, pH, etc.).

If the solvent is water, the extract is filtered if necessary, concentrated by evaporation or reverse osmosis and in some cases spray-dried. When the solvent is organic, the extract is distilled and the solvent is condensed and recycled. Recycling can be done simultaneously with extraction as shown in Figure 10.43; in the case of supercritical CO_2, the solvent returns to the gaseous state by expansion and then into a liquid state by mechanical compression (see Volume 2). Solid/liquid extraction can be facilitated by the microwave treatment of the solid/liquid mixture.

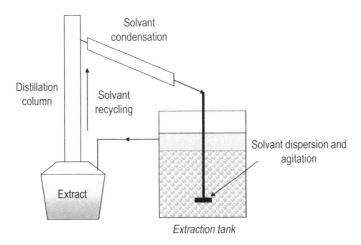

Figure 10.43. *Extraction with solvent recycling*

10.4.2.2. *Liquid/liquid extraction*

This technique is used to extract volatile aromatic compounds from a liquid. Since the liquid is usually aqueous, the solvent used is organic and immiscible in water to facilitate the separation of the two phases. The extraction rate depends on the contact surface area and the partition coefficients between the two phases. The extract obtained by the solvent is called oleoresin.

Extraction can be done in batch by creating an emulsion of the organic phase in the aqueous continuous phase, followed by phase separation to recover the organic phase. Extraction can also be carried out continuously in packed or tray columns. Filling columns with Raschig rings (pieces of tube used in large numbers as a packed bed within columns) increases the contact surface area between the two liquid phases: the heavy phase is added at the top and the light phase (usually the organic phase) at the bottom (Figure 10.44). The phase separation of the two liquids is sometimes difficult to achieve due to the presence of amphiphilic molecules, which can stabilize emulsions.

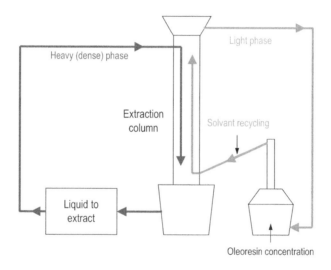

Figure 10.44. *Continuous liquid/liquid extractor*

10.4.2.3. *Distillation*

Since aromatic molecules are volatile, they can be extracted by distillation. Direct heating of the tank generally results in the thermal degradation of the raw material, which affects the quality of the flavoring. To limit thermal degradation, water vapor is added to the tank releasing latent heat by condensation. The vapor and the volatile molecules condense and are collected in an essencier (decanting flask) (Figure 10.45). Extraction by distillation can be carried out under vacuum to reduce the temperature in the extraction tank. When the volatile aromatic fraction is immiscible in water, two phases are obtained: the aqueous phase and the aromatic fraction known as essence. The latter can be heavier (onion essence) or lighter (citrus essence) than the aqueous phase, which is much larger in volume than essence and can be recycled in the extraction tank. When the aromatic fraction is soluble in water, flavors can be solvent-extracted as described earlier.

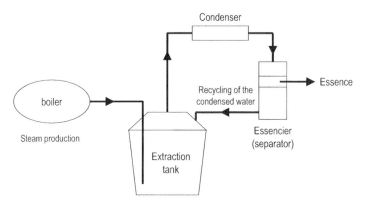

Figure 10.45. *Steam distillation*

10.4.2.4. Expression

Some essences can be obtained solely by expression, which is the case with citrus essence in particular. These fruits contain large quantities of hydrophobic molecules located inside oil vesicles. The principle of extraction of essential oils is based on rupturing these pockets of essence by mechanical means (pressure, incision or abrasion) at low temperatures. Essential oils are separated from the aqueous phase by decantation or centrifugation.

10.4.3. Formulation

Aromatic extracts, whether essences or oleoresins, cannot be used as such due to their hydrophobicity, volatility and the low concentrations used in food formulation. Moreover, the aromatic base must be adapted to the product and the food production technology (resistance to heat, acid, oxygen, etc.). The flavoring must be diluted to the desired concentration in organic phase (oil, fat, alcohol, etc.) or a solid phase (sugar, maltodextrin, cyclodextrin, starch, protein, salts, etc.). Diluting flavorings in a solid phase can be achieved by spraying an alcohol solution containing the flavoring onto the solid support and then removing the solvent. Alternatively, this can be done by co-spraying the dissolved flavoring and support (gum arabic, maltodextrin, cyclodextrin).

In the spraying method, the flavor is adsorbed on the surface of the support, which has two drawbacks: loss by evaporation and degradation upon contact with oxygen. On the other hand, in the co-spraying method, the aromatic fraction is located in the center of the powder granule because, during drying, water transfer from the center of the droplet to the solid/air interface is faster than the equivalent transfer of aromatic molecules, resulting in better retention and less exposure to air. In the case of cyclodextrins, hydrophobic substances are included within these cyclic molecules containing 6, 7 or 8 glucose units. Cyclodextrins are obtained from the hydrolysis of starch and bridged at the ends by a bond between carbons 1 and 4 (Figure 10.46). They are ring-like structures with a hydrophilic exterior and a hydrophobic interior, making them able to entrap aromatic molecules. This inclusion has the dual advantage of stabilizing the flavor during processing and storage and delaying its release in the mouth, which increases the length of olfactory perception.

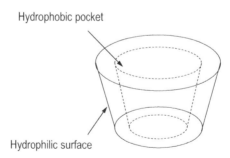

Figure 10.46. *Structure of β-cyclodextrin*

Bibliography

[ABE 97] ABECASSIS J., CHAURAND M. "Appréciation de la valeur d'utilisation du blé dur en semoulerie et en pastification", in GODON B., LOISEL W. (eds), *Guide d'Analyses dans les Industries des Céréales*, Lavoisier, Paris, 1997.

[ALL 99] ALLOSIO-OUARNIER N., QUEMENER B., BERTRAND D. *et al.*, "Application of high performance anion exchange chromatography to the study of carbohydrate changes in barley during malting", *Journal of the Institute of Brewing*, vol. 106, pp. 45–52, 1999.

[ALU 98] ALUKO R.E., KEERATIURAI M., MINE Y., "Competitive adsorption between egg yolk lipoproteins and whey proteins on oil-in-water interfaces", *Colloids and Surfaces B: Biointerfaces*, vol. 10, pp. 385–393, 1998.

[ANT 97] ANTON M., GANDEMER G., "Composition, solubility and emulsifying properties of granules and plasma of hen egg yolk", *Journal of Food Science*, vol. 62, no. 3, pp. 484–487, 1997.

[ANT 00] ANTON M., LE DENMAT M., GANDEMER G., "Thermostability of hen egg yolk granules: contribution of granular structure", *Journal of Food Science*, vol. 65, no. 4, pp. 581–584, 2000.

[AUB 85] AUBERT C., *Les aliments fermentés traditionnels*, Terre Vivante, Paris, 1985.

[BAK 41] BAKER J.C., MIZE M.D., "The origin of the gas cell in bread dough", *Cereal Chemistry*, vol. 18, pp. 19–34, 1941.

[BAM 85] BAMFORTH C.W., "The foaming properties of beer", *Journal of the Institute of Brewing*, vol. 91, pp. 370–383, 1985.

[BAR 03] BARON F., NAU F., GUÉRIN-DUBIARD C. et al., "Effect of dry heating on the microbiological quality, functional properties, and natural bacteriostatic ability of egg white after reconstitution", *Journal of Food Protection*, vol. 66, pp. 825–832.

[BEG 96] BEGUIN G., CLAEYS C., DELAVET C., "Procédé et dispositif de découpe de légumes et de fruits", Brevet français, Patent no 2732922, 1996.

[BEG 98] BEGUIN G., VAROQUAUX P., "Procédé de lavage et de désinfection de feuilles de légumes tels que des salades", Patent no FR 2750576, 1998.

[BER 85] BEROT S., DAVIN A., "Technologie d'extraction et de purification des matières protéiques végétales", in *Protéines Végétales*, Lavoisier, Paris, pp. 335–472, 1985.

[BEV 02] BEVERIDGE T., "Opalescent and cloudy fruit juices: formation and particle stability", *Critical Reviews in Food Science and Nutrition*, vol. 42, pp. 317–337, 2002.

[BRA 01] BRACCINI I., PEREZ S., "Molecular basis of Ca2+-induced gelation in alginates and pectins: the egg-box model revisited", *Biomacromolecules*, vol. 2, no. 4, pp. 1089–1096, 2001.

[BRI 02] BRIGGS D.E., "Malt modification. A century of evolving views", *Journal of the Institute of Brewing*, vol. 108, pp. 395–405, 2002.

[BRI 02] BRITTEN M., "Ingrédients laitiers", in VIGNOLA C.L, *Science et Technologie du Lait – Transformation du Lait*, Presses Internationales Polytechniques, Montréal, pp. 471–526, 2002.

[BRU 06] BRUMMELL D.A., "Cell wall disassembly in ripening fruit", *Functional Plant Biology*, vol. 33, pp. 103–119, 2006.

[CHA 89] CHAMBROY Y., "Physiologie et température des produits frais découpés", *Revue Générale du froid* 3, Colloque AFF, Avignon, 8–10 November 1989.

[CHA 99] CHANDRA G.S., PROUDLOVE M.O., BAXTER E.D., "The structure of barley endosperm. An important determinant of malt modification", *Journal of the Science of Food and Agriculture*, vol. 79, pp. 37–46, 1999.

[CHA 05] CHARLES F., SANCHEZ J., GONTARD N., "Modeling of active modified atmosphere packaging of endives exposed to several postharvest temperatures", *Journal of Food Science*, vol. 8, pp. 443–449, 2005.

[CHE 92] CHEFTEL J.C., CHEFTEL H., "Graisses et huiles", in *Introduction à la Biochimie et à la Technologie des Aliments*, vol. 1, Lavoisier, Paris, 1992.

[CHR 95] CHRISTIE W.W., "Composition and structure of milk lipids", in FOX P. F. (ed.), *Advanced Dairy Chemistry Lipids*, 2nd ed., vol. 2, Springer, 1995.

[COL 86] COLONNA P., ROUAU X., "L'amidon, utilisations industrielles", *Industries des céréales*, vol. 41, pp. 7–11, 1986.

[CRA 83] CRANDALL P.G., MATTHEWS R.F., BAKER R.A., "Citrus beverage clouding aspects: review and status", *Food Technology*, vol. 37, pp.106–109, 1983.

[CRO 02] CROGUENNEC T., NAU F., BRULÉ G., "Influence of pH and salts on egg white gelation", *Journal of Food Science*, vol. 67, no. 2, pp. 608–614, 2002.

[DAL 04] DALGLEISH D.G., SPAGNUOLO P. A., GOFF H. D., "A possible structure of the casein micelle based on high-resolution field-emission scanning electron microscopy", *International Dairy Journal*, pp. 1025–1031, 2004.

[DAM 98] DAMODARAN S., ANAND K., RAZUMOVSKY L., "Competitive adsorption of egg white proteins at the air-water interface: direct evidence for electrostatic complex formation between lysozyme and other egg proteins at the interface", *Journal of Agricultural and Food Chemistry*, vol. 46, pp. 872–876, 1998.

[DEE 01] DE ELL J.R., KHANIZADEH S., SAAD F. *et al.*, "Factors affecting apple fruit firmness: a review", *Journal of the American Pomological Society*, vol. 55, pp. 8–27, 2001.

[DEN 00] DENES J..M., BARON A., RENARD C. *et al.*, "Different action patterns for apple pectin methylesterase at pH 7.0 and 4.5", *Carbohydrate Research*, vol. 327, no. 4, pp. 385–393, 2000.

[DIN 99] DINSDALE M.G., LLOYD D., MCINTYRE P. *et al.*, "Yeast vitality during cider fermentation: assessment by energy metabolism", *Yeast*, vol. 15, pp. 285–293, 1999.

[DIX 00] DIXON J., HEWETT E.W., "Factors affecting apple aroma/flavour volatile concentration: a review", *New Zealand Journal of Crop and Horticultural Science*, vol. 28, pp. 155–173, 2000.

[DOI 93] DOI E., "Gels and gelling of globular proteins", *Trends Food Science & Technology*, vol. 4, pp. 1–5, 1993.

[DON 75] DONOVAN J.W., MAPES C. J., DAVIS J. *et al.*, "A differential scanning calorimetric study of the stability of egg white to heat denaturation", *Journal of the Science of Food Chemistry*, vol. 26, pp. 73, 1975.

[DRA 79] DRAPRON R., GENOT C., "Les lipides des céréales", *Ind. Alim. Agric.*, vol. 12, pp. 1257–1273, 1979.

[DUR 99] DURAND P., *Technologies des Produits et des Salaisons*, Lavoisier, Paris, 1999.

[DYE 93] DYER-HURDON J.N., NNANNA I.A., "Cholesterol content and functionality of plasma and granules fractionated from egg yolk", *Journal of Food Science*, vol. 58, pp. 1277–1281, 1993.

[END 65] ENDO A., "Studies on pectolytic enzymes of molds: Part XIII. Clarification of apple juice by the joint action of purified pectolytic enzymes", *Agricultural and Biological Chemistry*, vol. 29, pp. 129–136, 1965.

[EVA 99] EVANS D.E., SHEEHAN M.C., STEWART D.C., "The impact of malt derived proteins on beer foam quality: Part II: The influence of malt foam-positive proteins and non -starch polysaccharides on beer foam quality", *Journal of the Institute of Brewing*, vol. 105, pp. 171–177, 1999.

[FAU 85a] FAUQUANT J., VIECO E., BRULÉ G. *et al.*, "Clarification des lactosérums doux par agrégation thermocalcique de la matière grasse résiduelle", *Lait*, vol. 65, pp. 1–20, 1985a.

[FAU 85b] FAUQUANT J., PIERRE A., BRULÉ G., "Clarification du lactosérum acide de caséinerie", *Tech Lait*, vol. 1003, pp. 37–39, 1985b.

[FEI 00] FEILLET P., *Le Grain de Blé: Composition et Utilisation*, INRA Éditions, Paris, 2000.

[FIL 99] FILLAUDEAU L., BLANPAIN-AVET P., "Applications en brasserie de la microfiltration tangentielle", *Techniques de l'ingenieur*, vol. 3260, pp. 1–13, 1999.

[FLA 98] FLANZY C., *Œnologie: Fondements Scientifiques et Technologiques*, Tec & Doc Lavoisier, Paris, 1998.

[FLE 11] FLEURENT E., *Le Pain de Froment*, Ed Gauthier-Villars, Paris, 1911.

[FON 89] FONG C.H., HASEGAWA S., HERMAN Z. *et al.*, "Limonoid glucosides in commercial citrus juices", *Journal of Food Science*, vol. 54, pp. 1505–1506, 1989.

[FRE 90] FRENTZ J.C., ZERT P., *L'Encyclopédie de la Charcuterie*, 3rd édition, Soussana,1990.

[GAN 34] GANE R., "Production of ethylene by some fruits", *Nature*, vol. 134, no. 3400, p. 1008, 1934.

[GAN 35] GANE R., "The formation of ethylene by plant tissue and its significance in the ripening of fruits", *Journal of Pomplogy and Horticultural Science*, vol. 13, pp. 351–358, 1935.

[GAU 05] GAUCHERON F., "The minerals of milk", *Reproduction Nutrition Development*, vol. 45, pp. 473–483, 2005.

[GIO 01] GIOVANNONI J., "Molecular biology of fruit maturation and ripening", *Annual Review of Plant Physiology and Plant Molecular Biology*, vol. 52, pp. 725–749, 2001.

[GIR 88] GIRARD J.P., *Technologie de la Viande et des Produits Carnés*, Lavoisier, Paris, 1988.

[GOF 97] GOFF H.D., "Colloidal aspect of ice cream: a review", *International Dairy Journal*, vol. 7, pp. 363–373, 1997.

[GON 94] GONTARD N., GUILBERT S., "Bio-packaging: technology and properties of edible and/or biodegradable material of agricultural origin", in MATHLOUTHI M. *Food Packaging and Preservation*, Blackie Academic & Professional, Glasgow, pp. 159–181, 1994.

[GRA 73] GRANT G..T., MORRIS E. R., REES D.A. *et al.*, "Biological interactions between polysaccharides and divalent cations: the egg-box model", *FEBS Letters*, vol. 32, no. 1, pp. 195–198, 1973.

[GRO 96] GROSJEAN F., BARRIER-GUILLOT B., "Les polysaccharides non amylacés des céréales", *Industries des céréales*, vol. 98, pp. 13–33, 1996.

[GU 04] GU L.W., KELM M.A., HAMMERSTONE J.F. *et al.*, "Concentrations of proanthocyanidins in common foods and estimations of normal consumption", *Journal of Nutrition*, vol. 134, pp. 613–617, 2004.

[GUE 06] GUÉRIN-DUBIARD C., PASCO M., MOLLÉ D. *et al.*, "Proteomic analysis of hen egg white", *Journal of Agricultural and Food Chemistry*, vol. 54, pp. 3901–3910, 2006.

[GUN 98] GUNSTONE F.D., "Movements towards tailor-made fats", *Progress in Lipid Research*, vol. 37, no. 5, pp. 277–305, 1998.

[HA 05] HA M.A., VIETOR R.J., JARDINE G.D. *et al.*, "Conformation and mobility of the arabinan and galactan side-chains of pectin", *Phytochemistry*, vol. 66, pp. 1817–1824, 2005.

[HAM 95] HAMM W., "Trends in edible oil fractionation", *Trends Food Science & Technology*, vol. 6, pp. 121–126, 1995.

[HAL 92] HALL G.M., AHMAD N.H., "Surimi and fish mince products", *Fish Processing Technology*, Blackie Academic & Professionnal, New York, pp. 72–89, 1992.

[HEE 85] HEERTJE I., VISSER J., SMITS P., "Structure formation in acid milk gels", *Food Microstructure*, vol. 4, pp. 267–277, 1985.

[HEN 88] HENRY R.J., "The carbohydrates of barley grains. A review", *Journal of the Institute of Brewing,* vol. 94, pp. 71–78, 1988.

[HER 93] HEREDIA A., GUILLEN R., JIMENEZ A. *et al.*, "Review: plant cell wall structure", *Revista Española de Ciencia y Tecnología de Alimentos*, vol. 33, pp. 113–131, 1993.

[HOL 96] HOLT C., HORNE D.S., "The hairy casein micelle: evolution of the concept and its implication for dairy technology", *Netherlands Milk Dairy Journal*, vol. 50, pp. 85–111, 1996.

[HOR 02] HORNE D.S., "Casein structure, self-assembly and gelation", *Current Opinion in Colloid & Interface Science,* vol. 7, pp. 456–461, 2002.

[HUL 71] HULME A.C., RHODES M.J.C., "Pome fruits", *The Biochemistry of Fruits and their Products*; ACH eds, pp. 333–369, 1971.

[HUS 88] HUSS H.H., "Le poisson frais: qualité et altérations de la qualité", *Collection FAO Pêche,* no. 29, p. 132, 1988.

[IRE 14] IRELAND H.S., GUNASEELAN K., MUDDUMAGE R. *et al.*, "Ethylene regulates apple (malus x domestica) fruit softening through a dose x time-dependent mechanism and through differential sensitivities and dependencies of cell wall-modifying genes", *Plant and Cell Physiology,* vol. 55, no. 5, pp. 1005–1016, 2014.

[JAC 78] JACKSON G., WAINWRIGHT T., "Melanoidins and beer foam", *Journal of the American Society of Brewing Chemistry,* vol. 36, pp. 192–195, 1978.

[JAY 05] JAYANI R.S., SAXENA S., GUPTA R., "Microbial pectinolytic enzymes: a review", *Process Biochemistry*, vol. 40, pp. 2931–2944, 2005.

[JEA 00] JEANTET R., CROGUENNEC T., MAHAUT M. *et al.*, *Les Produits Industriels Laitiers,* Lavoisier, Paris, 2000.

[JEA 16a] JEANTET R., CROGUENNEC T., SCHUCK P., BRULÉ G., *Handbook of Food Science and Technology 1 – Food Alteration and Food Quality*, ISTE, London and John Wiley & Sons, New York, 2016.

[JEA 16b] JEANTET R., CROGUENNEC T., SCHUCK P., BRULÉ G., *Handbook of Food Science and Technology 2 – Food Process Engineering and Packaging*, ISTE, London and John Wiley & Sons, New York, 2016.

[JOH 02] JOHNSTON J. W., HEWETT E.W., HERTOG M. L., "Postharvest softening of apple (Malus domestica) fruit: a review", *New Zealand Journal of Crop and Horticultural Science,* vol. 30, pp. 145–160, 2002.

[JOU 58] JOUBERT F.J., COOK W. H., "Preparation and characterization of phosvitin from hen egg yolk", *Canadian Journal of Biochemistry and Physiology*, vol. 36, pp. 399–408, 1958.

[KAD 86] KADER A.A., "Biochemical and physiological basis for effects of controlled and modified atmospheres on fruits and vegetables", *Food Technology*, vol. 40, pp. 99–104, 1986.

[KAD 89] KADER A.A., ZAGORY D., KERBEL E.L., "Modified atmosphere packaging of fruits and vegetables", *Critical Reviews in Food Science and Nutrition*, vol. 28, no. 1, pp. 1–30, 1989.

[KAL 02] KALLAY T., "Genetic determination of maturation processes in climacteric fruits – a review", *Acta Alimentaria*, vol. 31, pp. 169–177, 2002.

[KAS 01] KASHYAP D.R., VOHRA P.K., CHOPRA S. et al., "Applications of pectinases in the commercial sector: a review", *Bioresource Technology*, vol. 77, pp. 215–227, 2001.

[KAT 89] KATO A., IBRAHIM H.R., WATANABE H. et al., "New approach to improve the gelling and surface functional properties of dried egg white by heating in dry state", *Journal of Agricultural and Food Chemistry*, vol. 37, pp. 433–437, 1989.

[KAT 90a] KATO A., IBRAHIM H.R., WATANABE H. et al., "Structural and gelling properties of dry-heating egg white proteins", *Journal of Agricultural and Food Chemistry*, vol. 38, pp. 32–37, 1990a.

[KAT 90b] KATO A., IBRAHIM H.R., TAKAGI T. et al., "Excellent gelation of egg white preheated in the dry state is due to the decreasing degree of aggregation", *Journal of Agricultural and Food Chemistry*, vol. 38, pp. 1868–1872, 1990b.

[KID 33] KIDD F., WEST C., "The effects of ethylene and apple vapours on ripening of fruits", in *Great Britain Department of Science and of Industry Research Report, Food Investigation Board for 1932*, pp. 55–58, 1933.

[KIO 89] KIOSSEOGLOU V.D., "Egg yolk", in CHARALAMBOUS G., DOXASTAKIS G. (eds.), *Food Emulsifiers: Chemistry, Technology, Functional Properties and Applications*, Elsevier, London, pp. 63–85, 1989.

[KIT 88] KITABATAKE N., SHIMIZU A., DOI E., "Preparation of transparent egg white gel with salt by two-step heating method", *Journal of Food Science*, vol. 53, pp. 735–738, 1988.

[KNO 89] KNOCKAERT C., "Les marinades des produits de la mer", Collection Valorisation des produits de la mer, IFREMER, Brest, 1989.

[KOR 06] KORHONEN H., PIHLANTO A., "Bioactive peptides: production and functionality", *International Dairy Journal*, vol. 16, pp. 945–960, 2006.

[KUC 93] KUCZINSKI A., VAROQUAUX P., SOUTY M., "Reflectance spectra of 'ready-to-use' apple products for determination of enzymatic browning", *International Agrophysics*, vol. 7, pp. 85–92, 1993.

[COU 84] COURTINE R. (ed.), Larousse Gastronomique, Larousse, Paris, 1984.

[LEA 03] LEA A.G.H., DRILLEAU J.F., "Cidermaking", in LEA AGH, PIGGOTT J. R. (eds.), *Fermented Beverage Production* (2nd edition), Kluwer Academic/Plenum Publishers, 2003.

[LEC 03] LECHEVALIER V., CROGUENNEC T., PEZENNEC S. *et al.*, "Ovalbumin, ovotransferrin, lysozyme: three model proteins for structural modifications at the air-water interface", *Journal of Agricultural and Food Chemistry*, vol. 51, pp. 6354–6361, 2003.

[LEC 05] LECHEVALIER V., CROGUENNEC T., PEZENNEC S. *et al.*, "Evidence for synergy in the denaturation at the air-water interface of ovalbumin, ovotransferrin and lysozyme in ternary mixture", *Food Chemistry*, vol. 92, pp. 79–87, 2005.

[LED 99] LE DENMAT M., ANTON M., GANDEMER G., "Protein denaturation and emulsifying properties of plasma and granules of egg yolk as related to heat treatment", *Journal of Food Science*, vol. 64, no. 2, pp. 194–197, 1999.

[LED 00] LE DENMAT M., ANTON M., BEAUMAL V., "Characterisation of emulsion properties and of interface composition in oil-in-water emulsions prepared with hen egg yolk, plasma and granules", *Food Hydrocolloids*, vol. 14, pp. 539–549, 2000.

[LEI 85] LEISTNER L., "Allegemeines über Rohwurst une Rohschinken", *Mikrobiologie, und Qualität von Rohwurst und Rohschinken*, Bundesanstlt für Fleischforschung, Kulmbach, Allemagne, 1985.

[LEQ 06] LE QUÉRÉ J.M., HUSSON F., RENARD C.M. *et al.*, "French cider characterization by sensory, technological and chemical evaluations", *Lebensmittel-Wissenschaft & Technologie*, vol. 39, pp. 1033–1044, 2006.

[LIN 94] LINDEN G., LORIENT D., *Biochimie Agro Industrielle: Valorisation Alimentaire de la Production Agricole,* Masson, Paris, 1994.

[LIC 89] LI-CHAN E., NAKAI S., "Biochemical basis for the properties of egg white", *Critical Reviews in Poultry Biology*, vol. 2, pp. 21–58, 1989.

[LIN 94] LINDEN G., LORIENT D., "Biochimie agro-industrielle: valorisation alimentaire de la production agricole", Masson, p. 368, 1994.

[LUC 98] LUCEY J.A., SINGH H., "Formation and physical properties of acid milk gel: a review", *Food Research International*, vol. 30, no. 7, pp. 529–542, 1998.

[LUS 95] LUSK L. T., GOLDSTEIN H., RYDER D., "Independent role of beer proteins, melanoidins and polysaccharides in foam formation", *Journal of the American Society of Brewing Chemistry,* vol. 53, pp. 93–103, 1995.

[MAH 00] MAHAUT M., JEANTET R., BRULÉ G. *et al.*, *Les Produits Laitiers Industriels*, Lavoisier, Paris, 2000.

[MAI 12] MAILLARD L.C., "Action des acides aminés sur les sucres; formation de mélanoïdines par voie métabolique", *Comptes Rendus Hebdomadaires de l'Académie des Sciences de Paris,* vol. 154, pp. 743–752, 1912.

[MAN 98] MANGAS J.J., MORENO J., PICINELLI A. *et al.*, "Characterization of cider apple fruits according to their degree of ripening. A chemometric approach", *Journal of Agricultural and Food Chemistry*, vol. 46, pp. 4174–4178, 1998.

[MAR 02] MARTINET V., BEAUMAL V., DALAGALARRONDO M. *et al.*, "Emulsifying properties and adsorption behavior of egg yolk lipoproteins (LDL and HDL) in o/w emulsions", *Recent Research Development in Agricultural & Food Chemistry*, vol. 37, pp. 103–116, 2002.

[MAR 03] MARTINET V., SAULNIER P., BEAUMAL V. *et al.*, "Surface properties of hen egg yolk low-density lipoproteins spread at the air-water interface", *Colloids and Surfaces B: Biointerfaces*, vol. 31, pp. 185–194, 2003.

[MAS 91] MASTERS K., *Spray drying*, Longman Scientific & Technical and John Wiley & Sons Inc, Essex, 1991

[MAU 69] MAUBOIS J.L., MOCQUOT G., VASSAL L., "Procédé de traitement du lait et de sous produits laitiers", Patent no FR2052121, 1969.

[MAU 32] MAURISIO, *Histoire de l'Alimentation Végétale Depuis la Préhistoire Jusqu'à nos Jours*, Payot, Paris, p. 663, 1932.

[MCM 84] MCMAHON D.J., BROWN R.J., "Composition, structure and integrity of casein micelles: a review", *Journal of Dairy Science*, vol. 67, pp. 499–512, 1984.

[MEC 49] MECHAM D., OLCOTT H., "Phosvitin, the principal phosphoprotein of egg yolk", *Journal of the American Chemical Society*, vol. 71, pp. 3822–3833, 1949.

[MIC 01] MICHALSKI M. C., MICHEL F., SAINMONT D. *et al.*, "Apparent ζ-potential as a tool to assess mechanical damages to the milk fat globule membrane", *Colloids and Interface B: Biointerfaces,* vol, 23, pp. 23–30, 2001.

[MIE 91] MIETTON B., in *Les Bactéries Lactiques*, vol. II, édition Lorica, Uriage, France, pp. 55–133, 1991.

[MIL 98] MILLET J., "A propos de la méthode Nizo", *Revue des ENIL,* vol. 215, pp. 19–22, 1998.

[MIN 00] MINE Y., KEERATIURAI M., "Selective displacement of caseinate proteins by hens egg yolk lipoproteins at oil-in-water interfaces", *Colloids and Surfaces B: Biointerfaces*, vol. 18, pp. 1–11, 2000.

[MIZ 85] MIZUTANI R., NAKAMURA R., "Physical state of the dispersed phases of emulsions prepared with egg yolk low density lipoprotein and bovine serum albumin", *Journal of Food Science*, vol. 50, pp. 1621–1623, 1985.

[MOK 61] MOK C. C., MARTIN W. G., COMMON R. H., "A comparison of phosvitins prepared from hen's serum and from hen's egg yolk", *Canadian Journal of Biochemistry Physiology*, vol. 39, pp. 109–117, 1961.

[NAK 77] NAKAMURA R., HAYAKAWA R., SATO Y., "Isolation and fractionation of the protein moiety of egg yolk low density lipoprotein", *Poultry Science*, vol. 56, pp.1148–1152, 1977.

[NGU 94] NGUYEN-THE C., CARLIN F., "The microbiology of minimally processed fresh fruits and vegetables", *Critical Reviews in Food Science and Nutrition*, vol. 34, no. 4, pp. 371–401, 1994.

[OHA 87] OHATA K., KAMADA K., "Studies on foam stability of beer", *Proceedings of the European Brewery Convention Congress*, Madrid, 1987.

[ONE 04] O'NEILL M.A., ISHII T., ALBERSHEIM P. *et al.*, "Rhamnogalacturonan II: structure and function of a borate cross-linked cell wall pectic polysaccharide", *Annual Review of Plant Biology*, vol. 55, pp. 109–139, 2004.

[ORT 09] ORTIZ-BASURTO R.I., WILLIAMS P., BELLEVILLE M.P. *et al.*, "Presence of rhamnogalacturonan II in the juices produced by enzymatic liquefaction of Agave pulquero stem (Agave mapisaga)", *Carbohydrate Polymers,* vol. 77, no. 4, pp. 870–875, 2009.

[OZD 05] OZDEMIR I., MONNET F., GOUBLE B., "Simple determination of the O_2 and CO_2 permeances og microperforated pouches for modified atmosphere packaging of respiring foods", *Postharvest Biology and Technology,* vol. 36, pp. 209–213, 2005.

[PEP 96] PEPPELENBOS H.W., VANTLEVEN J., "Evaluation of four types of inhibition for modelling the influence of carbon dioxide on oxygen consumption of fruits and vegetables", *Postharvest Biology and Technology,* vol. 7, nos. 1–2, pp. 27–40, 1996.

[PER 03] PEREZ S., RODRIGUEZ-CARVAJAL M.A., DOCO, T., "A complex plant cell wall polysaccharide: rhamnogalacturonan II. A structure in quest of a function", *Biochimie,* vol. 85, nos. 1–2, pp. 109–121, 2003.

[PET 96] PETRICH-MURRAY H., DUCROO P., "Les pentosanes en panification", *Industries des Céréales,* vol. 97, pp. 13–17, 1996.

[PIC 87] PICLET G., "Le poisson aliment: composition – intérêt nutritionnel", *Critical Nutrition and Dietetics*, vol. 22, no. 4, pp. 317–336, 1987.

[PIS 81] PISECKY J., "Technology of skimmed milk", *Journal Society Dairy Technology*, vol. 34, pp. 57–67, 1981.

[PLA 97] PLANCHOT V., COLONNA P., SAULNIER L., "Dosage des glucides et des amylases", in GODON B., LOISEL W. (eds.), *Guide Pratique d'Analyses dans les Industries des Céréales*, Lavoisier, Paris, 1997.

[POP 82] POPPE J., VINCENT R., "La gélatine alimentaire. Dans: protéines animales – extraits, concentrés et isolats en alimentation humaine", *Technique et Documentation*, Lavoisier, Paris, pp. 228–255, 1982.

[POW 86] POWRIE W.D., NAKAI S., "The chemistry of eggs and egg products", in STADEMAN W.J,. COTTERILL O.J. (eds.), *Egg Science and Technology*, The Avi Publishing Company, Inc. pp. 97–139, 1986.

[REN 11] RENARD C.M.G.C., LE QUERE J.M., BAUDUIN R. *et al*., "Modulating polyphenolic composition and organoleptic properties of apple juices by manipulating pressing conditions", *Food Chemistry*, vol. 124, no. 1, pp. 117–125, 2011.

[REN 98] RENARD C., THÉRY S., "Détermination des méthodes physico-chimiques pour prédire la qualité biscuitière et boulangères des blés français", *Industries des Céréales*, vol. 109, pp. 31–36, 1998.

[REN 51] RENAUDIN C., *La Fabrication Industrielle des Pâtes Alimentaires*, Dunod, Paris, 1951.

[ROU 02] ROUSSEL P., CHIRON H., *Les Pains Français; Évolution, Qualité, Production*, Mae-Erti, Vesoul, 2002.

[ROU 96] ROUAU X., "Les hémicellulases en panification", *Industries des Céréales*, vol. 96, pp. 13–19, 1996.

[STA 86] STARON T., *L'encyclopédie nutritionnelle de l'homme*, INRA, 1986.

[SAL 97] SALTVEIT M. E., "Physical and physiological changes in minimally processed fruits and vegetables", in TOMAS-BARBERAN F.A. (ed.), *Phytochemistry of Fruits and Vegetables*, Oxford University Press, pp. 205–220, 1997.

[SAL 00] SALTVEIT M. E., "Wound induced changes in phenolic metabolism and tissue browning are altered by heat shock", *Postharvest Biology and Technology*, vol. 21, no. 1, pp. 61–69, 2000.

[SAU 88] SAUVEUR B., "Structure, composition et valeur nutritionnelle de l'œuf", *Reproduction des Volailles et Production d'Oeufs*, INRA, chap. 14, pp. 347–436, 1988.

[SCH 94] SCHUCK P., PIOT M., MÉJEAN S. *et al.*, "Déshydratation des laits enrichis en caséine micellaire par microfiltration; comparaison des propriétés des poudres obtenues avec celles d'une poudre de lait ultra-propre", *Lait*, vol. 74, pp. 47–63, 1994.

[SEG 67] SEGEL E., GLENISTER P.R., KOEPPL K.G., "Beer foam", *Technical Quaterly Master Brewers Association of the Americas*, vol. 4, pp. 104–113, 1967.

[SER 01] SERVENTI S., SABBAN F., *Les Pâtes, Histoire d'une Culture Universelle*, Actes Sud, Arles, 2001.

[SHE 79] SHENTON A.J., Membrane composition and performance of food emulsions, PhD Thesis, University of London, UK, 1979.

[SHE 74] SHEWAN J.M., "The biodeterioration of certain proteinaceous foodstuffs at chill temperatures", in SPENCER B. (ed.), *Industrial Aspects of Biochemistry*, North Holland Publishing Co, 1974.

[SIE 06] SIEBERT K.J., "Haze formation in beverages", *Lebensmittel-Wissenschaft & Technologie*, vol. 39, pp. 987–994, 2006.

[SIN 03] SINGH S.V., JAIN R. K., GUPTA A. *et al.*, "Debittering of citrus juices – A review", *Journal of Food Science and Technology*, vol. 40, pp. 247–253, 2003.

[SOL 05] SOLIGNAT G., "Produits de charcuterie. Saison humide: lardons, jambon cuit", *Tech. Ing.*, vol. 6, no. 504, pp. 1–10, 2005.

[SOM 04] SOMERVILLE C., BAUER S., BRININSTOOL G. *et al.*, "Toward a systems approach to understanding plant cell walls", *Science*, vol. 306, pp. 2206–2210, 2004.

[SPA 92] SPANOS G.A., WROLSTAD R. E., "Phenolics of apple, pear and white grape juices and their changes with processing and storage – a review", *Journal of Agricultural and Food Chemistry*, vol. 40, pp. 1478–1487, 1992.

[STE 91] STEVENS L., "Egg white proteins", *Comp. Biochem. Physiol.*, vol. 100B, pp. 1–9, 1991.

[TAI 05] TAILLANDIER P., BONNET J., *Le Vin: Composition et Transformations Chimiques*, Editions Tec&Doc, Lavoisier, Paris, 2005.

[TAM 02] TAMIME A.Y., "Fermented milks: a historical food with modern applications: a review", *European Journal of Clinical Nutrition*, vol. 00, pp. 1–14, 2002.

[TAM 99] TAMIME A.Y., ROBINSON A.K., "Background to manufacturing practice", in TAMIME A.Y., ROBINSON A.K. (eds.), *Yoghurt Science and Technology*, Woodhead Publishing Limited, Cambridge, pp. 11–128, 1999.

[THA 94] THAPON J.L., BOURGEOIS C.M., *L'Oeuf et les Ovoproduits*, Lavoisier, Paris, 1994.

[TOM 01] TOMAS-BARBERAN F.A., ESPIN J.C., "Phenolic compounds and related enzymes as determinants of quality in fruits and vegetables", *Journal of the Science of Food and Agriculture*, vol. 81, no. 9, pp. 853–876, 2001.

[TOT 02] TOTOSAUS A., MONTEJANO J.G., SALAZAR J.A. et al., "A review of physical and chemical protein-gel induction", *International Journal of Food Science and Technology*, vol. 37, pp. 589–601, 2002.

[VAR 99] VAROQUAUX P., GOUBLE B., BARRON C. et al., "Respiratory parameters and sugar catabolism of mushroom (*Agaricus bisporus* Lange)", *Postharvest Biology and Technology*, vol. 16, no. 1, pp. 51–61, 1999.

[VAR 02] VAROQUAUX P., MAZOLLIER J., "Overview of the European fresh-cut produce industry", in LAMINKARA O. (ed.), *Fresh-Cut Fruits and Vegetables, Science, Technology, and Market*, CRC Press, Boca Raton, pp. 21–43, 2002.

[VAN 01] VAN AKEN G.A., "Aeration of emulsions by whipping", *Colloïds and Surface A: Physicochemical and Engineering Aspects*, vol. 190, pp. 333–354, 2001.

[VER 99] VERCAUTEREN R., RAPAILLE A., "Productions industrielles des glucides alimentaires, propriétés technologiques et utilisations alimentaires", *Dossier Scientifiques de l'Institut Français pour la Nutrition*, vol. 11, p. 37, 1999.

[VIS 88] VISSER R.A., VAN DEN BOS M. J., FERGUSON W. P., "Lactose and its chemical derivatives", *Bulletin International Dairy Federation*, vol. 233, pp. 33–44, 1988.

[WIL 90] WILLM C., "Farines d'antan, farines d'aujourd'hui; comparaison des farines de meules et des farines de cylindres", *Industries des Céréales*, vol. 66, pp. 7–16, 1990.

[YAM 67] YAMASAKI M., KATO A., CHU S.Y. et al., "Pectic enzymes in the clarification of apple juice. Part II – The mechanism of clarification", *Agricultural and Biological Chemistry*, vol. 31, pp. 552–560, 1967.

[YOR 03] YORUK R., MARSHALL M.R., "Physicochemical properties and function of plant polyphenol oxidase: a review", *Journal of Food Biochemistry*, vol. 27, pp. 361–422, 2003.

List of Authors

Marc ANTON
INRA
UR 1268 Biopolymères
Interactions Assemblages
Nantes
France

Alain BARON
INRA
UR 1268 Biopolymères
Interactions Assemblages
Nantes
France

Gérard BRULÉ
Agrocampus Ouest
UMR 1253 Science et
Technologie du lait et de l'œuf
Rennes
France

Florence CHARLES
University of Avignon
Pôle Agroscience
France

Hubert CHIRON
INRA
UR 1268 Biopolymères
Interactions Assemblages
Nantes
France

Thomas CROGUENNEC
Agrocampus Ouest
UMR 1253 Science et
Technologie du lait et de l'œuf
Rennes
France

Catherine GUÉRIN
Agrocampus Ouest
UMR 1253 Science et
Technologie du lait et de l'œuf
Rennes
France

Romain JEANTET
Agrocampus Ouest
UMR 1253 Science et
Technologie du lait et de l'œuf
Rennes
France

Valérie LECHEVALIER
Agrocampus Ouest
UMR 1253 Science et
Technologie du lait et de l'œuf
Rennes
France

Françoise NAU
Agrocampus Ouest
UMR 1253 Science et
Technologie du lait et de l'œuf
Rennes
France

Ludivine PERROCHEAU
Saint-Antoine Hospital
INSERM U592
Paris
France

Jean-Michel LE QUÉRÉ
UR 1268 Biopolymères
Interactions Assemblages
Nantes
France

Philippe ROUSSEL
Pierre et Marie Curie University
Paris
France

Pierre SCHUCK
INRA
UMR 1253 Science et
Technologie du lait et de l'œuf
Rennes
France

Mohammad TURK
École Supérieure Chimie
Organique & Minérale
Compiègne
France

Patrick VAROQUAUX
INRA
UMR 408 Sécurité et qualité des
produits d'origine végétale
Avignon
France

Index

A, B, C

activity
  water, 71, 106, 141, 172, 241, 361
  respiratory, 298, 299, 301, 315, 317
bacteria
  acetic, 268-270, 289, 290
  lactic, 9, 22, 29, 42, 56, 61, 108, 148, 171, 189, 269, 285, 339, 340, 368
  propionic, 55, 56
centrifugation, 45, 49, 119, 139, 172, 224, 227, 255, 258, 261, 267, 268, 271, 272, 292, 303, 313, 333, 342, 343, 345, 350, 351, 358, 359, 365–367, 370, 371, 378–380, 381, 382, 400
cheese
  draining, 40, 44,
  fresh, 49
  hard, 39, 42, 43, 44, 45, 47, 56, 56, 364
  soft, 42, 44, 47, 55, 57
chromatography, 143, 161, 181, 343–345, 347, 395

collagen, 67, 77, 83–88, 101, 105, 112, 348
cooking, 58, 87, 88, 97, 101, 103, 104, 105, 109
crystallization
  lactose, 365, 366
  sucrose, 356, 359
crushing, 173, 174, 175, 225, 249, 256, 279–282, 295, 303, 362, 363

D, E

decantation, 350, 351, 365, 366, 370, 371, 375. 379, 400
drying
  milk, 38, 336
  pasta, 169
emulsion
  egg, 115, 127
  milk, 17, 18
enzyme
  amylase, 156
  arabinanase, 244, 245
  galactosidase, 9, 29, 235, 236, 367, 368
  lipoxygenase, 170, 197, 198, 263, 278, 279, 379

lyase, 236, 243, 244, 259, 302
polygalacturonase, 235, 242, 243, 255
rennet, 22, 23, 38–45, 54, 57, 335, 341
evaporation
  under vacuum, 33, 34

F, G, H, L

fermentation
  alcoholic, 287
  beer, 220
  bread, 147, 160, 164, 168, 171, 184, 190
  cider, 265, 267
  malolactic, 281, 288
  milk, 29
  wine, 275, 279, 285-289
filtration
  diafiltration, 338, 345
  frontal, 228
  microfiltration, 28, 32-34, 38, 229, 259, 339, 343
  reverse osmosis, 30
  ultrafiltration, 32, 41, 48, 350
glass transition, 156, 325, 364
hydrogenation
  glucose, 84, 95, 208, 367
homogenization, 8, 18, 19, 25, 27, 28, 30, 44, 45, 58, 59, 62, 127
lactoglobulin, 13, 14, 336, 343–345
lipid
  fat globule, 4, 7, 343,
  fish, 66
  milk, 4, 5, 7, 27, 53
  oxidation, 27, 102, 141, 379, 385
  phospholipid, 5, 123, 124

lysozyme, 57, 120–123, 126, 127, 132, 136, 143, 346, 347

M, O, P

malting, 205, 207, 209, 210, 213–218, 221, 222
maturation
  meat, 87, 99, 100
  wine, 289
mill, 147, 150, 151, 154, 162, 172–179, 198, 212, 221, 223–226, 257, 258, 276, 293, 299, 349, 350, 354, 355, 370
mould, 44, 45, 53, 55–57, 104, 105, 111, 118, 167, 193, 205, 312, 362, 363, 367
mutage, 284, 288, 294
oxydation, 56, 79
  carotenoid, 383
  lipid, 27, 102, 141, 375
  myoglobin, 104
packaging
  modified atmosphere, 80, 170, 298, 299, 303, 305, 312, 314, 315, 317, 318
  permeability, 316, 320
pasteurization, 26, 27, 30, 38, 54, 58, 59, 139, 224, 229, 230, 258, 262, 263, 265, 273
polyphenol, 170, 198, 203, 208, 219, 220, 228, 229, 233, 245, 249, 254, 257, 259, 260, 264, 266, 278, 302, 359, 391
polysaccharides
  alginates, 329, 369
  carrageenan, 369
  pectin, 329, 369
  starch, 214
powder
  rehydration, 326
  spray drying, 34

precipitation
  alginate, 375, 369
  carrageenan, 369, 373, 374
  casein, 339
  plant protein, 349, 351
  whey protein, 337, 338, 339, 341, 342
  xanthan, 376, 377, 378
pressing, 45, 48, 167, 197, 198, 200, 201, 238, 250, 251, 253, 258, 260, 265, 266, 268, 279, 280, 281, 282, 284, 285, 292, 293, 294, 354, 355, 379, 380
properties
  foaming, 115, 116, 125, 126, 140
  gelling, 116, 131, 133, 134, 139, 140, 323, 335, 116, 131, 133, 134, 139, 140, 324, 335
  thickening, 158, 324

S, T

salt
  brines, 110
  fish, 70
  meat, 87
  nitrite, 104, 105
solvent extraction, 350, 378–380
transesterification, 364, 379, 384, 386–388

Printed and bound by CPI Group (UK) Ltd, Croydon, CR0 4YY
19/03/2023
03203240-0004